THE COMPLETE HANDBOOK OF PLUMBING

BY ROBERT E. MORGAN

&

RAMESH SINGHAL

TAB **TAB BOOKS Inc.**
BLUE RIDGE SUMMIT, PA. 17214

FIRST EDITION

FIRST PRINTING

Copyright © 1982 by TAB BOOKS Inc.

Printed in the United States of America

Reproduction or publication of the content in any manner, without express permission of the publisher, is prohibited. No liability is assumed with respect to the use of the information herein.

Library of Congress Cataloging in Publication Data

Morgan, Robert E.
    The complete handbook of plumbing.

    Includes index.
    1. Plumbing—Amateurs' manuals.   I. Title.
TH6124.M67   696'.$ 81-9243
ISBN 0-8306-0042-6   AACR2
ISBN 0-8306-1374-9 (pbk.)

Contents

To my wife Susan and children, Gabriel and Gillie.

Introduction

Not too many years ago the do-it-yourself enthusiast was looked upon by many of his fellow human beings as somewhat of a strange person who attempted to do for himself things that might be more expertly (and sometimes less expensively) accomplished by a professional tradesman. For some, do-it-yourself projects were considered to be luxury pursuits rather than necessities.

Today, nothing could be further from the truth. The average homeowner has found it necessary to take on many tasks that were formerly farmed out to professionals in fields such as plumbing. Why? For many of us, spiraling inflation has signaled that if you don't do it yourself, you can't afford to have it done any other way.

This book is written in a manner that should make the information practical to almost everyone. If you are accustomed to doing some of the plumbing work around your home, this text allows you to venture further while learning and increasing your knowledge in the field. For those of you who avoid this pursuit due to lack of knowledge, the detailed explanations of equipment, tools, and accessories are invaluable.

By using the information contained in this book coupled with practical home experiences, you find that many plumbing jobs are not so complex and can be handled with a minimum of tools and supplies. You won't be an experienced plumber simply by learning the materials presented here, but by practicing what you learn, you find that almost no job is beyond your means, both financial and experiential.

The Home Plumbing System and Its Operation

Although there have always been some homeowners who were able to keep their plumbing system properly maintained, these persons were considered to be a small minority until recent years. This is understandable in a way. To the average person, a plumbing system looks like a quite complicated series of pipes and subsystems, the repair of which requires a professional with years of experience. It is true that a basic understanding of a plumbing system is necessary in order to do even minor repairs, but every system is scientifically designed according to the basic laws of nature and can be subdivided into separate sections, each performing a specific function in the overall system.

More and more homeowners today are beginning to take on the responsibility for installing or modifying their plumbing systems. This should give you an accurate impression that it is not as complex an undertaking as it seems to be at first glance. No two plumbing systems are exactly alike in layout and design. Each of them does function in much the same manner and is composed of basically the same parts installed in different combination. Once a basic understanding of how a household system is put together and why it operates as it does is obtained, it will be a simple matter to diagnose and repair things which would have necessitated an emergency call to a professional plumber, perhaps in the middle of the night. If you wish to go further, the information provided in this chapter and following ones will enable you to design and install a complete house plumbing system.

A residental plumbing system is composed of a number of subsystems, each of which performs a necessary task. Some of these systems stand alone and are never interconnected. Others are interrelated and interconnected. In this chapter each section will be discussed in detail.

WATER SUPPLY SYSTEMS

The most obvious point to begin with is the source of water. For those living in areas not connected to a municipal system, Chapter 5 deals exclusively with wells and their installation. In urban areas, homeowners "tap into" *water mains*, which are owned by the municipality or possibly a private water company. These mains supply water, which is purified, treated, and raised to the required pressure. This water can be obtained in a number of ways and from a number of sources such as a river, reservoir, or possibly a deep well. In any case, the resulting water is pumped through a network of pipes to areas designated to receive the supply.

Municipal Supply

If it is necessary to connect into a municipal water system, the normal procedure would be to contact the utility's offices and request that this service be performed. As a customer of the utility, each user generally is billed for this connection, which is called *tapping in*. This is normally a one-time charge and is known as a *water tap fee*. Once connection has been made, the utility will determine charges for each individual customer based on usage.

Street mains are located under the pavement of the street. The piping for this main is normally about 3 inches in diameter, although sizes may vary considerably. The system will already be under a certain amount of pressure. The theory is that this pressure will provide the users of the system with not only an adequate supply of water, but sufficient pressure no matter how much demand is placed on the overall supply.

The decision as to what methods are used to effect the actual hookup to a municipal water system will vary, depending upon the particular rules and regulations of the water department involved. The whole connection may be done by a plumbing contractor supplied by the utility. In such a

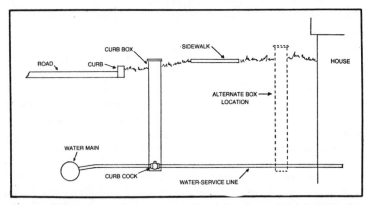

Fig. 1-1. Typical water service supply line from water main to just inside the house foundation.

situation, the plumber will run the pipe from the main to whatever point the customer desires. He may even, at the customer's request, do the final connection into the plumbing system itself inside the home. The customer will be billed accordingly if this is the case.

A more common situation would be for the water department to install a supply line running from the water main to the nearest point just outside, or sometimes just inside, the house foundation. The customer would then be responsible for the hookup to the interior plumbing system.

In smaller localities, it is not uncommon for the customer to be responsible for installing the supply line. The authorities may also require that the work be done by a licensed plumber rather than the homeowner. Regardless of who does the actual installation, the procedure itself is done in basically the same way.

The first step is to locate the main under the street. This information is obtained from the utility. A street cut is then made and a section of the pavement is removed. A trench is dug from the cut to the house. This trench should be wide enough to provide ample working space and deep enough so that it is below the frost line. In order to attach piping to the

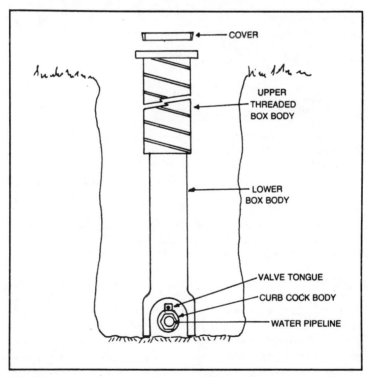

Fig. 1-2. Curb or stop box provides an additional shutoff point outside the home.

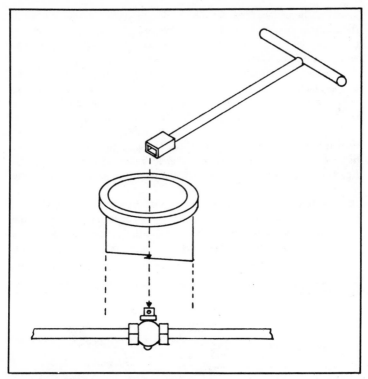

Fig. 1-3. A special key is needed to shut off the valve in the curb box.

water main, a process known as *wet tapping* must first be done. This is accomplished by bolting a saddle specially designed for this purpose to the water main itself. This saddle is sealed onto the pipe under pressure by means of a special gasket arrangement. A tapped hole is provided at the top of the saddle. The hole is positioned properly, and a valve is then threaded into the hole tightly. The valve is opened and a special tool with a drill bit attachment is attached through the open valve body. The bit is rotated to bore a hole through the wall of the main. Once this boring process is completed, the bit is removed. The result will be that even though the main is full of water under pressure, there will be no water leakage.

The piping can now be attached to the valve and run the length of the trench to the house, where it is connected to another valve. Local codes will normally dictate the exact location of this second valve. This valve will be located on municipal property, although this may not always be the case. The main purpose of the curb valve, as it is commonly known, is to provide an additional shutoff point outside the house in the event of a problem in the supply line. Even if this curb valve is not required by local authorities, the homeowner would be wise to install one anyway on his own property. A typical installation is pictured in Fig. 1-1.

Once the pipeline has been run into the residence, the curb valve is covered by a special adjustable length of large diameter pipe which is normally constructed of cast iron and has a removable cover. This is called the *stop box* or *curb box* (Fig. 1-2). The box is installed in such a manner that the cover protrudes at a point just above ground level. To facilitate operating this shutoff valve, a long rod inside the box extends from the top of the box to its attachment point on the valve top. Figure 1-3 shows a key which is used to turn the rod. These keys are normally possessed only by persons with the necessary authorizations to use them.

Although water meters will be discussed in detail later in this chapter, you should know that each locale will probably have different requirements for their installation. Some areas may not require a water meter, some may install them during the initial hookup, and some may even require that the homeowner purchase and install the meter himself. The exact location of the water meter may also be dictated by local regulations.

If no water meter is required, the water service line is run from the curb valve directly into the house and connected to a main shutoff valve just inside the wall of the foundation (Fig. 1-4). If the meter is required, there are a number of ways it can be installed. One way would be to run the water service line from the curb valve directly to the water meter. Another method is to install yet another valve of the same type as that used for the curb valve at a position in the line just on the street side of the meter. This valve is called a *meter valve* and provides for another shutoff valve if repairs are required to the meter itself (Fig. 1-5).

Private Supply

Not all water supplies provided to more than one home are municipally owned. Some are privately owned by individual water companies,

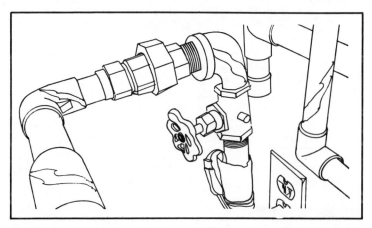

Fig. 1-4. A water supply line may be run directly into the house where it is connected to the main shutoff valve.

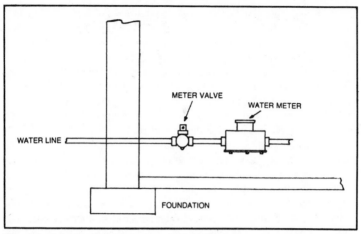

Fig. 1-5. If a water meter is required or desired, another valve may be installed on the street side of the meter.

and some are even owned and/or controlled by the homeowners as a group. A private water system is most often supplied by deep wells which are pressure fed through distribution mains by pumps, but they can also be operated by gravity feed from large storage tanks. A system such as this is commonly found in semirural subdivisions or housing developments where it would not be practical for each individual home to have its own separate water supply.

In order to tap into a privately owned water system, you must first obtain permission, usually in the form of a letter or similar document, authorizing such a connection. Again, as with municipal systems, the actual hookup may be done by employees of the company, or you may have to take care of the procedure by hiring an outside plumber, depending upon the rules of the company involved.

The actual procedures involved in tapping into a private supply are much the same as with a municipal system. Although the water mains may be a bit smaller, the water running through the lines is under pressure, and the wet tapping method described earlier is used. A stop valve is then installed at the main, along with a curb valve and stop box at the appropriate location in the line. The water service line is extended into the house and terminates in a main shutoff valve located just inside the walls of the building. Water meters will probably not be required in these types of systems; but if desired, the location will be the same as in a municipal system layout.

COLD WATER SYSTEM

The cold water system starts at the point where the water service line terminates inside the building. This may be at a main shutoff valve, a water pump, or a water meter, depending upon the individual system. The system is located entirely inside the building and serves to receive the supply of water from the street and distribute it to points such as sinks, toilets, showers, hot water heaters, etc. Figure 1-6 shows a typical water distribution system for a two-story home. The layout for a one-story or three-story home would be much the same, with appropriate modifications. As can be seen, the system consists of many pipes, fittings, valves, and other devices, which may seem a bit complicated at first. For the purposes of this discussion, I will begin at the distribution system's starting point where it attaches to the water supply line just inside the building.

Water Meters

As was mentioned previously, the installation of a water meter may not necessarily be required by local water authorities, but it does serve a

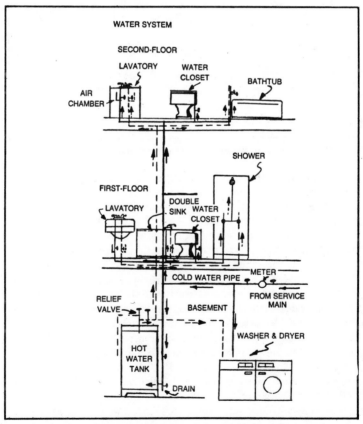

Fig. 1-6. A typical water distribution system for a two-story home.

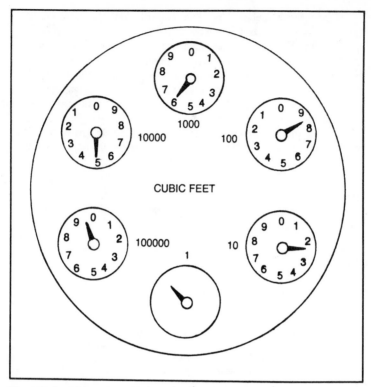

Fig. 1-7. A water meter which registers usage in cubic feet.

very useful purpose in the overall system. The sole function of a water meter is to provide a measurement of the quantity of water that is supplied to any given building. Most people are aware of the fact that there is a meter somewhere in their system, but few people really understand how to read one. Most modern water utilities now use some electronic method to "read" each individual water meter connected to their system. A bill is then sent to the customer identifying the amount of water used.

Learning how to read a water meter is very simple and can provide you with a great deal of useful information. Although water meters purchased in a wide variety of designs, shapes, and configurations, they are all read in much the same manner. Most are read in terms of cubic feet of water used. Figure 1-7 shows a typical water meter which registers using cubic feet. The fastest moving indicator is the 1-foot one. The other indicators read in multiples of 1, from 10 all the way up to 100,000. By noting the positions of all indicators and adding the figures shown, a total consumption figure can be arrived at. The only difficulty encountered with this type of meter is that most people don't think of water in terms of cubic feet. In order to determine the amount of gallons of water used, the cubic feet must be multiplied by 7.841, which is the number of gallons of water in 1 cubic foot. Water heaters currently being manufactured have been modified to enable the average homeowner to read his own meter in an easier manner. These newer meters look much like an odometer in the speedometer of an automobile. The dial reads in gallons for simplicity (Fig. 1-8). It is a simple matter for you to keep record of water usage either daily, weekly, or monthly in a small notebook for easy reference. This is also a good way to double-check the bills received from the utility itself.

A water meter can come in quite handy for making periodic checks on a home plumbing system. The way to use the meter in this manner is to first make sure all taps in the building are shut tightly and all water using devices are shut down. Then watch the meter for any indication of water usage. This type of periodic maintenance can save you some money. If there is a leak in the system which goes undetected for any length of time because it is located somewhere within the walls or floors, a good deal of damage can occur. Even if the leak is a small one, a gradual seeping of water can cause dry rot, ruin insulation, cause a freeze up, and possibly damage interior ceilings and/or walls.

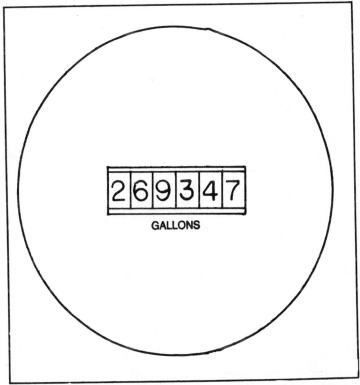

Fig. 1-8. A water meter which reads in gallons is much easier to measure.

In these inflationary times, everyone is looking for shortcuts and ways to cut down on living expenses. This is certainly understandable, and water meters can prove quite invaluable for this purpose. For instance, measuring water usage on a daily basis is an excellent way to find out which appliances and fixtures are energy efficient and which are not. Once it is known where the water goes after it enters the home and what most of this water is being used for, it is an easy matter to set "rationing" limits and try to stay within these limits. If everyone in the family is made aware of the ramifications of those 20-minute showers or running the dishwasher when it is only half full, they will undoubtedly join in the conservation effort.

Most people are aware that their water usage usually increases during the summer. This is due to such tasks as filling up the swimming pool, watering the garden or lawn, or even washing the car. By reading the water meter before and after these operations, you will be able to see how much each task costs and will probably look to either delegating a certain limit to each of them or finding a less expensive method of providing for the same end result.

A water meter reading can be an aid in a maintenance program in homes where a pump is involved. The amount of wear and tear on a water pump and its motor is directly related to a certain degree to the amount of water that travels through the system. With this knowledge, it is an easy matter to set up a periodic maintenance program based upon the number of gallons of water that are pumped. A program such as this can provide you with a degree of foresight in preventing unforeseen and possibly expensive breakdowns in a water supply system, especially in those areas where the water has a high quantity of minerals that can accumulate and cause malfunctions.

Distribution Lines

As can be seen in Fig. 1-6, once the supply line exits the water meter, it splits into two separate lines. One line leads to the hot water heater, which will be discussed later in this chapter. The other is called a *riser* and will be the cold water distribution line. This line continues full-size throughout the building, running near the points of greatest concentrated cold water usage until it reaches the next to last point, where the line is often reduced one size and carried in this pipeline to the final usage point. As can be seen in Fig. 1-6, this line is very seldom straight and will most often have numerous branches extending from the main line to the various fixtures requiring a cold water supply. Those fixtures which may require only hot water, such as some dishwashers, will be bypassed by the cold water lines. As the main line is tapped, the branches are reduced in size. In some cases, a reduced branch pipe is carried to a usage area, where it is again reduced in size and continued to another usage point. These branches are commonly referred to as *fixture branches*. Both the size and type of pipe to be used in this section of the system will be determined by demand and local regulations.

Pressure-Reducing Valves

Although low pressure is quite common, particularly in older homes, there may be some instances where high pressure is present in the system. The average residential system needs a pressure rating of approximately 40 to 50 pounds per square inch. A rating of 30 pounds or less would be considered to be insufficient, while anything over 70 pounds is too much pressure for the system and can result in problems.

It is not uncommon to find pressure in a municipal water main to be much higher than what is considered to be adequate for a home. Some may have as much as 200 pounds of pressure, and this will definitely damage a residential plumbing system. Fortunately, there is a device available

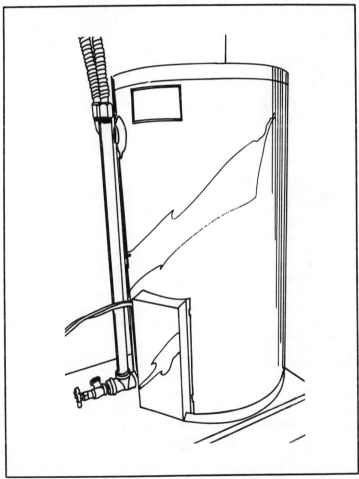

Fig. 1-9. A typical tank-type hot water heater with storage capacity.

which can alleviate some of this pressure. Called a *pressure-reducing valve*, the device is manufactured in many different pipe sizes and is preset at the factory to specific pressure ratings. Most valves are also made to be adjusted within a specified range by means of a screwdriver, so a certain amount of flexibility is available for individual needs. Once the valve is installed, it will automatically reduce the incoming water pressure. This type of valve can also be used not only to reduce pressure, but level it in situations where the water pressures in the main fluctuate to some degree. The valve will smooth out any ups and downs that occur above its set pressure level and deliver a continuously constant pressure.

A pressure-reducing valve is installed at the end of the water service line, on the output side of a main stop valve, water meter, or at any convenient location at the onset of the cold water distribution system. In situations where the main pressure is higher than the pressure rating of the pipe being used for the water service line, the valve may be installed in close proximity to the water main. This will prevent any damage from occurring to the water supply line.

HOT WATER SYSTEM

The portion of a plumbing system which delivers hot water to the required areas of the building is generally called the *domestic hot water system*. This is done to differentiate between the heating system and the hot water plumbing system. The two are completely different and will never interconnect in any manner.

During the heating process, the domestic hot water does not lose any of its original pressure. This pressure will be consistent with that of the cold water system. As seen in Fig. 1-6, the hot water lines are generally of the same size as those used for cold water. The lines run parallel and adjacent in the system. Similarly, branches extend from the main line out to those fixtures and/or appliances requiring hot water such as sinks, showers, bathtubs, washing machines, and dishwashers. Those fixtures not requiring hot water, such as toilets, are bypassed.

At its point of exit from the hot water heater, the temperature of the hot water may be quite hot, sometimes in excess of 180 degrees. Because of this, certain materials may not be used for the hot water piping. Any type of metal pipe can be employed to deliver hot water, and copper pipe is the one normally used today. Some types of plastic pipe, particularly the CPVC type, may be used. The CPVC pipe has the most heat-resistant properties, but it may be prohibited by local codes for use either completely or in the area closest to the hot water heater. In the latter situation a few lengths of copper pipe could be used for the line where it exits the hot water heater, and the CPVC pipe could be used for the rest of the run.

When installing the cold and hot water lines, it is important to allow for some distance between the two. If the two pipes are made of different metals, any contact may cause them to corrode very quickly once the system is in operation and the pipes get wet.

Types of Water Heaters

A hot water heater may be located almost anywhere in a building, but it is most often installed in the basement or on the first floor. The water can be heated by electricity, heat exchange, gas, or even the sun. Water heaters come in two basic designs: the *storage* or *tank* type, and the *instantaneous* or *tankless* type.

The tank type is the most commonly used heater and consists of a special water tank (Fig. 1-9). These tanks can be purchased in numerous designs and configurations with capacities ranging anywhere from 10 to 82 gallons. A hot water tank is a freestanding unit which is most often located directly adjacent to the point of greatest hot water usage to provide maximum operating efficiency. Inside the tank itself is a heating device, either a gas flame or electrical heating coils, depending upon the method to be used to heat the water. An electric hot water heater has a bit more flexibility with regard to its location. It can be installed almost anywhere in a home. However, gas heaters need fresh air inlets and a flue to provide for the proper discharge of exhaust fumes associated with gas. Most building codes have strict regulations regarding the installation of gas heaters. These codes will specify proper location, separating distances from any combustible materials, and possibly the materials to be used for construction of surrounding walls, ceilings, and floors.

The instantaneous type of water heater is used in conjunction with the home's heating system (Fig. 1-10). In this type of installation, the furnace or boiler is equipped with copper coils that are supplied with water. As the cold water travels through this coil, it is heated from the hot water system of the home. There is no storage of this water, and as long as great amounts

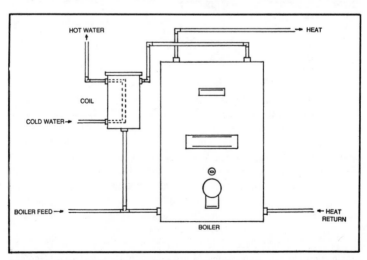

Fig. 1-10. An instantaneous or tankless hot water heater works in conjunction with the home heating system.

21

of water are not needed in a short period of time, this system will work quite adequately. Instantaneous heaters are rated in terms of gallons of hot water per minute, computed for a certain rise of temperature—generally 100°F. For example, if a unit is rated to produce 4 gallons per minute (gpm) and raise water temperature by 100°F, and the entering water is 50°F, the unit will deliver water at a rate of 4 gpm at 150°F. If the incoming water is of a cooler temperature, the water delivered will be correspondingly lower. Similarly, if the water is delivered at a rate higher than what is specified, the temperature will be correspondingly lower.

There is another type of water heater available for use in the home. Called the instant tankless heater, this unit is quite small and is designed to serve only one utilization point. The heater operates automatically when the water is turned on and can be powered either by gas or electricity. It is connected by means of a cold water supply pipe at its inlet and a short length of pipe or tubing running to the fixture at its outlet. A number of these units can be installed, with each serving an individual fixture. This eliminates the need for great lengths of pipe, water storage capacity, and tank space.

Selecting a Water Heater

The proper selection of a water heater is a simple task. The first thing to do is add up the number of fixtures in the building and, using information supplied by the manufacturers, determine their water consumption rates to arrive at an approximate requirement to provide hot water for them. If no such information is available, the following figures may be used as guidelines, keeping in mind that they may vary somewhat depending upon the equipment and individual preference with regard to temperature:

Shower:	4	to	10	gallons/use
Bathtub:	10	to	15	gallons/use
Automatic Washer:	12	to	25	gallons/use
Dishwasher:	6	to	15	gallons/use
Manual Dishwashing:	2	to	5	gallons/use

If you are planning to replace an existing hot water heater, it is a good idea to find out the capacities and specifications of the old one. This information is normally provided on a small plate attached somewhere near the base of the unit and gives capacity in gallons, recovery per hour, and Btu rating. If the old unit provided a sufficient amount of hot water during its operation, a new heater with the same capacity and recovery value should do the same. However, if this is not the case, or if an addition is planned which may require additional draw on the new heater, it may be desirable to purchase one with either a larger storage capacity or a higher gallons per hour recovery than the old heater.

Although most specifications regarding recovery rates are pretty much standardized, some manufacturers state this rate in a lesser temperature. For example, they may state the recovery rate in terms of 60 degrees rather than the usual 100 degrees. A tank with a 60-degree recovery rating may at first glance appear to be the better buy, but it will not perform as well. In any case, I suggest that all claims made by the manufacturer be closely scrutinized before a purchase is made. There are many heaters available on today's market at widely varying prices, and although those sold by the better known manufacturers may be a bit more expensive, their reputation speaks for itself. Make sure that the water heater comes with a warranty. Obviously, the longer the warranty, the better, and it can range anywhere from 7 ½ years to 15 years or better. A number of linings are also available, each with a different price tag. The most expensive types of water heaters have *galvanized tanks*. These will be adequate as long as the water supply is not high in chemical composition, which can corrode the interior of the tank. The other two choices are copper or glass. The copper interior is quite adequate and cheaper than the glass, but the glass is both cleaner and a bit more durable in the long run.

Water Heater Parts

All hot water heaters consist of basically the same parts. Figure 1-11 provides a cutaway drawing of a common gas water heater which is composed of:

■ An insulated water tank.

■ A source of energy, in this case, a gas burner.

■ Both cold and hot water connections at the top with a dip tube carrying water to the bottom of the tank. If the cold water inlet were at the bottom, this tube would not be necessary.

■ A thermostat control and, in the case of a gas heater, a *pilot light*.

■ Appropriate connections for either gas or electricity.

Although the cold water connection may be either at the top or the bottom of the tank itself, the hot water outlet must always be at the top since heat rises. In a gas water heater, there is a pilot light which burns continuously, with its flame playing on a device which is known as a *thermocouple*. If the pilot light happens to go out for any reason, this device will immediately signal the gas valve, which will automatically shut off the gas supply to the burner since there is no flame to burn it. Generally, pilot lights do not normally burn out if maintained properly. If lint is present in the orifice area, it may cause the pilot to go out. This can be prevented by periodically cleaning the area with a small brush or even an old toothbrush. A defective thermocouple might cause the pilot to malfunction. If this is the case, the device should be replaced. This is not a difficult installation procedure, and most manufacturers provide complete instructions which should be strictly adhered to.

Inside the flue running from the burner to the vent at the top of the tank are *baffles* which serve an efficient purpose in the system. The vents are designed to discharge the gases generated to some point outside the building for obvious safety reasons. The baffles capture some of the escaping heat and deliver it back into the tank for additional energy which is used to heat the water.

Another safety feature found on all water heaters is a relief valve located at the top of the tank. These can be purchased in two different designs: the *pressure* type and the *temperature-cum-pressure* type. Regardless of the type installed with a heater, they both perform the same useful purpose. If the temperature of the water at the top of the interior of the tank (where the hottest water is) rises to a level that might be dangerous, the valve will automatically open up and release cold water into the tank. The temperature setting of the valve would normally be 200 degrees or some lesser figure, depending upon the specifications of the type of heater being used. If for some reason the water heater does not have one of these valves, it is a good idea to either have one installed or install one yourself. High temperatures and/or pressures can build up to the point where an explosion might occur.

A third device that most water heaters come equipped with is a *drain* located near the bottom of the tank. The purpose of this drain is to discharge any overflow from the heater to either a house drain or a sump pump. Overflow sometimes occurs due to the expansion of the cold water as it is heated. The line running from this drain should not simply be run to a point outside the building, as freezing may occur and the purposes for which the drain is installed would be defeated.

Electric water heaters are equipped with a thermostat to control the temperature of the water. In order to set this thermostat, it may be necessary to turn the unit off and, sometimes, to remove a side panel to gain access to the dial. Some older units merely have thermostat settings for warm, normal, or hot. With these heaters, there will not be as specific a degree of flexibility. In years past, normal temperature settings ranged from 140°F to 160°F. However, as costs rise and people continue to look for ways to cut back on overall fuel consumption, it is not unusual to find some people setting their temperatures back to 120 degrees with satisfactory results. In buildings with long runs of hot water lines, there may be a loss in temperature during the time the water exits the tank and reaches its final destination. It may be necessary to adjust the thermostat upward a bit, adding 1 degree per linear foot of pipeline between the heater and the fixtures.

It is not uncommon for an electric heater to have two thermostats, one at the bottom and one at the top. The heating elements themselves can be inside the tank, or they can be wrapped around on the outer surface of the tank. The number of elements on the tank will correspond to the number of thermostats. The principle behind this double installation is quite simple. As the hot water near the top of the tank exits for distribution to the building's fixtures, the temperature inside the tank begins to drop and the upper thermostat kicks in its element. When the hot water reaches the desired setting, the element shuts off. At the same time the lower element starts, due to the fact that the water in the lower portion of the tank still

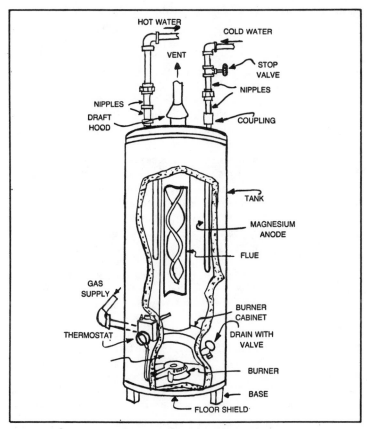

Fig. 1-11. Cutaway drawing of a gas hot water heater.

needs to be heated. In order to prevent the lower element from kicking in while the upper one is on, both are interlocked. Only one will be operating at any given time. As with the upper element, once the water in the lower portion of the tank reaches the desired temperature, the lower element automatically shuts off. Generally, most water heaters manufactured by the more reputable and reliable companies are equipped with some sort of safety interlock device which automatically shuts down both elements in the event water reaches a dangerously hot temperature. If this does occur, it will be necessary to reset the device. This is done by first shutting off all power. The access panel is then removed from the tank to reveal a red button which is simply pushed to reset the interlock.

Installing a Water Heater

Using the gas water heater in Fig. 1-11 as an example, the steps involved in actual installation will be discussed. Before attempting to install a heater, it is wise to check with the local authorities to make sure the code permits homeowners to install their own heaters. Some may require that only licensed plumbers do the installation.

The actual installation consists of making the proper connections to the cold water and hot water distribution lines and to the gas line or electric outlet. For the gas water heater shown in Fig. 1-11, the following pipes and fittings will be required: one coupling, three nipples, two unions, one gate valve, one tee, and a 48-inch length of ¾-inch galvanized pipe. Also, if the gas supply line is very old or rusted, some additional pipe lengths and fittings may be required. The flue line will consist of a galvanized furnace pipe and an elbow of 3 inches for a 30-gallon recovery heater, or of 4 inches for a 40 to 50-gallon recovery heater.

Once the proper tools and materials have been assembled, the first step is to remove the old water heater, where applicable. This is done by closing the gate valve on the cold water line at the top and attaching a garden hose to the drain valve located toward the bottom. Let the water run out until the tank is empty. A way to speed up this process would be to remove the union on the hot water line. This will allow air to enter, causing the water to run out much quicker. Once the tank is empty, the hot and cold water lines can be disconnected and the gas line shut off at the valve. The union on this line is also disconnected. Now the flue pipes are removed, and the water heater is completely disconnected.

The new water heater is placed in its proper position and connected to the hot and cold water lines (Fig. 1-11). All the fittings in this installation should be sealed with an appropriate pipe joint compound, as all the joints should be completely watertight. It is considered a good practice to have a *sediment trap* in the gas line. This trap will prevent rust and scale from traveling to the thermostat and depositing there. If there was no such trap in the system with the old water heater, one can be constructed quite easily using a ½-inch cap, a ½ by 6-inch nipple and a ½-inch tee. Once the trap is in place, the gas line and flue line are connected. No specific instructions are provided here, as most heaters come with their own installation procedures. These may vary to a degree. In any case, follow the instructions provided, and the heater will perform satisfactorily.

It is not difficult to light the burner on a gas heater. First, turn the pilot knob and the thermostat to their off positions. Turn the pilot knob to the point marked pilot. Light a match and hold it to the pilot, keeping the red button depressed for about 30 seconds. Release the red button. If the pilot light remains on, turn the pilot knob to the on position. If it goes out, repeat the process of lighting it again until it stays lit. Once the pilot remains lit, the thermostat is set to the desired temperature, causing the burner to ignite. The heater is now functioning.

If greater heating capacity is desired, an alternative to purchasing a larger tank and discarding a perfectly operational, but not adequate, tank would be to connect two tanks in series (Fig. 1-12). In such a system, the tanks can be of different sizes, as long as they are interchanged yearly.

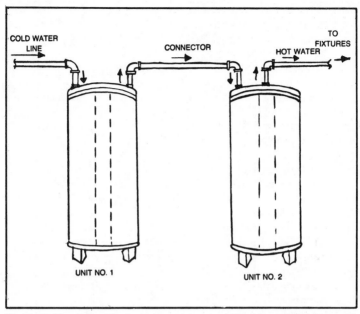

Fig. 1-12. Two hot water heaters may be connected in series to increase hot water supply.

Since only one of the tanks will be receiving water from the cold water system, it will be working harder to keep the water at the desired temperature. If the two tanks are of the same capacity and recover rate, it may be possible to connect them in parallel.

Most readers will undoubtedly have had the experience of having to wait as long as a few minutes after turning on the shower for the water to get hot. This may or may not be considered an inconvenience, but there is a solution available which, if used during the original construction of a house, will provide for almost instantaneous hot water at even the farthest point on a run. This type of installation is called the *upfeed and gravity system* (Fig. 1-13). A riser pipe delivers the hot water to the fixtures in the normal manner, but a gravity return line is also installed which returns the cooler water to the tank. This provides for an almost constant circulation of water both to and from the water heater. Not only will this system assure instant hot water, but a degree of waste is prevented, since the fixture will not be running those extra minutes while waiting for the water to get hot.

If attempting to alleviate this problem on an already existing structure, the pipes can be wrapped with an insulating tape that uses aluminum foil. This will prevent some degree of heat loss. This method may not be possible if the pipes are not accessible, and the only alternative may be to adjust the thermostat on the water heater up somewhat. This may provide some, but not complete, relief.

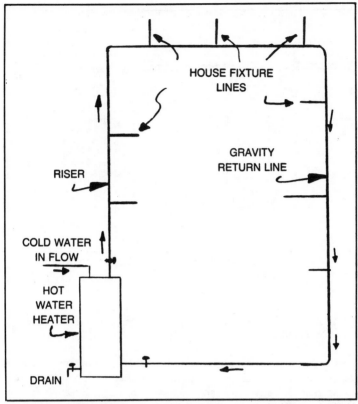

Fig. 1-13. An upfeed and gravity type system provides an almost continuous circulation of water.

As was stated previously, almost all new heaters are covered by some form of warranty. Some of the problems which may cause malfunctioning are not very difficult to repair, and for those do-it-yourselfers who wish to, Table 1-1 is a chart of the most common causes of malfunctions and how to repair them. This is basically a troubleshooting guide for the more minor problems which may occur. If following these suggestions does not solve the problem, it is recommended that a professional be contacted. In any case, you should not attempt to repair the burner on a gas heater. This takes both special repair equipment and great care in order to prevent either an explosion or a fire hazard. A professional should perform any work to be done on this portion of a gas hot water heater.

If the water from the fixtures ever becomes discolored, this is usually caused by sediment deposits building up in the tank. This can be prevented by periodically shutting off the water supply and opening the drain valve at the bottom of the tank. Check the flow. If it is clear, close the valve and

Table 1-1. Troubleshooting Chart for Hot Water Heaters.

MALFUNCTION	PROBABLE CAUSE	SUGGESTED REPAIRS
Abnormally long recovery time	Calcium deposits in the heating element	Take out the heating element and remove the calcium with vinegar and a brush
Water collects under the tank	1. Condensation may be forming on the heater surface 2. Tank may have developed a hole 3. Leakage at plumbing connections 4. Element may be leaking	1. It is nothing to worry about. Collect drip in a pan. 2. Start looking for a new heater—stop the leak temporarily with a boiler plug. 3. Repair as necessary 4. Replace the element.
No hot water reaching the fixtures	1. The circuit breaker may have tripped or a fuse may have blown out 2. Thermostat may be defective 3. Calcium build-up on heating element (electric heater) 4. Upper heating element burned out 5. Thermocouple malfunctioning (gas heater)	1. Reset the circuit breaker or replace the fuse. If it goes out again, take the help of a professional. 2. Replace the thermostat 3. Heating element should be removed and cleaned. 4. Replace upper heating element. 5. Check pilot light, clean the orifice. If pilot does not stay lit, thermocouple is defective; replace it.
Water at fixtures not hot enough	1. The temperature setting at the thermostat may be low 2. Capacity of water heater tank insufficient. 3. Lower heating element may have burned out (electric heater) 4. Dip tube may have developed a hole.	1. Set within the normal range of 140° to 150° F. 2. Get a unit of larger capacity or add another one in series. 3. Replace defective element 4. Remove dip tube and install a new one.
Steam coming out with hot water	1. Thermostat not functioning due to stuck contact points (electric heater) 2. Burner does not shut off 3. Thermostat set at a very high temperature.	1. Replace shorted or burned out terminals 2. Set thermostat lower; if it still fails to shut off, replace thermostat. 3. Set thermostat at normal setting of 140° to 150° F.

restore the water supply. In most cases, the water will be a bit dirty in the beginning, but it will clear up in a few minutes. Once clean water is observed, shut off the drain valve and restart the heater. This procedure should be done at regular intervals, perhaps once every other month until it is determined what amount of time is required to suit individual conditions. The amount of deposit will vary with the amount of sediment in the water and may even fluctuate from season to season.

THE DRAIN-WASTE-VENT SYSTEM

The basic purpose of this portion of a plumbing system is to collect waste from the fixtures and transfer it safely outside of the residence, either to a septic system or to the street sewers. It is completely separate from the rest of the plumbing in the home, although its lines may run in close proximity with either the hot water or cold water lines. The regulations and guidelines laid down in plumbing codes are quite specific and strict with regard to this portion of the system. This is due to safety for the persons in the residence, since improper drainage and venting may result in contaminated water supplies. A drain-waste-vent (DWV) system is composed of three subsystems: one for drainage, one for waste, and one for venting. Although each subsystem is interconnected, they all have a different function in the overall system.

Drainage

A typical DWV system is shown in Fig. 1-14. The drainage system consists of a series of pipes, fittings, and traps which carry used water and human waste away by gravity to the appropriate sewage system. A pipe that carries nonhuman liquid waste is called a *waste pipe*. If a pipe carries human wastes made up of both solids and liquids, it is a *soil pipe*. Pipes that are installed horizontally are often called *laterals*, while the vertical pipes are *stacks*. A lateral or a stack that carries human waste and nonhuman waste together is also a *soil pipe*. Often two or more fixtures will drain into one line, which in turn may connect to yet another line and eventually go into a larger, central line. These subsystems are called *branches*. A principal branch of a stack may be called a *primary branch*, while others may be *horizontal branches* or *secondary branches*, depending upon how they are arranged. A stack that receives nonhuman liquid waste (sometimes called gray water) and/or human waste as well is a *soil stack*. If a stack receives only nonhuman liquid waste, it is a *waste stack*. A principal pipe that receives waste from several smaller ones is often called a *main*.

This may all seem a bit confusing, but it's really not. For the purposes of this discussion, let's look at a house with one bath and one kitchen sink. When used water is dumped into a sink, it travels through a basket strainer and into a short tailpiece or tailpipe. The same is true of a lavatory basin, except there is no strainer. This short length of pipe carries the waste water directly into the trap and on into the drainage line. The section of

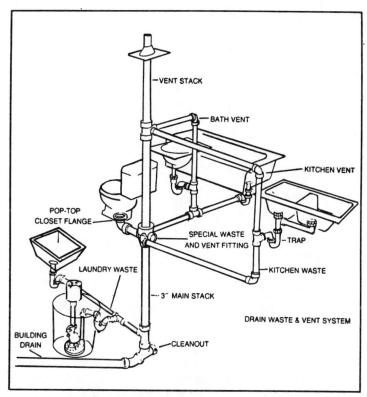

Fig. 1-14. A compact drain-waste-vent system.

drainage line between the trap and the nearest other drainpipe is called a *fixture drain*. The fixture drain may also be designated by its particular function; i.e., a sink drain or a lavatory drain. In any case, this pipe then connects to a larger one, which in a simple system would be a soil stack. If these fixtures are located some distance from the soil stack, they are connected to a branch line or to a waste stack. If the connection is made through a branch line, that branch will eventually connect with either a soil stack or a waste stack. Either or both of these stacks then drop down to a horizontal line at the lowest point of the drainage system a large-diameter pipe known as a *house drain* or a *building drain*. This pipe is the ultimate collector of all drainage from the system. It exits from the foundation at some convenient point and terminates approximately 5 feet from the building, at which point the sewage system takes over.

The tub or shower is connected in much the same manner. Used water flows through a drain, into a short length of pipe and to a trap, which is located as close as is conveniently possible to the fixture. The water then flows through the fixture drain and into the soil stack or other larger diameter pipe.

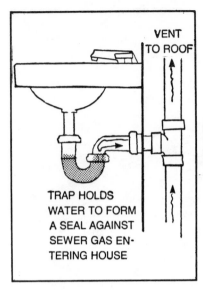

Fig. 1-15. The purpose of a trap is to seal the drainage system off from the interior atmosphere of the home.

A toilet installation is somewhat different but works on the same principle. Since it is already equipped with a built-in trap, a second one will not be necessary. A floor-mounted toilet is affixed to a special fitting called a *closet* or *floor flange*, which in turn is connected to a closet bend and a short length of soil pipe. The soil pipe runs directly into the soil stack. A wall-mounted toilet drains through the rear and must be mounted directly to a special fitting that is installed in the soil stack. In both cases, the sewage drops directly down the soil stack and into the house drain.

The basic principles outlined here will remain the same in a system made up of more fixtures, with the addition of more branches being the only real difference. In situations where there are a number of fixtures in close proximity to each other, a common variation would be to connect all of them to a single trap and fixture drain. This is known as a continuous waste pipe, and is done more for convenience than anything else. The system would work equally well with each fixture having its own individual trap and drain. A continuous waste pipe, if installed properly, will save on both materials and labor.

Traps

The primary function of a trap is to prevent sewer gas entry into the house, as well as to aid in proper drainage from the fixtures. A trap is absolutely essential on every single fixture in a system (Fig. 1-15). Traps are generally located as close as is conveniently possible to the fixture. In most cases, this point is at the end of a very short length of pipe that carries the drain water from the fixture. There are many different configurations of traps that are used to fulfill various specific installation requirements such as the running trap, drum trap, P-trap, S-trap, and so on. But each serves the same basic purpose. The trap is always filled with water, and the water serves to seal the drainage system off from the building's interior atmosphere. Sewer systems are not only loaded with all manner of lethal bacteria; they are also charged with a variety of gases under low pressure. These gases can be extremely poisonous, highly explosive, or very corrosive, not to mention quite offensive to the senses. Since there is really no way to overcome these characteristics, this gas must be prevented from entering the building. The small plug of water in each trap does this with great effectiveness and in a quite ingenious manner. Because water always seeks its own level, the water quantity in the trap remains constant. It cannot drain away under normal operating conditions and does not stagnate because as the drain is used, the water is continuously replaced. No valves are present in this portion of the system, so human error is never a problem. The only thing that may be required is an occasional cleaning if the trap becomes clogged.

Cleanouts

Figure 1-16 shows a *cleanout*, which is a specialized fitting that provides for a method of cleaning out the drainage lines if they become clogged. Some of the better-grade traps sold on today's market come equipped with a cleanout plug. With these devices, the cleaning process is more simplified. All that is required is to unscrew the plug and drain the trap. As can be seen, these plugs serve a very important function in the drainage system and should be installed on all the main portions of the piping. They are not generally necessary on the smaller sections because

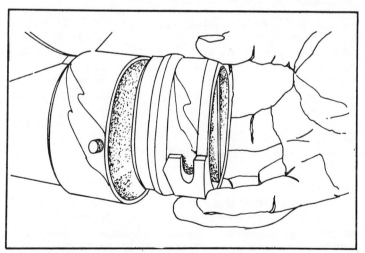

Fig. 1-16. A cleanout should be installed on every drain to provide a means of access for clearing blockages.

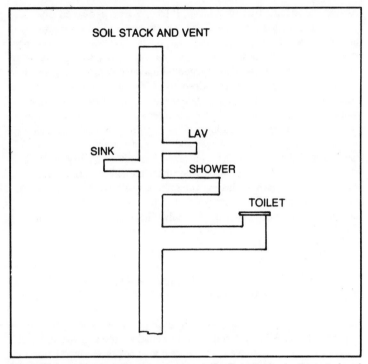

Fig. 1-17. In a direct venting arrangement, the fixtures are both vented and drained through the same pipeline.

it is a relatively simple procedure to enter these lines with an auger through the drains of the fixtures themselves. Cleanouts are usually installed at the point where a soil or waste stack connects to the building drain. In systems requiring large numbers of lines, such as a three-story structure, cleanout plugs might also be installed at the end of each long branch line where it connects to a stack. In a less complex system, one cleanout installed at the point where the stack connects to the house drain will probably be sufficient. As a general rule, a cleanout should be installed wherever a stack changes direction by 45 degrees or more, or where there is more than 45 feet of pipeline from the last cleanout.

Venting System

The venting portion of the drainage system is vital to the proper operation of the entire installation. If no venting is provided for, the system will not work properly, if at all, and contamination is a danger. The venting system is not really separate from the drainage system. The two are actually interconnected and frequently use the same pipes. The principles behind the venting system are quite simple.

The presence of an unrestricted air flow through the vent pipes allows waste liquids to drain away quickly and efficiently through the drainage pipes. If the drainage pipe system were completely closed, the flow would be slow and sluggish, and solids would not be carried along easily. This can be likened to trying to pour liquid from a can with only one opening. The flow would be quite slow. If another opening is made in the can, the flow is speeded up considerably. This free air also serves to prevent water from being siphoned from the traps in the system. A strong flow of water through a pipe creates a vacuum behind the water. The suction created

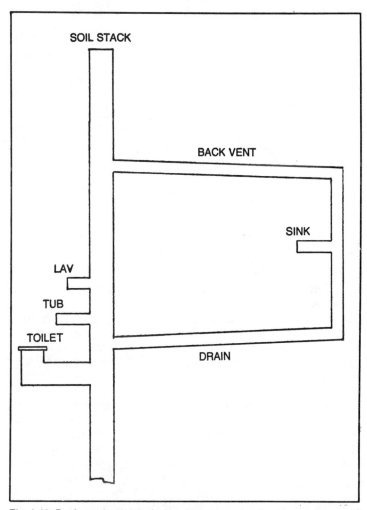

Fig. 1-18. Back venting is used when fixtures are too far apart to be vented and drained through the same stack.

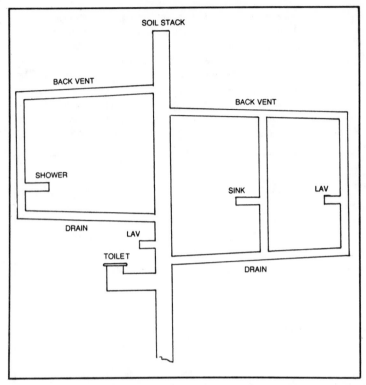

Fig. 1-19. Two or more fixtures may be back vented together.

could easily empty a trap and pull the trap water down the drainpipe. As was stated previously, the trap prevents gases from entering the drainage system and escaping into the house. Sewer gases are always present in the sewage lines, and they can easily travel back along the pipes. If the system is vented properly, any gases that do accumulate are exhausted outside the building where they can do no harm. This constant flow of fresh air through the drainage system also prevents the growth of slime and algae. Not only can sewer gases be quite poisonous, they can also cause damage to the piping itself through corrosion.

The installation of the venting system is quite important. In a simple system where the fixtures themselves are close enough together, they can be both drained and vented in the same pipe. This method is shown in Fig. 1-17 is called *direct venting*. It can also be said to be a form of wet-venting, which occurs when any portion of a waste pipe serves as a vent for other fixtures attached to the same pipe. This situation occurs with all fixtures, even if only for a very short distance.

If the system is more extensive and the fixtures are located quite distant from each other and/or the soil stack, the fixtures cannot be properly drained and vented simultaneously through the soil stack. In a situation such as this, the venting system will require additional piping to allow for the system to function properly. This can be done in a number of ways.

The most common method is called *back venting* and is shown in Fig. 1-18. As can be seen, a pipe is connected close to the fixture drain. This piping runs up and back to the nearest soil or waste stack. Another method would be to back vent two or more fixings together (Fig. 1-19). A fixture can also be vented individually with a separate pipe extending directly through the roof to the outside of the building (Fig. 1-20). Two or more fixtures can also be piped to a common vent pipe, which is then connected either to a stack or run through the roof.

A vent pipe which rises directly through the roof is called a *vent stack*. It might also be necessary to include a relief vent. This is a length of pipe which runs from one waste or soil stack to another to provide for air circulation between the two. A section of vent piping that picks up two or more fixtures and eventually terminates at a stack is called a *branch bent*.

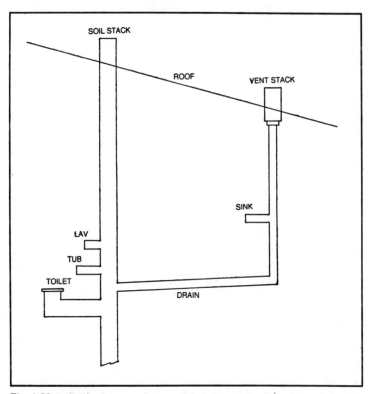

Fig. 1-20. Individual venting for one fixture located a distance away from a group of fixtures.

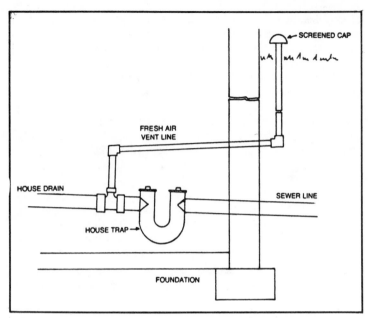

Fig. 1-21. A fresh-air vent installed in the basement will provide further circulation for the system.

The most common mistakes which occur in this portion of the drainage system are attributed to the lack of enough venting. Each fixture trap must be adequately vented in some way, with the vent being of sufficient diameter and in the proper place to allow for efficient operation. It is not uncommon for a home to be equipped with a double-venting system, in which at least two vent pipes run through the roof. This system can consist of a soil stack and a waste stack, or a soil stack and a vent stack. Similarly, more than two stacks can be used.

Another type of venting which is being used more and more in drainage systems today is called the *fresh-air vent*. This vent is normally installed at some point in the house drain before it exits the building. If a house trap is installed, the fresh-air vent is installed just on the house side of the trap. Unlike other vents already discussed, this one does not extend upward to the roof. Instead, the pipe rises a bit and exits through a wall, where it terminates in a screened cap just above ground level. A typical fresh-air vent is shown in Fig. 1-21. This type of vent serves a very useful purpose in the overall system, and although not required by most plumbing codes, it is generally installed. The admittance of fresh air at such a low point in the drainage system allows excellent circulation throughout the lines. Also, should blockage occur in the sewage line, this vent acts as a safety valve. The backed up waste will flow out of the vent rather than continuing upward to the fixtures in the building itself. This type of blockage will become apparent immediately if a fresh-air vent is provided for in the system, thus preventing any damage from occurring. Blockages that might occur on the house side of this vent will be treated in the normal manner. Although the installation of a fresh-air vent may require a little extra work and some additional parts, the advantages of having one far outweigh both the time and money involved.

SUMMARY

The information covered in this chapter should give you a basic understanding of a whole house plumbing system and how it works. Each individual system has its own special characteristics and requirements. The main purposes of each of the separate subsystems will remain the same, regardless of its particular design configurations. Some systems will require extensive piping and a wide variety of components; others will be very simple, with a limited number of additional attachments. Generally speaking, local plumbing codes will dictate specifications, materials and their sizes, and the proper placement of the various traps, vents, and cleanout plugs. Although these authorities serve the primary function of regulating the plumbing industry, they can sometimes be a source of invaluable aid to the do-it-yourselfer as well. If a good working relationship is established, they may be willing to share a bit of their knowledge and experience acquired over the years.

Understanding how a plumbing system operates can save you many dollars and prevent those times of frustration and worry when something goes wrong and outside help is needed. Identifying and dealing with a plumbing emergency is not as difficult as many think it is, and the feeling of accomplishment at being able to repair a leaky faucet or plumb a whole bathroom is quite satisfying.

Although not everyone is meant to be a home plumber, it is well within the scope of the average person to be able to learn the basic fundamentals of plumbing. It is up to you to decide just how far you want to go in this field. A whole house plumbing system is most decidedly a real chore requiring a good deal of time, materials, and study. It can and is being done all the time, however, by people all over the country. All it takes is a desire to learn and some patience, and the result will be a fully operational plumbing system that has the stamp of a professional on it.

2 Plumbing Components

Since its inception, plumbing has been greatly improved. Existing systems, installed many years ago, are continuously modernized. You will find that there is a great degree of standardization in component parts. You may experience some confusion at the outset of your project upon learning of the dozens of types of pipes and other components available to choose from. In remodeling or making repairs to an existing system, it will probably be necessary to limit your choice to the original system's basic type of pipes; when starting from scratch, there are numerous considerations which can affect your decision. Plumbing codes, which differ from locality to locality, may dictate certain materials, and cost will play a large part. The amount of specialization required in the design of each individual systems may present yet another factor to be considered, because some types of piping offer less variety than others. The water and soil characteristics in your area may tell you something about the effects their properties might have on a particular type of pipe. Since this is something over which you have little or no control, this may be the final deciding factor in pipe selection.

As can be seen, there is a great deal of planning involved in the initial stages of any plumbing project. Fortunately, there is also enough information available on the subject to give the neophyte emple materials to study. A trip to your local plumbing supply house will also be helpful. Most retail merchants in this field will be able to provide some assistance in the proper material selection for the project or projects being planned.

In this chapter is a complete description of each of the types of pipe commonly used in plumbing today and a discussion of each material's advantages, disadvantages, and properties. With this information in hand, it should be a simple task to make the necessary decisions regarding material selection best suited for the projects being attempted.

COPPER PIPE

Rigid copper pipe has been around for many years. It is approved in almost every local plumbing code for home installation. Much can be said for copper pipe. It is very durable, quite lightweight, easy to work with, and provides a great deal of variety regarding thickness and fitting selection. Copper pipe normally has a very smooth inner surface which offers little resistance to the flow of liquids. Due to this factor, a pipe of the same size made of another material would have a smaller carrying capacity. Thus, a smaller size copper pipe would deliver the same quantity of flow as a larger size pipe of cast iron, clay, or concrete, whose surfaces are not as smooth.

Rigid copper pipe normally can be purchased in four different thicknesses. For underground work which dictates the use of quite heavy materials, *type k* would be chosen for its durability and thickness. *Type L* is a medium-weight copper pipe and is normally used mainly for interior pipelines. The remaining two types, *type M* and *type DWV*, are very lightweight, and some plumbing codes do not allow for their use. Type M is used in much the same manner as type L, and type DWV is primarily used for vent, waste, and drainpipes. The two thinner pipes are made from a thin copper sheet, which accounts for their lightness. Although plumbing codes and regulations will be discussed in another chapter, I would like to mention here that the rules imposed by each locality, although they do differ, have determined that certain diameters and strengths of pipe work best under set conditions. DWV pipe can sometimes be used for the applications listed here, but only if its diameter exceeds a specified measurement which provides the best possible ease of flow through the system.

The copper pipe just discussed is rigid, but it can also be purchased as flexible copper tubing. This tubing is normally sold in coils of varying lengths, from 15 to 60 feet. Flexible copper pipe quite obviously can provide a great deal of convenience and ease in assembly. It bends quite easily and can be wound off its coil and around corners or obstructions. Passing it through walls or ceilings is a simple task. It is similar to working with electric cable and eliminates the need to drill holes through ceilings and/or plastered walls. Because of the pipe's ability to be wound through walls, etc., there will be less, if any, joints, which gives the added advantage of simplifying the work. The danger of leaks at joints can be completely eliminated. Some plumbers simply unroll a whole coil and use one piece of tubing to run the entire system from the bottom up. If this is done, there will be only two joints, one where the tubing connects to the water supply, and one at each fixture or faucet. The only precaution in working with this type of tubing is to be careful not to bend it anywhere to a degree where it will cause a kink. This will restrict the opening of the tube

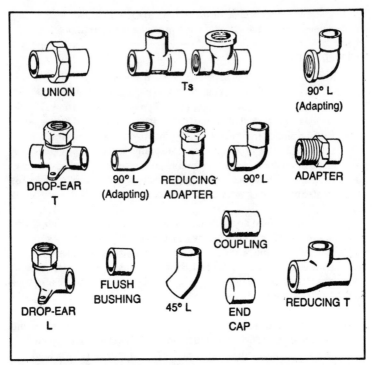

Fig. 2-1. A variety of copper fittings.

and can reduce the overall flow capacity. When making gradual bends and/or curves for a line, a *tube bender* is the best method. It can be done by simply bending it over your knees carefully.

There is quite a variation in fittings for copper pipe. Figure 2-1 shows an assortment of the *solder*, or *sweat-type*, fittings which are most commonly employed in general plumbing work. Adapters such as those shown are used to connect larger pipes to smaller ones and are measured in both inside and outside diameter for ease of selection. When a turn is needed anywhere in a line, an *elbow* is used. Elbows come in degrees of turn. For example, if the turn is beneath a sink and a 90-degree angle is required, an elbow with a 90-degree turn would be chosen. Elbows are available in 90, 45, 22½, 11¼, and 5⅝-degree turns, so there should be no difficulty in finding the angle required in a system. When connecting different sections of a system together, such as a main pipe to a branch pipe at a right angle, a tee is used. Similarly, a Y is used to connect this same branch to the main line if the angle is other than a right angle. When four pipes must be connected at right angles to each other, a *cross* is used. These are the most commonly required fittings, but if individual circumstances dictate any other types of connections, check with your local plumbing supplier for specialized fittings which are also available.

Soldered fittings can be used with both rigid or flexible copper pipe but another type of fitting, the *flare* or *compression* type, has been developed solely for use with flexible copper tubing. Flare fittings will normally be used primarily for air piping and fuel-oil lines; thus, their use for the applications discussed is minimal. It is not recommended that they be used in general plumbing work. Some samples of flare fittings are shown in Fig. 2-2.

SOLDERING COPPER JOINTS

Soldering is a simple task, and a brief explanation is given here to acquaint you with the proper technique. After the pipe has been cut, smooth both the inside and outside edges with a reamer or file. Remove any dirt or pieces of metal with a clean cloth or a piece of steel wool, but be careful not to cause too much abrasion. This can possibly result in increasing the gap between the pipe and the fitting beyond its limit, causing the joint to leak. Apply a thin coat of soldering flux to both the *inside* of the fitting and the *outside* of the pipe. Insert the pipe in the fitting and twist it around a few times to allow the flux to spread evenly. Remove all excess flux from the outside of the fitting. Adjust the propane torch to the desired flame following the manufacturer's directions. Heat the connection and the

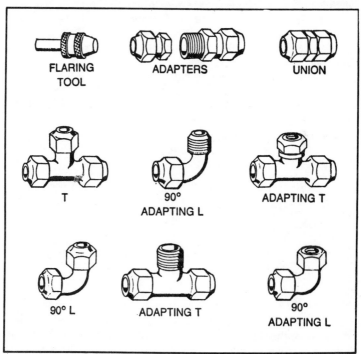

Fig. 2-2. Some of the more common flare fittings and a flaring tool.

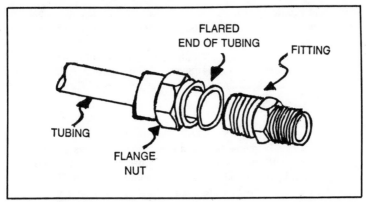

FLARED
END OF TUBING

FITTING

TUBING

FLANGE
NUT

Fig. 2-3. After the tubing has been flared, it is pressed onto the fitting and the nut is tightened.

pipe *evenly* on all sides. When the fitting is sufficiently heated and ready to be soldered, the flux will begin to bubble out. Remove the torch and touch the end of a length of solder to the edge of the joint. If the temperature is correct, the solder will melt and be immediately drawn into the joint. If this does not occur, repeat the process with the propane torch again until the molten solder is drawn in. Let the solder flow until a gleam of solder can be seen all around the seam, indicating the joint is full. Wipe any excess solder off with a cloth, being careful not to touch it with your hands, as it is very hot.

Let the joint cool for a few minutes. Do not ever attempt to cool a joint down with cold water. This can crack a fitting. After the joint has been allowed to cool down somewhat, it can be tested. If a leak is discovered, it is a simple procedure to reheat the fitting and apply more solder until the joint is full.

Here are a few simple procedures to follow in soldering:

■ When working near wood or any other combustible substance, a piece of asbestos or metal should be positioned between the two.

■ When working near another soldered fitting, keep the completed joint cool by placing a cool, wet rag over it.

■ Always apply the propane torch to the fitting and pipe, *not* to the solder.

■ During the cooling process, the fitting and pipe must be kept absolutely motionless in order to make for a solid joint.

FLARED JOINTS

A *flared joint* is made with a flaring tool, the use of which will be discussed in the next chapter. Once the flange nut is removed from the fitting, slip it over the tubing and clamp the tubing into the flaring tool. The cone of the tool is then screwed down into the end of the tube. The end will flare out until it has filled the recess in the flange nut. The flared end is then pressed onto the end of the fitting, and the connection is made by tightening the nut over the threads. This procedure is shown in Fig. 2-3.

Another method quite similar to flared fittings is a *compression connection*, except that no tools are required. The connection is made by removing the flange nut from the fitting and slipping it over the end of the copper tubing, along with the compression ring which matches the tube's diameter. The tube is then inserted into the fitting and tightened by hand. Using two open-end wrenches, hold the fitting in place while turning the flange nut. This process will seal the ring and squeeze it against the tubing. A compression fitting is shown in Fig. 2-4.

GALVANIZED STEEL PIPE

Galvanized pipe can be made of either wrought iron or steel, and until recently, was the sole type of pipe used in home systems. Wrought iron pipe will usually be found in older homes and is no longer manufactured today. Although steel pipe is still used in home plumbing systems, it has been largely replaced by its copper and plastic counterparts. Today, its most popular use is for hot and cold water supplies, heating systems, gas and air distribution, and venting. Where soil and water supplies in an area may have corrosive properties, steel is the ideal choice. In localities where lime and scale deposits are found in the soil, this same steel is not recommended.

Because of its rough surface and the types of fittings used, flow will be more restrictive than with the smoother copper pipes. The equipment that will be required to cut, thread, and repair steel pipe can also make its use more expensive and time-consuming. In spite of these disadvantages, steel pipe is quite strong and durable and is ideal for those areas in the basement where it may be exposed to possible damage from outside sources. In areas such as these, plastic or copper pipes could be easily damaged.

Steel pipe is either plain or galvanized. Galvanized simply means that the steel has been coated with zinc. It can be purchased in varying lengths

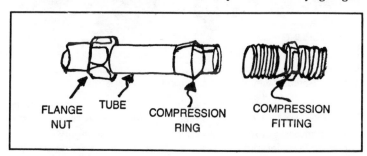

Fig. 2-4. A compression fitting is similar to a flared fitting, except that no tools are required.

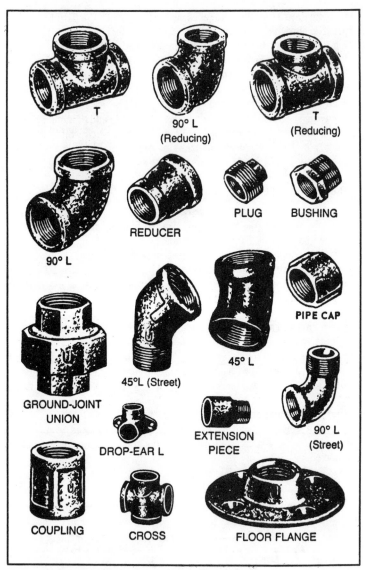

Fig. 2-5. Some common malleable iron fittings.

and can be either threaded or unthreaded. There are normally three strengths: *standard* (s), *extra strong* (x), and *double extra strong* (xx). Another advantage to using steel pipe is its cost, which can often be one of the major considerations in planning a system.

Fittings for steel pipes are made from malleable iron, which can be bent and/or shaped to a certain extent without causing any damage. When doing any repairs on an old plumbing system, it is recommended that the existing pipes be measured exactly. A plumbing supplier will be able to locate the proper fitting easily in this manner. When referring to steel pipe, always ask for pipe, not a tube. Copper pipe is commonly called tube in the plumbing business.

Figure 2-5 shows some common malleable iron fittings. These fittings can be either ordinary (pressure) or drainage (recessed). The only real difference is that drainage fittings are smoother on the inside, so that the pipe screws in and fills the recess completely. This does not allow for any areas where solid matter can lodge and obstruct the line. As with copper fittings, each type of fitting is specially suited to perform a certain function in the system. A *coupling* is more or less a permanent fitting that should be used only in situations where the pipes will not need to be taken apart. It is possible to disassemble pipes joined with a coupling, but this connection should be considered to be permanent except when necessary repairs or alterations are done. When it is known beforehand that a certain amount of disassembly may be needed from time to time, a *union* should be used instead of a coupling. A *reducing coupling* is used when connecting pipes of different sizes, and it can be compared to the adapter used in copper connections. *Elbows* are also used in much the same manner as described in the previous section on copper fittings, but there is less of a selection in the degree of angle. Iron elbow fittings are most commonly available in either 45 or 90-degree angles and are used to change the direction of a line by one of these two amounts. When connecting short sections of pipe to a galvanized pipeline, a *nipple* is used. These are available in a wide variety of lengths and pipe sizes. A *cap* is used at the end of a line to close it off. To close the female ends of fittings, a *plug* is used. A *cross* is a fitting with four openings at 90-degree angles to each other and is used to connect four pipes at right angles. A *tee* is used to joint one pipe to two others at a 90-degree angle. Similarly, a *reducing tee* would be used to join three pipes of differing sizes together. When it is necessary to reduce the size of a coupling or fitting, a *bushing* can be inserted to eliminate the need for replacement of a complete joint.

When measuring threaded pipe, you will have a number of methods to choose from. In each option, though, the length of the pipe going into the fitting and the fitting dimensions will have to be considered to obtain the length of pipe needed. Figure 2-6 shows the various methods that are considered to be most commonly used. The first is the end-to-end measure, which is the total length of the pipe, including the portion which goes inside the fittings at both ends. The end-to-center measure is used for length of pipe which will have a fitting at only one end. The pipe length will be equal to the distance from the far end to the center of the fitting, minus the dimension from the near end of the pipe to the center of the fitting, plus the length of pipe going inside the fitting. The face-to-end method is also used for a pipe length with a fitting at only one end, except that the pipe length is equal to the measure from one end to the fitting plus the length of

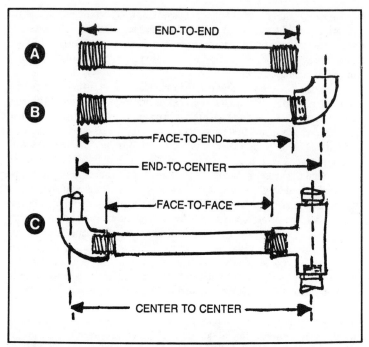

Fig. 2-6. Various methods used to measure threaded pipe.

overlap, or the length of pipe going inside the fitting. In the face-to-face measure, which is used for a length of pipe having fittings at both ends, the pipe length is equal to the distance from one fitting to the other plus twice the length of overlap. Alternately, the center-to-center measure can be used when there will be fittings at both ends. The pipe length will be the measurement from the center of one fitting to the center of the other, minus the sum of the distances from where the pipe enters the fitting to the fitting's center, plus twice the length of thread engagement. Some of these methods may seem a bit complicated at first, but once the desired selection is made suitable to the work being done, you will find it to be a relatively easy task.

Before beginning any job, it will be necessary to calculate the amount of pipe that will be needed. There are two ways this can be done. The first is the *dry-run* method and involves a minimum of mathematical calculating. To arrive at the length of pipe, measure the distance between the connection where your new line will start and the first fitting. For this step, it may be helpful to have another person assist in holding this first fitting in place. Now, measure the exact distance between these points and add the screw-in distances at both ends. Continue with this process, measuring each distance, until you arrive at the entire length of pipe needed for the whole job. If there will be differing sizes of pipe used in the project, be sure

1	2	3	4	5	6
Pipe label and size	Distance between center (inches)	Fitting numbers	Center to face of fitting, in inches (subtract)	Allowance for threads, in inches (add)	Total length of pipe required (inches)
A 3/4″	30″	1 - 2	1 5/16 + 1 5/16	9/16 + 9/16	28 1/2 ″
B 3/4″	50″	2 - 3	1 5/16 + 1 5/16	9/16 + 9/16	48 1/2″
C 3/4 ″	24″	3 - 4	1 5/16 + 1 1/16	9/16 + 9/16	22 3/4″
D 1/2″	20″	4 - 5	1 1/16 + 1 5/16	1/2 + 1/2	18 5/8″

to make a note of this in your calculations, breaking down each pipe size separately.

The second method for determining the total amount of pipe that will be required for a job is a bit more complicated. Figure 2-7 shows the layout of a system starting from a tee (1), going through two elbows (2) and (3), then through a reducer (4), and ending at a cross (5). Size of the pipe is ¾ inch all the way through, with the exception of the last run between (4) and (5), which is ½ inch. Table 2-1 gives the computations. Column 1 gives the sizes of pipes and column 2 gives the distance between the centers of the fittings. The third column identifies the fittings at both ends of the pipe, and column 4 is the center-to-face distance of each fitting, which can be

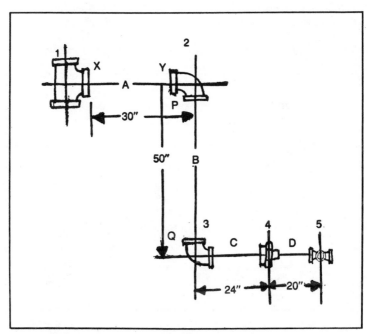

Fig. 2-7. Simple layout of a plumbing system.

Table 2-2. Computation Chart for Measuring Pipes for Offsets.

Degree of offset	Value of B when A = 1	Value of C when A = 1	Value of A when B = 1
60°	0.577	1.155	1.732
45°	1.000	1.414	1.000
30°	1.732	2.000	0.577
22 1/2°	2.414	2.613	0.414
11 1/4°	5.027	5.126	0.199
5 5/8°	10.168	10.217	0.098

measured out for each fitting or obtained from the individual manufacturer's specifications. The fifth column gives the distance the pipe will go inside the fittings on either side, and column 6 is the length of each run. Therefore, column 6 equals column 2 *minus* column 4 *plus* column 5. The total length of ¾-inch pipe needed for the example given in Fig. 2-6 will be 28½ inches + 48½ inches + 22¾ inches = 8 feet 3¾ inches. The amount of ½-inch pipe needed will be 18⅝ inches. These are exact calculations, and it is suggested that a little more than the required amount be purchased to allow for any possible waste. When measuring pipes for offsets, such as in Fig. 2-8, the process is a little bit different. Using Table 2-2 for computation purposes, the actual calculation is given here. A 1-inch pipeline has an offset at 30 degrees and dimension A is 36 inches. Fittings 1 and 2 are 30-degree elbows. In order to calculate the length of threaded steel pipe

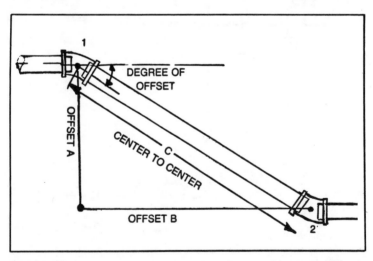

Fig. 2-8. Measuring method for pipes for offsets is done in a different manner.

that will be required between fittings, refer to Fig. 2-10. The value of C for a 30-degree offset equals 2 when A = 1. It is given that A is 36 inches. Hence, C = 2 × 36 = 72 inches. The standard length of elbow for 1-inch line is 2¼ inches, end-to-end of fitting. Hence, end-to-center length = ½ × 2¼ = 1⅛ inches. For the two elbows, the total length = 1⅛ + 1⅛ = 2¼ inches. The length of thread engagement for a 1-inch pipe is 11/16 inches. The total engagement length = 2 × 11/16 = 1⅜ inches. Length of pipe required = 72 inches − 2¼ inches + 1⅜ inches + 71⅛ inches.

A portion of this discussion on steel pipe is devoted to the actual cutting process of steel pipe because improper methods used here can result in the wasting of pipe and future blockage. The procedures given should be carefully adhered to in order to avoid a loss of time, high material costs, and future maintenance and repair costs.

The first thing to do if you are planning to do all the necessary cutting is to check the general condition of the equipment to be used. The pipe cutter should be checked to make sure it has no burrs or blunt edges on the cutting wheels. Once it has been determined that the equipment is in good working condition, calculate the lengths of pipe that will be required from fitting to fitting and mark the spots clearly to indicate where the pipes are to be cut. Set up the vise and lock the pipe in place tightly with the first mark about 6 to 9 inches away. Select a suitable cutter, a single-wheel or three-wheel type, depending upon the size of pipe to be cut. Turn its handle clockwise to open the jaws, and place it around the pipe from underneath, so that the cutting wheel is exactly over the mark on the pipe. When using a three-wheel cutter, the three wheels should stay level and perpendicular to the center line of the pipe. Turn the handle clockwise to close the jaws tightly. To get a bite on the pipe, turn the handle another quarter turn. Apply a little thread cutting oil to both the pipe and the wheels of the cutter. Rotate the cutter completely around the pipe to start the actual cutting process. Turn the handle a quarter turn and again rotate the cutter one complete revolution. Continue with this process until the pipe is fully cut. The threading of pipe is covered in the chapter on tools.

Once the pipe has been cut and threaded, it is advisable to screw the fitting on one end of the pipe while it is still clamped in the vise. The whole assembly should then be removed and taken to the installation where it is to be fitted and a wrench used to screw it in place. Before joining the pipe, the male threads on the pipe and the female threads on the fitting should be cleaned thoroughly with a hard brush. Coat the male threads with pipe joint compound. *Do not* coat the female threads on the fitting. Using hand pressure alone, screw the fitting on the pipe in three revolutions and then tighten with an appropriate wrench for the type of pipe being used. A second wrench may be used to hold the pipe in place during this process. If the threading has been done over the required length of pipe, two or three threads should be visible on the outside of the completed joint.

When it is not desirable to use a fitting for a minor change in direction in a line, it is possible to bend some smaller sizes of pipe. Table 2-3 shows

Table 2-3. Minimum Allowable Radii of Bends for Various Sizes of Pipes.

Diameter of pipe in inches	Minimum recommended radii in inches
1/8	1 1/4
1/4	1 1/4
3/8	1 7/8
1/2	2 1/2
3/4	3 3/4
1	5
1 1/4	6 1/4
1 1/2	7 1/2
2	10
2 1/2	12 1/2
3	15
3 1/2	17 1/2
4	20
5	30
6	42
8	45
10	50
12	60

the minimum allowable radii of bends for the various sizes of pipe. Whenever possible, it is recommended that larger radii than the ones given be used. The size of the pipe will determine the method selected for the actual bending process. Smaller pipes can be bent without heating with a *hickey*. To use this tool, take a small length of pipe and insert it into the side inlet of a tee fitting one or two times larger than the pipe to be bent. Push the hickey over the pipe and apply leverage until the pipe is bent to the required angle.

For larger size pipes, bench tools and heat must be used. Figure 2-9 shows a bending block, which can be made simply by drilling holes in a steel plate which is then fastened to the bench. Steel pins are then placed in the holes. The pipe is placed on the block between the pins so that it goes on the inside of one pin and the outside of the other. As leverage is exerted, the pipe bends. If necessary, heat can be applied to the inside of the bend as the pipe is shifted forward to make a smooth bend. The steel pins are not permanently installed in any particular holes to allow for them to be switched from hole to hole, depending upon the angle of bend required.

PIPES FOR DRAINAGE

Waste pipes are those lines in the plumbing system which carry the waste water from tubs, lavatories, washbasins, washers, etc., to the house drain. The drain provides for the transmission of this water out to the street sewer. This waste water will also contain some solids, and to avoid the settling of solid particles at any point in the system, any sharp changes in direction of flow must be avoided. This is done with specialized fittings. Such as *ridges* and *pipe shoulders*. The insides of these fittings are very smooth to prevent undue obstruction to the flow inside the pipe. Also, when joining a vertical line with a horizontal one, the latter is given a small, almost imperceptible pitch, such as ¼ inch per linear foot, to maintain a good flow velocity and prevent accumulation of solids. To make joints at these locations, threaded or machined right-angle elbows are used. These elbows are designed to provide a change of direction which is slightly over 90 degrees. This will provide sufficient pitch on the horizontal line to prevent obstruction.

Drainage pipes are commonly available in light copper, cast iron, threaded galvanized steel, clay tile, cement or concrete tile, fiber, and plastic. Since a drainage pipe does not have to withstand the pressure a water pipe does, its walls do not have to be as thick. For example, when using copper pipe for drainage, a lighter variety with thinner walls can be used. This will make for easier handling and, where expense is a consideration, keep overall costs down.

Cast-Iron and Threaded Steel Drainage Pipes

Cast iron and steel are used together to make up the drainage lines in many instances. Cast-iron pipe is employed for lines in sewers, soil stacks, waste stacks, main drains, and underground installations. It is recommended for its strength, durability, and high resistance to corrosive effects. The threaded steel pipe is used in conjunction with the cast iron for the branch drains and vent lines. Copper tubes usually make up the branches between the fixture drains and the main drain. The lighter DWV variety is the suggested choice here. Steel pipe is not for underground installations due to its high corrosive properties regarding chemicals.

The type of cast-iron pipe used for drainage purposes comes either plain or coated with a special preservative designed to reduce corrosion.

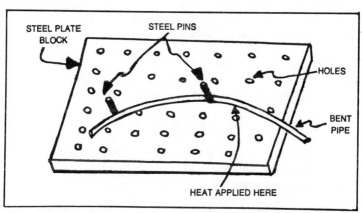

Fig. 2-9. Steel plate block used to bend larger size pipes.

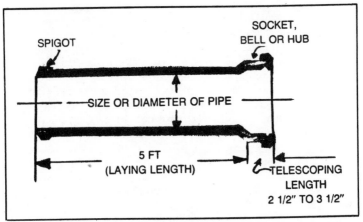

Fig. 2-10. Sample piece of cast-iron pipe.

Its useful life is estimated to be more than 100 years unless it is to be used with highly corrosive flows or in an area with a high concentration of chemicals in the soil. It is available in either service or extra heavy weight. For most common household plumbing, the service weight will be adequate. The extra heavy weight is normally used to provide extra safety when corrosive materials will be flowing through the pipes or when the pipe will be required to bear heavy loads, such as under a roadway that is heavily traveled. Tall office buildings or apartment complexes normally use the heavier weight due to the tall stacks that will be required. Cast-iron pipe normally comes in 5-foot lengths with a socket, hub, or bell at one end and spigot at the other end. Special pieces are available with a hub at both ends for situations where a piece shorter than 5 feet is required. Diameters range from 2 inches to 8 inches, measuring the inside of the pipe. A sample piece of cast-iron pipe is shown in Fig. 2-10. If you are purchasing a large supply of pipe at one time for a job that will take place over a long period, be sure to store it where it will be protected. This pipe is very brittle and can crack if mishandled or dropped.

Cast-Iron and Threaded Steel Pipe Fittings

There are an enormous amount of fittings available for use with cast-iron and threaded steel, and some of these are pictured in Fig. 2-11. See Table 2-4. It will not be necessary for you to know all the possible fittings, but it is recommended that you be familiar with those most commonly used to have an idea what is available and how to use them in a system. The example given here shows where some of these are used in the main stack, the secondary stack, the revent pipe, and the house drain. The sanitary type fittings must be used to make up all the drainage lines which will carry liquid flows. The straight cast-iron or ordinary threaded fittings are used on vent pipes that will carry only gases and no liquids.

54

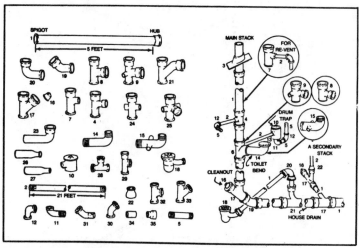

Fig. 2-11. Many fittings are available for use with cast-iron pipe.

Measuring cast-iron pipe is quite simple, since it is normally available in 5-foot lengths. As shown in Fig. 2-10, the total length of a section of pipe is the sum of the laying length (5 feet) and the telescoping length. The latter varies according to the diameter of the pipe being used. For example, if the pipe is 2 inches in diameter, the telescoping length will be 2½

Table 2-4. Parts List for Fig. 2-11.

1—Cast-iron pipe
2—Threaded steel pipe
3—Roof flashing
4—San. tapped tee
5—Threaded pipe nipple
6—San. tee with side   tap (RH)
7—Straight tapped tee
8—San. tee
9—San. tee with side tap (LH)
10—Drum trap
11—90°drain, street ell
12—90° drain, ell
13—Toilet floor flange
14—Toilet bend
15—Tapped toilet bend
16—Cleanout ferrule
17—San. 45° Y
18—Floor drain with trap

19—San. 1/8 bend
20—San. 1/4 bend
21—San. tap Y
22—Threaded spigot
23—San. long 1/4 bend
24—Straight tapped cross
25—San. tee, both sides tapped
26—Spigot-end increaser
27—Threaded-end increaser
28—Jiffy vent tee
29—Jiffy waste tee
30—45° drain. ell
31—45° drain. street ell
32—Drainage tee
33—Drainage Y
34—Coupling
35—Reducer

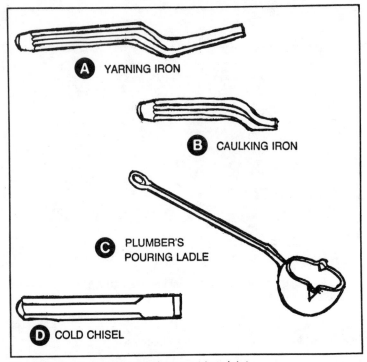

Fig. 2-12. Tools needed to make a cast-iron joint.

inches; for 3-inch pipe, it will be 2¾; and for 4, 5, and 6-inch pipe, it will be 3 inches.

To cut cast-iron pipe, the method will depend on the variety used. The service weight type is cut by making a 1/16-inch deep groove all around it with a hacksaw. The pipe is then placed over a thick plank of wood with the groove just overhanging the edge, and a few blows with a hammer will break off the grooved section. A pipe cutter cannot be used with cast-iron pipe.

To cut the heavier variety of cast-iron pipe, a hammer and chisel are used. In order to make the cut, mark the pipe to indicate where the cut is to be made. Next, place the pipe over a wooden board or mound of earth and score it around the marks with a small, pointed cold chisel. Make sure the chisel is not too sharp. Move the chisel along the mark a little at a time while tapping it lightly with a hammer at first. Continue the process while turning the pipe, increasing the intensity of blows steadily until the pipe breaks off. In cases where only a small length of pipe needs to be removed, this can be done using a hacksaw and an adjustable wrench. Use the hacksaw to cut a deep groove all around to a depth of not less than half the wall thickness of the pipe. Apply leverage with the adjustable wrench until the smaller section of the pipe is broken off.

Cast-Iron Pipe Joints

The technique used for making joints on cast-iron pipes is very different from the method described previously for joining steel pipes. It requires a few special tools. Figure 2-12 shows a *yarning iron, caulking iron, chisel, melting pot,* and *pouring ladle*. The basic procedure requires that the spigot end of one length of pipe be inserted into the bell end of the other length. The joint is then made by a filling of oakum and poured lead. A finished joint should have an equal thickness of the jointing material visible all around the joined pipes. The actual process is as follows:

■ Clean thoroughly the bell and spigot ends of the pipe, making sure no moisture and foreign matter remain. Slide the spigot of the upper pipe into the bell or socket of the lower, making certain that it is properly centered. If the upper pipe has a cut end, there will be no spigot.

■ Using the yarning iron, pack the hub with an oakum gasket or a thick layer of loose fibrous oakum, first wrapping it loosely around the plain or spigot end, and then packing it firmly with the yarning iron with a few blows of a hammer (Fig. 2-13). Do not use hard blows which may crack the pipe and/or fitting. Oakum is a fibrous material and is available at most plumbing supply houses. Pack in several layers until the annular space is completely filled, ¾ inch to 1 inch from the top of the hub. The oakum

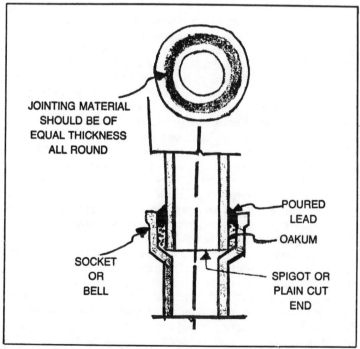

JOINTING MATERIAL SHOULD BE OF EQUAL THICKNESS ALL ROUND

POURED LEAD

OAKUM

SOCKET OR BELL

SPIGOT OR PLAIN CUT END

Fig. 2-13. Joining materials are packed firmly in place with the yarning iron.

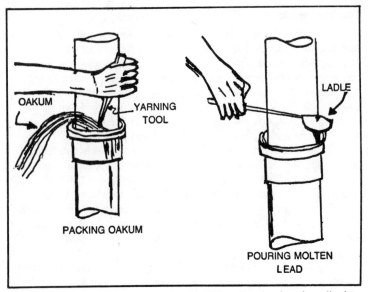

Fig. 2-14. The joint is filled completely with molten lead and caulked to complete the process.

should be thoroughly compressed so that it has an even top to receive the molten lead that will fill the remaining space.

■ Heat the lead in a melting pot until it melts and takes on a cherry red color. Preheat the pouring ladle and pour a small amount of lead on top of the oakum. Using light taps, caulk the area with a caulking iron. When this layer of lead has hardened, more molten lead can be poured into the joint until it is filled to a point a little above the top of the rim. Allow a few minutes for the lead to harden; then begin caulking the outer edge with an outside caulking iron, again using the hammer. Lead should grip the oakum and be packed tightly against both the pipes. If a fitting or a pipe is cracked during this process, it must be replaced. A finished cast-iron joint is shown in Fig. 2-14.

The diameter of the pipe being joined will dictate the amount of lead and oakum needed to make a good connection. These requirements and the ratio of lead in relationship to the amount of oakum are shown in Table 2-5.

In situations where the bell is pointing downwards instead of upwards, as in the example given here, the procedure will be a bit different. The first steps of aligning the socket inside the hub and caulking the oakum in the joint remain the same. After this portion of the procedure is completed, a *joint runner* such as the one shown in Fig. 2-15 is clamped around the pipe. A joint runner is a fat asbestos rope with a clamp attached. The ends of the runner are raised and clamped tightly. The runner is then slipped up the pipe until it is resting tightly against the joint. It may be necessary to tap it lightly with a hammer. Using putty, plaster, or fine clay

Pipe Diameter in inches	Amount of Lead per joint in lbs.	Amount of Oakum per joint in lbs.
2	1 - 2	0.20
3	2 - 3	0.30
4	3 - 4	0.45
5	4 - 5	0.55
6	4 1/2 · 6	0.65
7	5 - 7	0.75
8	6 - 8	0.85
10	7 - 10	0.95
12	9 - 12	1.25

soil, make a funnel in the raised end. Allow the funnel to dry thoroughly before continuing. If the molten lead is poured in too quickly, its contact with the material used to make the funnel may cause steam to rise. Once the funnel has dried, pour the molten lead in on top of the funnel, filling the

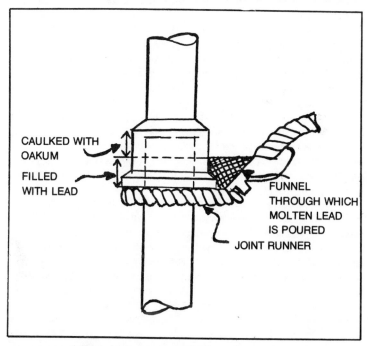

Fig. 2-15. The procedure is a bit different when the bell is pointing downwards.

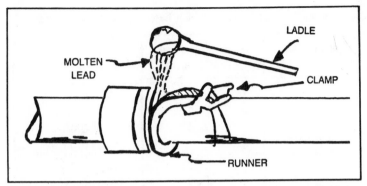

Fig. 2-16. In a horizontal joint, the molten lead will flow down and fill the joint.

joint completely. Once the joint has cooled down, the runner and clamp may be removed. Using a chisel, cut the lump of lead from the space outside the pipe joint and caulk the lead with the caulking irons, first on the outside and then on the inside. A horizontal joint is made in much the same manner, except that it will not be necessary to form a channel. The molten lead will simply flow down through the opening at the clamp and fill the joint (Fig. 2-16).

There may be some situations in which it will not be possible to use the molten lead method. In such a case, the joint can be made with lead wool or shredded lead, both of which will be available at most plumbing stores. This material is rolled into strands, each about ½ inch in diameter and about 2 feet long. The rolls are then inserted into the hub and caulking is done until the joint is completed.

Another method of joining cast-iron pipes is with a gasket joint. A gasket joint is usually used when the pipe has one plain end and one end with a hub. A neoprene gasket is placed inside the bell of one section, and the plain end of the second section is pushed in after a lubricant has been applied. This tight arrangement will provide a satisfactory waterproof joint.

In recent years, a newer and much more simple method has been introduced called the no-hub system. The ends of all the pipes and fittings in such a system are manufactured with plain ends and are joined with a neoprene sleeve which is held in place with a shield. Clamps are used to screw the shield down tightly. This produces a joint which can be assembled and/or disassembled quite simply. Before designing a system using this method, check the local plumbing codes for your particular area, as some codes prohibit its use in underground installations.

VITRIFIED CLAY PIPE

Vitrified clay pipe is most commonly used in the connection of the house drainage system to the septic tank and sewer. It is easily distinguished from other types of pipe due to its dark brown, glossy appearance. Clay pipe is quite brittle, making it inflexible. Movement of any sort will cause the pipes to break or the joints to leak. For this reason, plumbing codes most often prohibit its use inside any buildings.

Measuring clay pipe is done in much the same manner as for cast-iron pipe. The overall length is its laying length plus the telescoping length, which can varying from 1½ inches for 4-inch pipe to 4 inches for 36-inch pipe. One end is plain and the other end is usually bell-shaped. Cutting is very seldom required, because clay pipe comes in quite short lengths up to a maximum of 6 feet. However, when it is necessary to cut clay pipe, a hammer and brick chisel are used. Care must be taken during the cutting process and should be done in a gradual manner, as the pipe is brittle and can easily crack.

Cement mortar or bituminous compounds are used to make joints with clay pipe. The spigot or plain end is inserted into the bell end and aligned properly (Fig. 2-17). The hub is then packed with oakum to a depth of about ¾ inch. A yarning iron should not be used during this process. Instead, a thin, planed-down piece of wood can be used, as this will not cause any damage to the pipe itself. Fill the joint completely with the mortar or compound and pack it in tightly. It should then be finished with a beveled edge outside. Use mortar composed of 1 part cement and 2 parts clean sand. When joining a few short feet of pipe, these can be placed in a vertical position, the joint made, and the pipes then moved into the system as designed. When making joints in a horizontal position on the ground, spread paper under the pipe to prevent any loose earth or other materials from entering the lines.

CONCRETE PIPE

A brief description of the properties and ways of using concrete pipe is given here, although you will seldom be required to work with it.

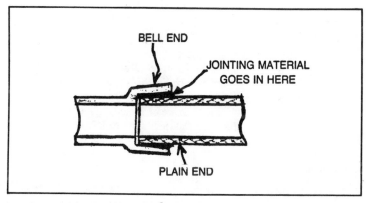

Fig. 2-17. Clay piping is joined by means of either cement mortar or bituminous compounds.

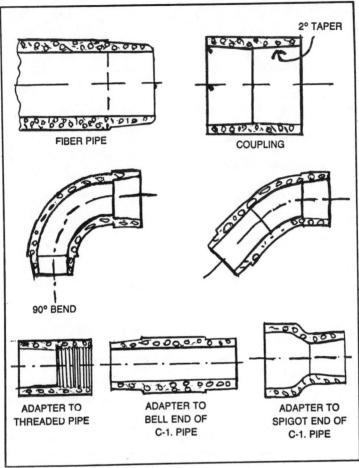

Fig. 2-18. Some common bituminized fiber pipe fittings.

Concrete pipe is primarily used for storm sewer and sanitary sewer mains and for large water supply mains in the street. Maintenance and repair of this portion of your water system is normally the responsibility of the local water authority. Sizes range from 4 to 108 inches. The pipe comes with or without steel reinforcement. The ends are bell-and-spigot or tongue-and-groove. It is very seldom, if ever cut, and joints are made in much the same way as for vitrified clay pipe.

BITUMINIZED FIBER PIPE

Like the vitrified clay pipe, *bituminized fiber pipe* is primarily used to convey waste water from a building to a septic tank or a sewer. It is composed of interwoven fibers soaked with bituminous compounds, which makes for a corrosion-resistant product that is very light and easy to work with. It comes in sizes from 2 to 6 inches and in length of 5 and 8 feet. Bituminous pipe is sometimes used in leaching fields to spread the overflow from the septic tank into porous soil where it is absorbed. When used in this manner, the pipes are laid at a very slight pitch, usually ¼ inch to ⅛ inch per foot. The 8-foot pipes are better suited for precision laying.

Figure 2-18 shows some of the commonly available fittings used with fiber pipes. Adapters can also be purchased when it is necessary to connect other types of pipe, such as cast-iron, to the fiber pipe.

Fiber pipe is cut using a crosscut or rip hacksaw. Its ends are tapered or tooled. Cuts produce plain ends which must be tapered using a fiber-pipe tapering tool before the pieces can be attached to fittings or to other pipe lengths. In order to make a joint, clear any grease or burrs from both the pipe and the fitting. Attach the two together. Normally, the fitting will slide quite easily up to within ¼ to ⅓ inch of the shoulder on the taper. Place a wooden block against the fitting if it is to be joined to an installed pipe. Alternately, place the block against the end of the pipe if it is to be joined to an already installed fitting on the line (Fig. 2-19). Keeping the block in place with one hand and the line braced, apply light taps on the block with a sledge. This will gradually bring the pipe and fitting together until the fitting is resting against the taper shoulder. The heat produce during this process will serve to fuse the materials of the pipe and fitting, thus producing a watertight joint.

PLASTIC PIPE

Plastic pipe and its components will be discussed in more detail in this text than were its counterparts, primarily because since its inception, it has become increasingly popular with the do-it-yourselfer. It has many

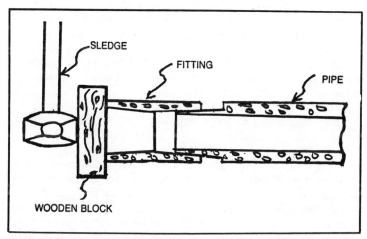

Fig. 2-19. Use a block of wood to assist in making a fiber joint.

advantages over the heavier pipes, the most important of which to the beginner will be the ease with which it can be installed and the smaller complement of tools required. A check should be made as to the plumbing codes in your area, as new rules and regulations regarding plastic pipe are still forthcoming even today, due to its newness. This test will point out both the advantages and disadvantages found in working with plastic pipe and the various fittings and types of plastic pipe that are available for home plumbing use.

Not only is plastic quite easy to work with, but it is available in varying flexibilities, ranging from limber to virtually limp. For example, polyvinyl chloride, or chlorinated polyvinyl chloride, is considered to be rigid as far as plastic pipe is concerned, but it has a certain amount of flexibility, certainly far more than a piece of steel pipe would have if used in its place. The PE (polyethylene) water supply comes in coils and can be unrolled in a relatively straight line, but it is flexible enough to round corners and be wiggled into place with little difficulty. Another type of plastic pipe, *polybutylene tubing,* is so flexible that it can be pulled through bored holes or wound through a building in much the same way as an electrical wire. This makes it the easiest piping to install. It works wonderfully well in those hard-to-reach places where a length of rigid pipe would make installation uncomfortable and tedious. Due to plastic pipe's flexibility, long runs can be had with few or even no fittings. This can save the home plumber a good deal of time and money.

Although the cost factor can at this time be listed as one of the advantages of plastic pipe, the majority of the materials from which plastic products are manufactured depend heavily on crude oil. We are now in an age where much emphasis is being placed on a need for less dependence upon overseas oil, and this may well mean that the cost of plastic plumbing components will rise at a somewhat greater rate than metallic piping and fixtures. Costs may vary from area to area, but even taking into consideration the current oil crisis, plastic is still less expensive on the retail level. The decreased labor costs due to the ease of its installation, as a general rule, keep the overall expense well below that of its metallic counterparts.

On the average, a plastic water pipe and fitting will weigh about one-tenth as much as black or galvanized steel pipe. This provides not only greater ease in handling for the home plumber, but also a lot less weight that must be supported by the house structure once the system is in operation. Plastic pipe weighs a little less than 1 ½ pounds per each 10-foot length, while galvanized pipe in a nominal ¾-inch size weighs anywhere from a little over 1 pound to as much as 1½ pounds per foot. When comparing the weight difference between plastic and cast iron, the 4-inch polyvinyl chloride (PVC) pipe weighs about 2 pounds per foot, while the same diameter of single-hub service-weight cast-iron pipe weighs about 8 pounds per foot. Even though the plastic pipe is only one-fourth as heavy as the cast-iron pipe, rather than one-tenth when compared with steel, the difference is even more marked when it comes to actually handling the material during installation.

Aside from all of its other advantages, plastic pipe is also very tough and sturdy. It can withstand an amount of rough abuse and mishandling that copper or even cast iron cannot. A hard blow with a hammer will seriously damage a piece of ½-inch copper tubing, but will have no effect at all on a piece of CPVC pipe. If PB (polybutylene) tubing is squashed flat, it will return to its former shape. If a length of cast iron is dropped, it will smash into pieces. A length of 4-inch drainpipe made of plastic will literally bounce. Because of the materials it is made of, plastic pipe is extremely corrosion-resistant and can be buried in damp acid soil or directly into cinders without fear of any damage. It can also be buried right next to a length of metallic pipe with no fear of any galvanic action taking place and corroding the pipes. Plastic is totally nonconductive of electric currents, does not rust, and is not bothered by direct burial in the earth. It will not be harmed by direct exposure to the air and the elements. Since the material is quite slick and very smooth, very little, if anything, can adhere to its inside walls. Minerals that may be present in the water supply such as lime, calcium, or even rust will very seldom be a problem. The same cannot be said of steel pipes, which can have such a substantial buildup of mineral deposits that they will become completely obstructed.

Flow Rate

Flow rate has been discussed previously in this chapter as a characteristic of pipe that is an important part of the overall system. The smoother the inside of a pipe, the higher rate of velocity that can be obtained. Because plastic pipe is quite smooth, it is sometimes possible to substitute a piece of plastic tubing which is smaller than what would be required if steel were used to do the same job. This flow rate, once established, will remain pretty much consistent throughout the life of the system, if installed properly, since very little, if any, buildup will occur.

Expansion Ability

One of the other characteristics of plastic pipe is its ability to expand to a degree. This makes it an excellent choice for installation of water supply lines where winters can be severe. When water begins to freeze inside the lines, this pushes the pipe's walls outward rather than rupturing them. However, the same cannot be said for the fittings should they freeze. The advantage of this is that flexible plastic water supply lines can be installed just below ground level or even on the ground surface. In the event of cold weather, they are simply opened. Although some portions of the line may be partly filled with water under no pressure, the pipe will not break. If metallic piping were used under these circumstances, it would have to be completely drained to avoid splitting.

Although this discussion pertains to the more flexible types of plastic pipe, even a fairly rigid plastic water supply pipe is able to tolerate a certain amount of freeze-up without danger of splitting. But here again, as long as the freeze does not occur at a fitting, the pipe will swell and return to its normal shape when thawed. The rigid pipe is a bit more fragile and will not withstand very cold temperatures for any extended period of time. It will, however, stand a much better chance of holding up during a short accidental freeze-up in a residential system than its metallic counter part.

Because of its composition, plastic piping has yet another advantage over other materials. It has good insulating properties, which can be a real money-saver in a home plumbing system. When used for hot water supply, a plastic pipeline will lose significantly less heat through its walls and into the air than a metallic pipeline. Thus, hot water will travel from the tank to the tap both faster and with less heat loss, resulting in a lower hot water usage overall. If you have ever had the experience of touching a hot water pipeline once installed and in operation, you will know that it will most often be quite hot to the touch. This is not the case with a system constructed of plastic. A plastic hot water pipeline's exterior walls will seldom become so hot as to cause a burn, so exposed pipes will never pose any safety hazard to adults, as well as young children.

Metallic pipes frequently sweat due to condensation when exposed to dampness and humidity. The insulating properties of plastic pipe work here in reverse regarding cold water pipelines. The interior water flowing through the pipelines is insulated from the outside air temperatures. This prevents mildew, dry rot, and other effects present in a metallic system which can cause extra work for the home plumber. Plastic is also a sound insulator, and the whole system will prove to be much quieter than a metallic system.

There are some disadvantages of plastic pipe which should be mentioned. None of the difficulties that may be encountered should be considered major, the most can be dealt with routinely and with a minimum amount of inconvenience to the home plumber.

Codes and Regulations

Plumbing codes and regulations will vary from area to area. In some instances, these codes will be quite similar, while great variation may be the case when comparing these regulations nationwide. Unfortunately, there are still many areas in this country where plastic plumbing components are prohibited completely. Some areas may allow the use of plastic pipe for such purposes as septic leaching field drainpipes, ordinary ground or rainwater drainage systems, etc. In other locales it can be installed for these purposes as well as residential drain, waste, and vent systems. Sometimes plastics can be used for the house water supply line, but not for water distribution within the house itself. In areas where there are no zoning laws, there will probably be no building and/or plumbing codes either. In this case, the choice of materials is left up to you. Perhaps the plumbing codes make no mention of plastic pipe. It would be to your advantage to apply for the necessary permits with regard to the use of plastic. This will involve working closely with the officials in adding to the codes to allow for plastic plumbing systems. Regardless of the situation in your area, it is recommended that anyone planning to use plastic pipe in any part of an intended system check with the local authorities before proceeding with its installation.

Lack of Professional Assistance

Another difficulty that may be encountered is the lack of experienced and knowledgeable assistance in certain areas of the country. There are a number of professional plumbers who steadfastly believe in the old tried and true materials and will refuse to install a system using plastic materials. If professional assistance is not required, this will not prove to be a problem, but it can be a consideration for those who do not plan to do a complete installation themselves. As plastic components gain a positive reputation, however, this situation will undoubtedly correct itself as attitudes change.

Minor Problems

The disadvantages discussed thus far really have nothing to do with the actual material or the system itself, but are really a reflection of the process of change. As to the plumbing system on the whole, there are some minor drawbacks or problems that may occur, but these can most often be traced to improper installation, misapplication of pipe or fittings, or poor workmanship. These situations can well apply to any system and, again, have no bearing on the material used. You should follow proper procedures during assembly regardless of what type of pipeline is used.

Disassembling and Thawing Difficulties

Plastic pipe, once installed, is not quite as simple to disassemble as would be the case with a metallic piping system. In a metallic system, additions, revisions, or corrections of mistakes are usually a simple matter of unscrewing or unsoldering the fittings, making the necessary changes or repairs, and assembling again. In a plastic system that uses welded joints, the fittings are not reusable. However, there are ways to alleviate this problem with the addition of couplings and/or other fittings. It is a much simpler task to repair or revise a metallic system than a plastic one.

In areas where winters are severe, it may be necessary to apply heat tape to the pipes to prevent freezing. If the system is metallic, it is a simple process to wrap these pipes with either tape or insulation, and the problem is solved. However, if plastic piping is used, normal heat tape cannot be used as any sort of permanent insulation. Over a period of time, heat from the tape can deform the plastic, and eventually the pipe or the fittings will be damaged. Also, if a heat-taped pipe is emptied of water, the tape will deform and melt the plastic in a short time and ruin that portion of the system completely. There is a specialized heat tape manufactured for use with plastic pipes, but it is quite expensive and difficult to find, since it is not stocked by most hardware and plumbing stores. This unavailability should be taken into consideration when designing a plumbing system, and alterations may be necessary to insure that the plastic piping is not installed where freezing may be a danger.

If the situation of frozen pipes does occur, the procedure of thawing the system is a bit more difficult than the method used to thaw metallic pipes. A dc welder cannot be clamped to a plastic pipe because the plastic is an insulator and will not allow any current to flow to effect a thaw. By the same token, if a torch or some other form of open-flame heat were used, the plastic would become deformed or melt. In this situation, a low-power heat tape is wrapped around the pipe, but only temporarily. The tape should not be left in place too long and should be watched closely to make sure deformation does not take place. An electric heater or sun lamp can also be used, but it should not be positioned too close to the pipe. Wrapping rags soaked in hot water around the pipeline will also do the job. All of these methods will work, but they will usually take a longer period of time than the thawing of metallic pipes.

The Right Pipe for the Job

Although the various types of plastic pipe will be discussed later in this chapter, it should be mentioned that care must be taken to use the proper type of pipe suited for the job being done. For example, not all kinds of plastic pipe are designed to carry hot water. CPVC or PB can carry hot water at maximum temperatures of 180°F, which is above the average homeowner's requirements. If the improper type of plastic pipe is used for hot water purposes, this will result in eventual pipe failure. By the same token, plastic piping should never be run close to high heat producing equipment or appliances.

When a plastic system is to be solvent-welded, close attention to the ambient temperature of the air in the working area is necessary. The system should be made up under dry conditions in air temperatures no lower than about 50°F to insure a tight seal. If there is no possible way to do the work at this specified temperature, there is a special welding solvent available that can be used in freezing temperatures, but only with some types of pipes and fittings.

Some incompatibility of pipes and fittings may be encountered when working with plastic due to a lesser degree of standardization from manufacturer to manufacturer. This applies to both the pipes and fittings themselves and also to the chemical compounds that make up the plastic material. The result of this lack of standardization can be joints that are too tight to fit properly, or too loose for a correct and full-strength welded joint. Plastic materials in parts of fittings that are incompatible will not weld properly. An incompatibility of the welding solvent with the plastic materials being employed will have the same result. The solution to these problems is simple enough and should be adhered to as strictly as possible.

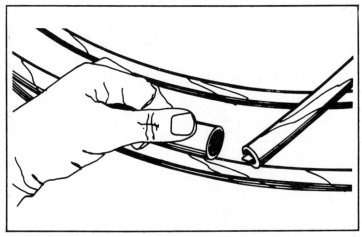

Fig. 2-20. Polyethylene or PE pipe is very flexible but also quite durable.

Always use pipes and fittings for a particular system that are manufactured by the same company, and always use the correct fittings and welding solvents designed for those materials. Any substitutions can lead to problems that could have been easily avoided.

All plumbing systems consist of a number of separate components, each being designed to do a particular job. To perform properly, the specific components must be used in the right places and be compatible with one another. In the early stages of development of plastic piping, hot water could not be carried, but this has long since been corrected. However, failure to use the proper type of pipe in the system will cause it to fail. The following discussion will give you an understanding of the various components of a plastic plumbing system and their intended uses.

POLYETHYLENE PIPE AND FITTINGS

Polyethylene or *PE pipe* is shown in Fig. 2-20 and is commonly used in the water supply portion of a plumbing system. PE pipe is flexible, quite tough, and is available in several sizes and grades. Its most common use is for a water supply for domestic purposes when tapping into a water main; use pipe in the ¾-inch or 1-inch diameter. The same pipe in larger sizes is widely used to supply water from wells and is strong enough to support the weight of a submersible pump, often without the use of a safety cable. It is also available in several pressure ratings. If PE is to be used to supply domestic water to a home, it must be approved by the National Sanitation Foundation for carrying drinking water. For normal irrigation purposes outdoors, this will not be necessary. It is commonly used in lawn sprinkling systems because it is so lightweight, inexpensive, and easy to handle. Normally available in rolls of standard lengths of 100, 150, and 220 feet, it can also be purchased through some suppliers by the foot.

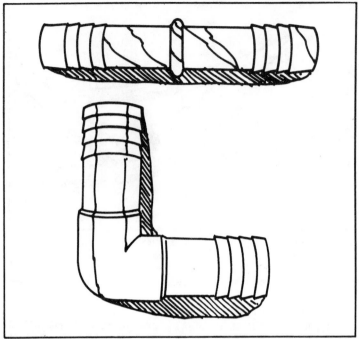

Fig. 2-21. Fittings for PE pipe are commonly made of either metal or molded nylon. Shown here are a straight coupling and a 90-degree elbow.

Only a limited selection of fittings are available for use with polyethylene pipe. These are normally made of either metal or molded nylon. The nylon is to be preferred, but because of its limited availability, the former may have to be used. A straight coupling is used to join two sections of pipe and a 90-degree elbow to make a turn at a right angle. See Fig. 2-21. Lesser degrees of turns can be made by bending the pipe in a wide arc due to its flexibility. A tee is used to attach branch and main lines, and adapters are available when it is necessary to join a plastic line to a metallic line (Fig. 2-22). To step up or step down the size of a pipe to join it to a pipe of another size, a reducer is used. These insertion fittings are slipped inside the pipe and are held in place not only by the rings on the fittings, but also by stainless steel clamps secured around the outside of the pipe. This is the most commonly used method to join polyethylene pipe, although it is not the only one. Other methods will not be discussed in this text, as they are more complex. The method discussed presents few problems for the home plumber.

POLYVINYL CHLORIDE PIPE AND FITTINGS

Polyvinyl chloride or *PVC pipe* is commonly used for outdoor cold water supplies and is shown in Fig. 2-23. Although PVC pipe is quite

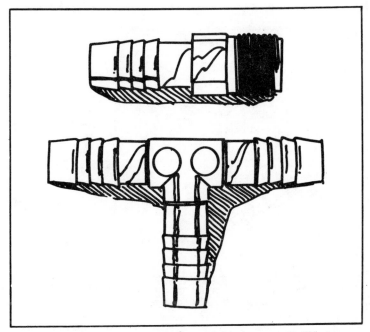

Fig. 2-22. Tees and adapters are also available for use with PE pipe.

limber and able to follow slight directional changes or irregular trench-bottom contours, it is considered one of the more rigid types of plastic pipe. It is normally packaged in bundles of 10-foot lengths and diameters ranging from ½ inch to 1 inch, with ¾ inch and 1 inch being the most popular for residential use. As was the case with PE pipe, approval must be obtained from the National Safety Foundation before being used to carry

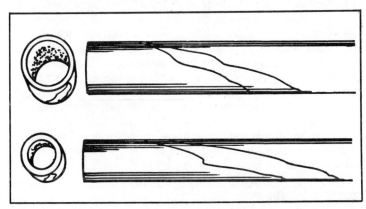

Fig. 2-23. Polyvinyl chloride or PVC pipe, although quite limber, is considered to be one of the more rigid types of plastic pipe.

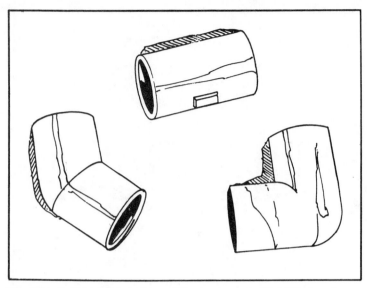

Fig. 2-24. Solvent-welded slip-on fittings are used with PVC pipe. A coupling, a 90-degree elbow, and a 45-degree elbow are shown.

drinking water. Although pressure ratings vary depending upon the diameter being used, PVC is capable of handling the normal pressures that would be placed upon it in a domestic water supply. It should only be used for cold water, *never* for hot water, and is used to supply water from water mains, nonsubmersible pumps, reservoirs, and cisterns. It can be used in the same manner for irrigation purposes as the PE pipe, although it is less practical. PVC pipe should not be used in areas where freezing may occur or for drainage purposes.

PVC pipe is most commonly joined to its fittings through the use of solvent-welded slip-on fittings. As with PE pipe, there is a limited selection available, but the choice is quite ample enough to make the necessary connections in a system. Lengths of pipe are joined with a coupling, and directional changes are made with either 90-degree or 45-degree elbows. These are shown in Fig. 2-24 and are used in the same manner as with PE and other types of pipe to join differing pipes, as well as adapters. Bushings are used to step up or down the size of one pipe for connection to a pipe of a differing size. Figure 2-25 shows some common PVC fittings. With the exception of the threaded portions of both male and female adapters, all the fittings are connected by means of a solvent designed especially for use with PVC pipe.

POLYBUTYLENE TUBING

Polybutylene or *PB tubing* is shown in Fig. 2-26 and is commonly used for water supply purposes. It is normally called tubing rather than pipe and is available in several sizes, but the ¾-inch size or the 1-inch size is the most common. PB pipe is quite flexible and is available in coils ranging from 25-foot to 100-foot lengths. Because of its flexibility, it can easily be laid in narrow trenches or in hard-to-reach areas and is quite resistant to corrosion and heat.

For normal water supply purposes, PB is jointed using the same type of insertion fittings as those used with PE pipe. Stainless steel clamps are used to secure the pipe to the fittings. Special transition fittings are used to secure the pipe to the fittings. Special transition fittings are available to allow for changing from PB tubing to other lines of pipe or fittings. Again, there are a limited number of fittings available, and a check with your local plumbing supplier is recommended before beginning any project involving PB tubing and fittings.

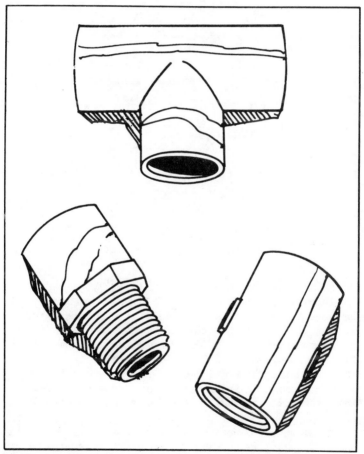

Fig. 2-25. Some common PVC fittings: tee, adapter, and bushing.

Fig. 2-26. Polybutylene or PB tubing is so flexible that it can be coiled or wound through hard-to-reach places, such as narrow trenches.

CHLORINATED POLYVINYL CHLORIDE PIPE AND FITTINGS

Chlorinated polyvinyl chloride or *CPVC* pipe is a rigid type of plastic piping that is commonly used for residential hot and cold water distribution systems. CPVC pipe has been designed especially to withstand constant hot water temperatures up to 180°F for indefinite periods of time (Fig. 2-27). It is a newer version of the PVC pipe which has been improved to allow for its use with hot water distribution systems. CPVC pipe is available in standard 10-foot lengths and diameters of ½ inch and ¾ inch. Other diameters are manufactured, but are not as readily available as those mentioned. Because of its improved characteristics and properties, it is possible to design a complete internal water distribution system using only CPVC piping. Its durability is proven by its ability to withstand normal household water pressures at maximum temperature ratings.

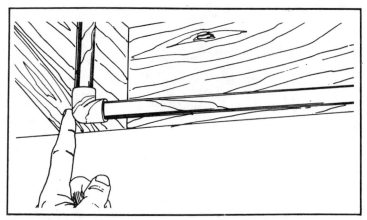

Fig. 2-27. CPVC pipe is designed to withstand high temperatures up to 180 degrees and is commonly used in water distribution systems.

Unlike the other types of plastic pipe already discussed, there is a good variety of fittings available for use with CPVC pipe. When it is necessary to join two straight lengths, slip-on coupling is used. If it is desirable to test a portion of the system before it is complete, caps are available to temporarily close the pipe or to prevent dirt from entering the lines while awaiting installation of fixtures or appliances. Figure 2-28 shows a coupling (left) and a cap (right). These caps can also be used as permanent fixtures, as shown in Fig. 2-29, to fabricate water hammer arresters or air chambers. Reducing bushings are used to allow for connecting two pipes of different diameters, and directional changes are made, again, with different degrees of elbows. A few of the more common elbows that are available are shown in Fig. 2-30. Another type of elbow, the 90-degree street elbow has one fitting socket at one end. The other end is designed to slip inside the socket of another fitting and is the same size as the pipe itself. Tees are used to connect branch and main lines and come in several different configurations. They can be ½-inch nominal at each socket (½ inch by ½ inch by ½ inch), ¾ inch at each end and ½ inch at the tee for branch socket (¾ inch by ¾ inch by ½ inch), or ¾ inch at one end and ½ inch at the other and at the tee or branch socket (¾ inch by ½ inch by ½ inch). These typical tee arrangements are shown in Fig. 2-31. Unions are also available which allow for direct welding to CPVC pipe to enable future

Fig. 2-28. A coupling is used to joint two lengths of CPVC; a cap is used to seal off a line, possibly for testing.

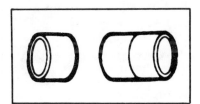

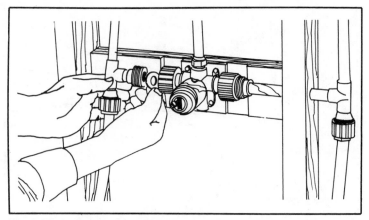

Fig. 2-29. A cap may be installed also as a permanent fixture and will perform much the same as an air chamber.

repairs without cutting. Stop valves made of plastic are available and can be set directly in the pipeline for additional flexibility in the system. These are shown in Fig. 2-32. A stop valve will provide a proper joint that can compensate for differences in coefficients of expansion between two dissimilar materials.

Figure 2-33 shows a plastic boiler drain which can be welded directly to CPVC pipe to allow for connections for appliances or even a garden hose. Figure 2-34 shows a wing elbow equipped with mounting ears and screw holes. This is of use in the transition from vertical shower supply riser to the shower arm, or with the boiler drain. Another fitting that can be used with plastic piping systems is the *escutcheon*, which is a round plastic trim plate primarily used to cover ragged holes made in walls through which pipes are run. These escutcheons are available as either a piece which slips over the pipe and rests against the wall or as a screw-in piece which is actually welded to the pipe to prevent movement. Special transition adapters are also available which are used to connect PB tubing to CPVC piping. These adapters can be angled or straight and can even include a stop valve. As can be seen, due to the increased use of this type of

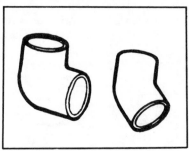

Fig. 2-30. CPVC 45 and 90-degree elbows.

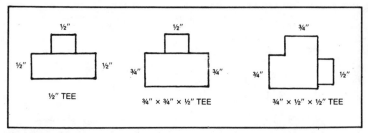

Fig. 2-31. Some of the more common tee arrangements that can be had with CPVC pipe.

plastic piping, a wide assortment of interesting fittings are available to provide both ease and versatility in an installation.

SMALL DIAMETER PB TUBING

Small diameter PB tubing is the same as the PB tubing discussed earlier. It is ideal for piping domestic water distribution systems. The ½-inch size is most commonly used to connect fixtures and/or appliances that use large quantities of water such as showers, washing machines, or dishwashers. For toilet tanks and lavatories that use a lower amount of water, the ¼-inch diameter tubing works nicely. The entire distribution portion of the system can be made up solely of PB tubing, using the various sizes and making the connections with special adapters that are available. PB tubing is very flexible, making it an easy task to run it under floors and through partitions. This will also preclude the necessity for elbows, which would be used to make changes in directions. For example, the ¼-inch tubing can be turned to a minimum radius of approximately 2 inches, the ½ inch to 4 inches, and the ¾ inch to 6 inches.

Connections to fixtures, appliances, or joints to CPVC or metallic pipe are made with the appropriate special fittings, which can sometimes

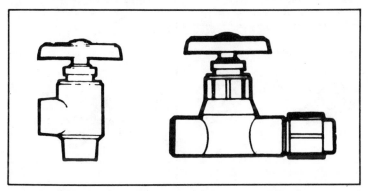

Fig. 2-32. Stop valves are also available made of plastic (courtesy Genova, Inc.).

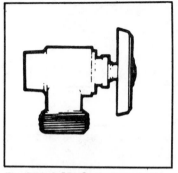

Fig. 2-33. A CPVC all-plastic boiler drain (courtesy Genova, Inc.).

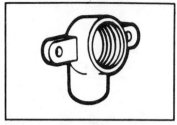

Fig. 2-34. The CPVC wing elbow attaches directly to the wall. It can be used to terminate a CPVC water line in a boiler drain, sill cock, or other valve (courtesy Genova, Inc.).

be used in combination with another fitting, a process which is called "cobbling up." Armed with the knowledge of what fittings are available and how they are used properly, this is a simple task. For example, if a coupling is to be made between two lengths of PB tubing, a straight adapter or transition fitting, PB at one end and CPVC at the other, is attached to each length of piping. A short stub of CPVC pipe is then solvent-welded between the two adapters (Fig. 2-35). If it is necessary to use a tee for a connection on a branch line, three adapters, three stubs of CPVC, and a CPVC tee fitting would be used. It can be seen that by using the various adapters, nipples, and other fittings, almost any combination needed to change directions, connect branches, and adapt the piping for connection to another part of the system can be had.

DRAIN-WASTE-VENT PIPING

Plastic piping for drain-waste-vent (DWV) systems is most commonly the polyvinyl chloride or PVC pipe already discussed previously, except that pipes of different grades and weights are specially designed for

Fig. 2-35. One way to make up a coupling for PB lines is with a short piece of CPVC and two transition adapters (courtesy Genova, Inc.).

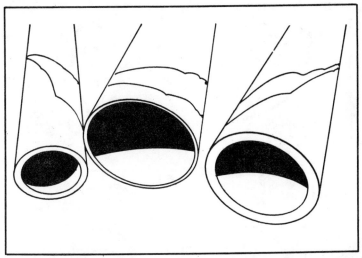

Fig. 2-36. Polyvinyl chloride or PVC pipe is most commonly used for drain-waste vent systems.

this portion of the system. PVC pipe for DWV purposes is quite tough, waterproof, and has a high resistance to corrosion, which is a must for piping used in this manner. Figure 2-36 shows some of the common diameters of PVC pipe which are available: namely 1½, 2, 3, 4, and 6 inches. This type of piping is not quite as flexible as some of the other plastic pipes already discussed. It comes in lengths ranging from 5 to 20 feet. Installation is still a simple task, though, and it is worth noting that the pipe is highly resistant to even the most powerful drain cleaners due to the thickness of the walls and their durability. As far as the selection of the proper diameter of PVC to be used for a given portion of the system, the 1½-inch size is usually the best choice for sinks, washbasins, and other low-volume appliances and fixtures. Two-inch PVC piping is most often used in shower and other drains, and the 3-inch diameter is best suited for the waste and vent stacks, the actual waste piping, and some drains. The 4-inch piping can sometimes be substituted for the 3-inch installations, but it will cost a little more.

Fittings for use with PVC-DWV piping are available in a wide variety. When connecting two straight lengths of pipe of the same size. normal

Fig. 2-37. A PVC-DWV reducing coupling to change drainpipe sizes (courtesy Genova, Inc.).

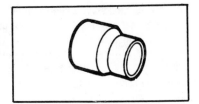

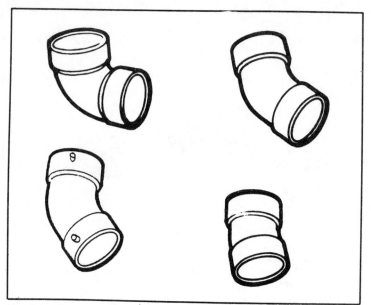

Fig. 2-38. The PVC-DWV 90, 60, 45, and 22½-degree elbows (courtesy Genova, Inc.).

couplings are employed. Reducing couplings are used in a similar manner to connect two sections of piping of differing sizes. Similarly, reducing bushings are available which perform the same function under varying circumstances. A reducing coupling is shown in Fig. 2-37. Male and female adapters are available which are used to connect the PVC piping to threaded pipe, and a different adapter can be purchased when it is necessary to join the pipe to a cast-iron system. Because of its rigidity, elbows will have to be used to make the necessary changes in direction. These are available in a number of angles such as 90 degree, 60 degree, 45 degree, and 22½ degree (Fig. 2-38). Elbows can be had either with hubs or sockets at each end or in the standard street form with a hub on one end and the other end pipe-size.

When it is necessary to connect the branch lines to main lines, a variety of fittings designed to provide proper drainage flow are available in many combinations. The first is called a *wye*, and it is used to connect a branch pipe into a horizontal drainpipe or to insert a cleanout plug in a drainpipe. Similarly, a reducing wye can be used when the main line is larger than the branch line. A wye is pictured in Fig. 2-39. Wyes are most commonly available with a 45-degree directional turn. If a 90-degree angle is desired in the same situation, a sanitary tee, or a reducing sanitary tee such as the one shown in Fig. 2-40, will be used. Both the wye and the tee are also manufactured in double form to enable the joining of two branches to the main line from opposite directions. A double tee with double side

Fig. 2-39. A wye is used to connect a branch pipe into a horizontal drainpipe (courtesy Genova, Inc.).

Fig. 2-40. A sanitary tee is used to change direction and size of piping (courtesy Genova, Inc.).

inlets provides four connections for branch lines, two of smaller size than the main line and two of the same size. Waste and vent tees are also available with either double side inlets or single side inlets to provide further flexibility in connecting 1½-inch or 2-inch branch vent or waste piping to a vertical drainpipe, as well as horizontal piping of the same size.

When installing two bathrooms back to back, special waste and vent fittings are available to make for a compact system. Likewise, using

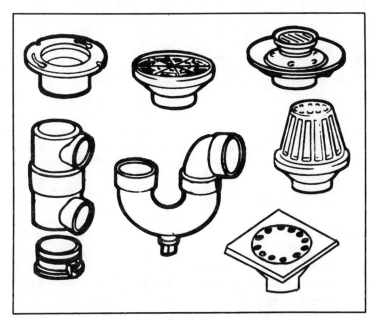

Fig. 2-41. Some of the specialized fittings that are available for use with PVC pipe (courtesy Genova, Inc.).

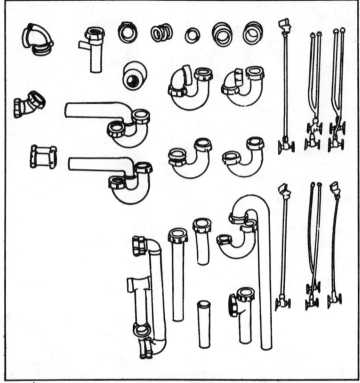

Fig. 2-42. Polypropylene pipe (PP) is used primarily in chemical and waste piping. Shown here are some of the many products made from this material.

combination fittings can also provide a great deal of flexibility in the installation of this type of piping. Some other specialized fittings which are available for use with PVC-DWV piping are closet flanges for toilets, floor and shower drains, special capped roof drain assemblies, drum traps, P-traps, and bell traps with replacement grates. These are shown in Fig. 2-41.

ABS PIPE AND FITTINGS

ABS pipe (acrylonitrile-butadiene-styrene) is another choice that can be used in place of PVC in drain-waste-vent systems. Although it has many of the same properties as PVC piping and has gained widespread approval, it is not quite as durable. ABS piping is a bit more susceptible to mechanical damage and is less heat-resistant. Certain caustics and chemicals can cause reactions that may lead to both pipe and fitting damage. It can also burn under certain conditions. Although the range and type of fittings available for use with ABS piping is not as extensive as with PVC pipe, the do-it-yourselfer should have no problem installing a complete residential system. There is a full complement of tees, couplings, elbows, traps, wyes, cleanouts, bushings, and adapters from which to choose. As with PVC, all of these fittings are of the slip-on, solvent-welded variety. Even though the two types of piping will fit together, they should not be considered interchangeable. Due to their different appearances, there should be no problem distinguishing one from the other in a system. PVC is normally buff-colored or beige, while ABS is black. It is important to note that although ABS piping may not be quite as durable as its PVC counterpart, it is being installed in many systems today under the right circumstances with success.

POLYPROPYLENE PIPE

Considered to be a newcomer in the plastic piping field, *polypropylene pipe* (PP) has already made a name for itself. It is virtually indestructible in most circumstances and is used primarily in chemical and waste piping. Some of the wide assortment of products made of this material are shown in Fig. 2-42. PP pipe can be used for fixture drainpipes and traps of the sort for installation under lavatories, sinks, and in other situations where the system is prone to failure from extended use and abuse. This type of piping is highly recommended for use when making repairs on a failing existing system, as well as for new installations in areas where durability is a must.

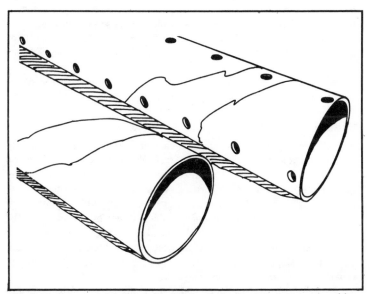

Fig. 2-43. Sewer and drain piping made of plastic materials can be either plain or perforated.

SEWER AND DRAIN PIPING

There are a number of choices available for use in underground installations of waste and drainage systems in the plastic field. Although the pipe selected must have a certain degree of rigidity and those available to choose from are classed as such, all are somewhat flexible. The four commonly used possibilities are PE, ABS, PVC and *styrene rubber* (SR), of which the ABS and PVC are most readily available. In addition, there are some newer varieties of specialized PE drainpipe that are corrugated and flexible. Both the PVC and ABS piping come in standard lengths which vary from manufacturer to manufacturer and can be either solid or perforated (Fig. 2-43). Although the 4-inch diameter is the most commonly used, a number of choices can be had, depending upon the area. It may be necessary to special order any diameter other than the 4-inch size, as it is the usual choice for residental applications and may be the only one available in stock.

Fittings for solid and perforated pipe are different, so be sure to be specific when ordering. Although the range is relatively small, the nature of sewer and drainage systems is such that specialized or complex fittings will not be necessary. Those few fittings that are available will be quite sufficient in any case. Elbows with 45-degree, 90-degree, and standard 22½-degree directional turns in either the short or long form are available. The actual pipe lengths are made with one belled end and are simply fitted together end to end. When cut pieces or pipe sections without belled ends are to be joined, ordinary couplings are used. In situations where a horizontal branch line is connected to a main sewer line, a wye or sanitary tee is used. Pipes are closed off with caps. A cross is used where it is necessary to join four drain lines, such as in seepage beds and evaporation fields. There are a number of adapters available, as well as reducing couplings, to join varying sizes of lines and connect the plastic pipe to lines made of other materials. All fittings are the slip-on , solvent-welded type to provide for a tight joint where necessary.

SUMMARY

The information provided in this chapter has been given to introduce you to the vast amount of options available when selecting pipe and fittings. Each situation and installation will have its own individual requirements, based upon the purposes for which it is to be used and the preferences of the builder. In most situations, there will be a number of routes that can be chosen to produce the desired results, you are cautioned at this point to proceed slowly and thoroughly. Investigate and research all of the options in order to make the best choices suitable for your particular situation. Making the right decision here can save numerous hours of repair and maintenance in the long run.

Always be sure to contact the local plumbing authorities *before* beginning an installation of any kind, large or small. These officials will be most helpful, in most cases, in explaining the rules and regulations guiding such installations. They can tell you what types of pipe are approved for your area, and if there is any doubt as to which is best suited in each case, they may even be able to advise and provide invaluable assistance in the proper selection. A soil test will usually be required before a major installation for a new residential system is begun. These officials can make recommendations as to the proper location for a drainage system for the best possible results.

Any installation will require a certain amount of money, both in labor and in materials. Take the time at the onset of the project to plan the system carefully. Contact the necessary local officials. Check prices at more than one supply house before proceeding. Time spent on these matters now will prove worthwhile in the years to come.

Rules and Regulations

The need for plumbing codes, which provide specifics regarding the proper installation of plumbing systems, has long been recognized and dealt with. Not only are there national code books, but each locality has its own code, which may or may not conform exactly to those developed nationally. There are a number of organizations in the United States which publish standardized rules and regulations in the event that no local codes exist in rural areas. It should be pointed out here that for the most part, these codes consist of tried and true methods regarding the actual installation, materials, techniques, designs, and practices that are known to be safe, effective, and adequate for intended uses. It is imperative that you investigate thoroughly any and all aspects of these codes before going ahead with a plumbing installation and, indeed, even before working out a design for one. It may be found that there is very little uniformity from one area to another. It is up to each person to adhere to those rules and regulations that apply to his particular area.

BOCA

One organization existing in this country which has developed a standardized book which may be used by localities either partially or completely is the Building Officials and Code Administrators International, Inc. (BOCA). Founded in 1915, BOCA is a nonprofit service organization dedicated to professional code administration and enforcement for the protection of public health, safety, and welfare. BOCA's objectives span both public and professional interests, and the organization's primary activities include the following:

—To serve the public's need for sound and progressive construction regulation through promulgation of the BOCA Basic Code series of model regulatory construction codes. The Basic Codes are performance-oriented model codes responsive to the latest advancements in construction technology.

—To serve governmental units, code administration personnel, and related building industry professionals by providing authoritative technical, educational, and informational services relating to all specialty areas of code administration and enforcement.

The nation's oldest professional association for regulatory code officials, BOCA currently serves a membership that includes both regulatory officials and a wide variety of private sector building and construction professionals. This broad membership base of professional participation assists in maintaining the Basic Codes as responsive documents published and promulgated in the public interest.

The BOCA complete model code service program is dedicated to the improvement of building regulations and the effective administration, organization, and methods of enforcement of these regulations by professionally staffed state and local government units. The BOCA Basic Codes are maintained in their current, responsive state through a democratic public hearing and revision procedure which allows all interested parties the opportunity to both propose changes to code provisions and testify regarding such change proposals. Change proposals to the BOCA Basic Codes are either accepted or rejected by vote of the organization's active membership, which consists of practicing regulatory code officials. Voting on change proposals is conducted at the organization's annual conference, at which time final testimony is heard. Public hearings on proposed code changes are held prior to the conference at the annual BOCA mid-winter meeting. All of the Basic Codes are completely revised and published in new additions every three years. This procedure is maintained for its responsiveness to our rapidly advancing building technology, and for its ability to retain code content in the hands of professional regulatory code officials and above the reach of various special interests. The BOCA Basic Codes are designed to protect public health, safety, and welfare through efficient and effective use of available materials and current building technology.

In addition to the Basic Code series, the BOCA organization publishes a variety of other publications useful to professional building departments and code personnel. These include the monthly *Building Official and Code Administrator magazine*, a wide and complete variety of building department forms and permits, textbooks and handbooks regarding code administration and enforcement, and code agency organization recommendations. Along with the monthly magazine, the only monthly magazine available to professional code officials in the United States, BOCA membership benefits include a biweekly bulletin, copies of all research reports issued by BOCA's Research and Evaluation Service, and draft copies of all proposed code changes and new code publications.

This organization serves the very useful purpose of providing a great degree of standardization to the plumbing industry as a whole. This assistance is passed along to the individual homeowner by local officials who depend upon this source in establishing their own regulations. The BOCA Basic Codes are designed for adoption by state or local governments by reference only. Jurisdictions adopting them may make necessary additions, deletions, and amendments in their adopting document. The organization requests that when a jurisdiction adopts one or more of the Basic Codes, a copy of the adopting document is sent to their headquarters in Chicago, Illinois.

The Basic Plumbing Code establishes minimum plumbing standards in terms of performance objectives, implemented by specific requirements, rather than in rigid specifications. This makes possible the acceptance of new materials which can be evaluated under nationally recognized standards, without the necessity of adopting cumbersome amendments for each variable condition. In addition, by presenting the purposes to be accomplished rather than the method to be followed, the designer is allowed the widest possible freedom, and the development of new and innovative plumbing systems is not hindered.

BASIC PRINCIPLES OF THE CODE

This code is founded upon certain basic principles of environmental sanitation and safety through properly designed, acceptably installed, and adequately maintained plumbing systems. Some of the details of plumbing construction may vary, but the basic sanitary and safety principles desirable and necessary to protect the health of the people are the same everywhere. As interpretations may be required, and as unforeseen situations arise which are not specially covered in this code, the 23 principles which follow shall be used to define the intent.

■ **All Occupied Premises Shall Have Potable Water.** All premises intended for human habitation, occupancy, or use shall be provided with a supply of potable water. Such a water supply shall not be connected with unsafe water sources; nor shall it be subject to the hazards of backflow or back siphonage.

■ **Adequate Water Required.** Plumbing fixtures, devices, and appurtenances shall be supplied with water in sufficient volume and at pressures adequate to enable them to function properly, and without undue noise under normal conditions of use.

■ **Hot Water Required.** Hot water shall be supplied to all plumbing fixtures which normally need or require hot water for their proper use and function.

■ **Water Conservation.** Plumbing shall be designed and adjusted to use the minimum quantity of water consistent with proper performance and cleaning.

■ **Dangers of Explosion or Overheating**. Devices for heating and storing water shall be so designed and installed as to guard against dangers from explosion or overheating.

■ **Use Public Water and Sewer Where Available**. Every building with installed plumbing fixtures and intended for human habitation, occupancy, or use, and located within 200 feet of a street, alley, easement, or adjacent or abutting property in which there is a public water supply and sewer service, shall have a connection with the water supply and sewer.

■ **Required Plumbing Fixtures**. Each family dwelling unit shall have at least one water closet, one lavatory, one kitchen-type sink and one bathtub or shower to meet the basic requirements of sanitation and personal hygiene. All other structures for human occupancy or use shall be equipped with sufficient sanitary facilities as prescribed in this code, and in no case less than one water closet and one lavatory.

■ **Smooth Surfaces Required**. Plumbing fixtures shall be made of durable, smooth, nonabsorbent, and corrosion-resistant material, and shall be free from concealed fouling surfaces.

■ **Drainage System of Adequate Size**. The drainage system shall be designed, constructed, and maintained to guard against fouling, deposit of solids, and clogging, and with adequate cleanouts so arranged that the pipes may be readily cleaned.

■ **Durable Materials and Good Workmanship**. The piping of the plumbing system shall be of durable material, free from defective workmanship, and so designed and constructed as to give satisfactory service for its reasonable expected life.

■ **Liquid Sealed Traps Required**. Each fixture directly connected to the drainage system shall be equipped with a liquid-seal trap.

■ **Trap Seals Shall be Protected**. The drainage system shall be designed to provide an adequate circulation of air in all pipes with no danger of siphonage, aspiration, or forcing of trap seals under conditions of ordinary use.

■ **Exhaust Foul Air to Outside**. Each vent terminal shall extend to the outer air and be so installed as to minimize the possibilities of clogging and the return of foul air to the building.

■ **Test the Plumbing System**. The plumbing system shall be subjected to such tests as will effectively disclose all leaks and defects in the work or the material.

■ **Exclude Certain Substances from the Plumbing System**. A storm, surface or ground water, nor any substance which will clog or accentuate clogging of pipes, produce explosive mixtures, destroy the pipes or their joints, or interfere unduly with the sewage disposal process, shall not be allowed to enter the building drainage system.

■ **Prevent Contamination**. Proper protection shall be provided to prevent contamination of food, water, sterile goods, and similar materials from backflow of sewage. When necessary, the fixture, device, or appliance shall be connected indirectly with the building drainage system.

■ **Light, Heat and Ventilation.** A water closet, urinal, lavatory, bathtub, or shower shall not be located in a room or compartment which is not properly lighted, heated, and ventilated in accordance with accepted practice.

■ **Individual Sewage Disposal Systems.** If water closets or other plumbing fixtures are installed in buildings where there is no sewer within a reasonable distance, suitable provision shall be made for disposing of the building sewage by an approved method of sewage treatment and disposal.

■ **Prevent Sewer Flooding.** Where a plumbing drainage system is subject to backflow of sewage from the public sewer, suitable provision shall be made to prevent its overflow in the building.

■ **Proper Maintenance.** Plumbing systems shall be maintained in a safe and serviceable condition from the standpoint of both mechanics and health.

■ **Fixtures Shall Be Accessible.** All plumbing fixtures shall be so installed with regard to spacing as to be accessible for their intended use and cleansing.

■ **Structural Safety.** Plumbing shall be installed with due regard to preservation of the strength of structural members, and prevention of damage to walls and other surfaces through fixture usage.

■ **Protect Ground and Surface Water.** Sewage or other waste shall not be discharged into surface or subsurface water unless it has first been subjected to an approved form of treatment.

These 23 principles form the basis for any plumbing installation and should be considered as common sense rules which will provide for a system that will not be a health hazard and will function properly. Even if these principles are not provided for in your area, it is wise to adhere to them as they apply to your individual system.

PLASTIC COMPONENTS

You may find yourself in a situation where the local plumbing code disallows the use of plastic plumbing components or possibly allows restricted use for certain things such as ground water drainage, septic tank leaching fields, sewer lines, or whatever. Although this material has been widely approved on a national scale, there may simply be nothing in the code in your particular area that specifically permits the use of plastic materials. If you are completely sold on the use of plastic, there is a process by which it is possible to obtain a variance to the code, which will allow you to go ahead with plastic.

Obtaining a variance to the local building code provisions is not usually a terribly difficult chore, although it will require some thought, energy, and time. There will be certain specific procedures which must be followed. The actual details involved in this process may vary a bit from area to area, but the information can be obtained from either the local Planning and Zoning Commission or the Board of Appeals. During the appeal process, be sure to point out that plastic plumbing materials are widely accepted throughout the country by many local codes. The amount of time involved in this should be weighed against each individual's preference in regard to the use of plastic. As far as ease of installation for the do-it-yourselfer, plastic is probably the simplest to work with.

OTHER HOMEOWNERS' RIGHTS

There is one other point with regard to plumbing codes, and particularly the National Plumbing Code, that might be of interest. In many areas, a homeowner is prohibited by local codes from doing his own work. This is an infringement upon your rights as a property owner, particularly in the case of a single-family residence. The complete control of the premises should be yours alone, and in any event, a suitable series of inspections can be easily arranged in matters where health and safety might be affected. If there is anything that can be done to effectively fight such a stipulation, do so.

The *National Plumbing Code* is another organization which provides a degree of standardization for the plumbing industry. Its rules and regulations are pretty much the same as those given earlier from BOCA International. Section 1.10(c) of these codes states the following: "Any permit required by this code may be issued to any person to do any plumbing or drainage work regulated by this code in a single-family dwelling used exclusively for living purposes . . . in the event that any such person is the bona fide owner of any such dwelling and that the same are occupied by or designed to be occupied by said owner, provided said owner shall personally purchase all materials and shall personally perform all labor in connection therewith."

Therefore, the National Plumbing Code states that a homeowner can do his own plumbing work in his own home. With this in mind, there are a number of alternatives. The first is to ignore the code and just go ahead with the installation, but this is not really a very wise choice. The officials involved would no doubt be angered by this and may even state that the system must be taken apart and redone by a professional. The second alternative is to simply comply completely with all provisions as necessary. This may have the effect of almost doubling the cost of the installation, since it will be necessary to hire a licensed plumber. Hourly rates for plumbers are quite high, certainly higher than your own labor costs anyway. The third option is to file an appeal and seek a variance to whatever article or points you disagree with. This should be done with both dignity and intelligence. Regardless of the results of such an appeal, strict adherence to the decision is a must. The main thing to remember during this process is to keep an open mind. You will obviously be working quite closely with these officials during the whole installation, and it is important to treat them with respect. A good working relationship will prove invaluable.

SUBDIVISION CONTROLS

As more residential subdivisions and developed tracts of single-family homes spring up, the residents of these mini-communities are forming their own groups and preparing written controls to regulate certain aspects of their particular area. Such controls may come about due to either a lack of or possibly weak zoning, or maybe in addition to zoning administered at the state, county, or municipal level. Although these controls may vary from area to area and are entirely locally devised and written, they are effective and enforceable. If the rules and regulations have been properly written, they can be considered to be perfectly legal. Enforcement is usually accomplished at the first level by authority to turn off an individual's water supply if regulations are not compiled with. This is crude but also quite effective. The authority is backed up by recourse to obtaining aid from local law enforcement agencies, filing lawsuits, and by various other courses of legal action.

The control of these rules and regulations is usually maintained by a group of homeowners residing in the subdivision or development. They may be members of a board of directors, or perhaps members of an architectural control committee. The rules and regulations themselves often take the form of a basic set of rules which are subject to additions and changes as the need arises. These rules generally have little to do with the interior design or construction of the residence. Rather, they are concerned with such matters as the proximity of structures to property lines, the visual impact of various appurtenances such as fences, trash containers, etc. With respect to the home plumbing system, the items most regulated will probably be in regard to water supply and waste disposal.

Procedures will establish the method to follow in obtaining permission for a water tap and the actual location of the tap. Generally speaking, the process of tapping in a new water supply line will be required to make sure everything has been done properly. The rules may also specify restrictions relative to the placement of a septic tank with regard to the property line and the distance from other buildings. There may also be requirements as to the size of the tank and the leaching field. In any case, take the time to investigate what rules and regulations are in operation, and get a solid picture of how they might affect any plans you may have.

SUMMARY

As can be seen, there are many things that must be considered prior to the actual installation of a plumbing system. The information provided here may or may not apply to your particular area, but there will almost always be some sort of regulating body that you will be required to work closely with during the planning and installing of the system. There will be permits to obtain every step of the way, starting possibly with a permit to obtain access to a water supply, and continuing all the way to a permit to actually install the in-house portion of the piping. All of this may seem complicated and unnecessary, but remember that an improperly installed drainage system will not only affect you, but may also have adverse results to other adjoining systems as well. Safety features have been built into plumbing codes for many years which have proven time and time again to be in the best interest of those using the system. Do not ignore these regulations. They exist for good reasons, as a rule, and will continue to exist for a long time to come.

If a good working relationship is established with local officials right from the start, you will find that these people have a wealth of knowledge that they will be more willing to share than if they feel you are fighting against them. These officials will probably be your neighbors as well as law enforcers. Remember that along with their authority to approve or disapprove your system as designed, they also probably have years of experience. This knowledge will really come in handy, especially if they feel you are open and willing to "bend" if need be. The main thing to remember is that adherence to all rules and regulations is a necessary part of the design and installation of a home plumbing system.

4 Tools and Supplies

As with most do-it-yourself undertakings, it will be necessary to obtain an assortment of tools and supplies before beginning. If you are already somewhat of a general "repairman" around the home, you will probably have some of the more common tools needed for the simpler plumbing repairs explained in this text. When planning to either completely design a whole house plumbing system or make some extensive additions, you will probably have to make some additions to your toolbox. Also, many hardware stores offer rentals of the larger and more expensive tools; if cost is a factor, this may be the best route to pursue. I recommend that unless you are planning to continue your plumbing endeavor beyond your home, renting of the more specialized equipment is the most practical choice.

When purchasing equipment, keep in mind that the cheapest tools, although they may perform adequately, will tend to have a shorter life and a greater chance of malfunctioning. On the other hand, it is not necessary to purchase the highest quality available either. The medium-priced tools will suffice, and most hardware stores have inventory reduction sales periodically where real bargains can be had.

If it is not possible to rent the tools needed, another approach is suggested. Purchase the tools and specialized equipment, complete the project or projects, and then sell those for which you will have no further use. There is always a market for used plumbing equipment, and an advertisement in the local newspaper or possibly some of the "handyman" types of magazines will result in a number of prospective callers. If the tools were well-maintained while in your possession, it may be possible to recoup most of the original cost, with the difference amounting to less than the cost of rental.

Maintaining tools in an organized manner makes good sense, and it is stressed here as a further reminder. If you already have a home workshop, you probably have some sort of toolbox and an established work area. If this is not the case, purchase a container or build one which will house most of the smaller tools. Also needed will be some compartments of varying sizes in which to store washers, clamps, drill bits, and other supplies. Maintaining organization from the start will greatly aid in the efficient completion of any project, large or small, with a minimum of wasted time and effort.

SCREWDRIVERS

In many plumbing systems, clamps are used to hold the fittings together. In order to tighten these clamps, a variety of screwdrivers of different sizes and lengths will be necessary. Figure 4-1 shows some flathead screwdrivers. The longer ones will come in handy when work is being done in hard-to-reach places, something not uncommon in this type of work. Make sure the screwdrivers purchased have large, easy-to-grip handles. These will provide leverage which will be necessary when tightening the clamps. Don't be afraid to apply too much pressure to a clamp, as most of them are made of stainless steel and are quite durable. If one of these devices is stripped or a screw snapped, a hole can be drilled through its body lengthwise and the screw removed. Although flathead screwdrivers will be required in most plumbing applications, it is recommended that a few Phillips head screwdrivers be purchased. There will be times when their use will be required.

PLIERS

The type of pliers found in most homes is shown in Fig. 4-2. These are commonly called *household* or *gas pliers*. They are normally 6 to 8 inches in length and are useful when working on small fittings or for holding or unscrewing small sizes of pipe after they have been loosened with a

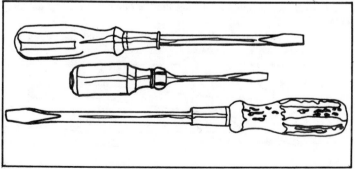

Fig. 4-1. Flathead screwdrivers in varying sizes will come in handy to tighten clamps which hold some fittings together.

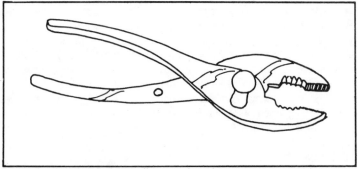

Fig. 4-2. Household or gas pliers are found in almost every home.

wrench. They are *not* meant to be substituted in place of wrenches. This practice can easily damage the pipes being worked on.

When working on pipes larger than 1½ inches in diameter, *adjustable slip-joint pliers*, shown in Fig. 4-3, will provide additional flexibility and ease in maintaining a firm grip on the pipe. These pliers have several channels cut across the side of the one arm, enabling their jaws to be set wide apart with no danger of loosening their grip under pressure.

For added convenience and freedom of movement, the *locking pliers*, or *visegrips*, shown in Fig. 4-4, can be locked into place. This leaves both hands free for other tasks once they are clamped in position. The jaw opening, which has teeth composed of deep grooves, is adjusted by means of a small screw at the end of the handle.

WRENCHES

The two types of wrenches most commonly used by plumbers for nuts and fittings are either *fixed* or *adjustable*, and some examples are shown in Fig. 4-5. The stillson or pipe wrench (top) is primarily used to either thread or unthread a section of pipe. These wrenches should be purchased in a number of sizes regarding length and jaw capacity and are normally used in pairs, one holding the fitting and the other unthreading the pipe section.

The adjustable *open-end* wrench (middle) is used for square and hexagonal nuts such as those found inside faucet and valve assemblies. A

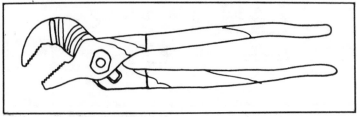

Fig. 4-3. Adjustable slip-joint pliers.

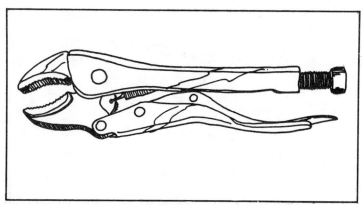

Fig. 4-4. Locking pliers or vise grips.

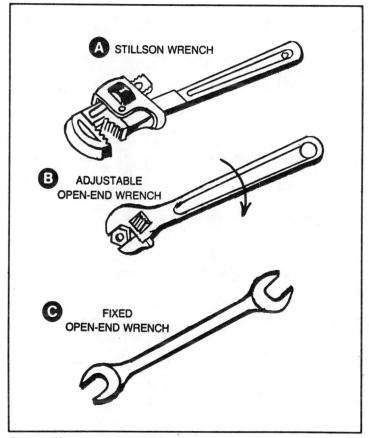

Fig. 4-5. Wrenches may be either fixed or adjustable.

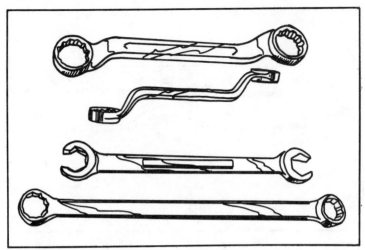

Fig. 4-6. Box-end wrenches are useful, as long as space is not a problem.

12-inch size wrench will be the one most commonly used, but a 6 or 8-inch size will come in handy when working on smaller pipes. These wrenches can be purchased in sets, with sizes ranging from as small as ¼ inch to 1¾ inch.

Fixed-end wrenches, although being more limited in their use, are also a useful addition to the home plumber's toolbox. They can also be purchased in sets, with sizes ranging from ⅜ inch to 1¼ inch. A fixed-end wrench is pictured at the bottom of Fig. 4-5.

Also available are *box-end* wrenches (Fig. 4-6). These can be quite useful. Because of their closed end, they can't be used unless the nut is in the open to provide space for the wrench to be slipped down over it. The same can be said of socket wrenches (Fig. 4-7)

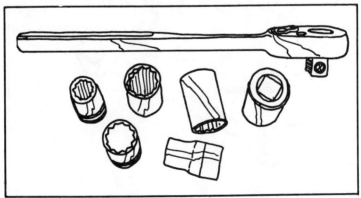

Fig. 4-7. Socket wrench.

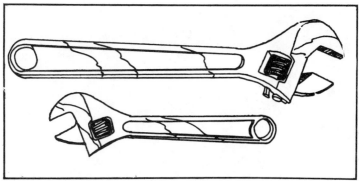

Fig. 4-8. Adjustable spud wrenches come in varying sizes.

In instances where flat-sided or delicate fittings such as brass or copper are used, it is recommended that you use an adjustable spud wrench (Fig. 4-8). Using a stillson or a pipe wrench on these fragile materials will cause damage. Another use for this wrench is for turning the large, flat-sided nuts found on drainpipes under a lavatory or kitchen sink. The closet spud wrench, a smaller version of the adjustable spud wrench, is perfect when working in close quarters, such as under a sink or lavatory.

Another wrench used for working on delicate fixtures made of brass, aluminum, plastic, or lead is the strap wrench (Fig. 4-9). It consists of a heavy webbed strap attached to a handle. The strap is looped around the pipe, passed through a slot in the handle, and drawn tight. When the handle is pulled, the loop is tightened, making a firm grip on the pipe and forcing it to turn.

The *chain wrench* shown in Fig. 4-10 is a heavy sprocket chain attached to a steel handle. Like the strap wrench, it grips the pipe and holds it firmly, but only turns in one direction. Due to its slimness and light weight, this wrench is ideal for those hard-to-reach places and is particularly suitable for large pipes such as those on cast-iron drains.

When working on faucet assemblies and/or spray attachments while fixed in place under a sink, the *basin wrench* shown in Fig. 4-11 will provide for easy access. It is also quite small and light, making what might be an unpleasant chore more simple.

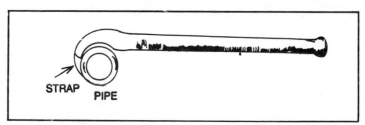

STRAP   PIPE

Fig. 4-9. A strap wrench is useful when working on fragile fixtures.

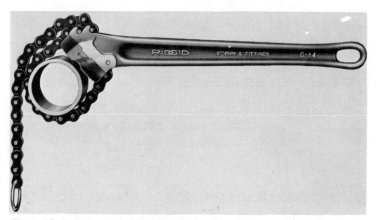

Fig. 4-10. A chain wrench is ideal for working in cramped quarters.

SAWS AND PIPE CUTTERS

The type of pipe you will be working with will dictate the amount of specialization needed in saws. When working with plastic materials that are easy to cut, a simple handsaw of whatever type you prefer will be sufficient. Make sure a fine-toothed blade is used rather than a coarse-toothed type. By doing so, the cut will be much less ragged and freer from burrs and chips that will have to be repaired before assembling the pipes. Some blades are specifically designed for woodworking, so make sure the saw is equipped with a metal-cutting blade. Some types of plastic pipe are rather abrasive, even though they may seem to be soft.

Fig. 4-11. A basin wrench is very lightweight and handy also for hard-to-reach areas.

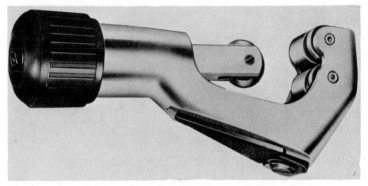

Fig. 4-12. A small tube cutter used for copper and plating tubing.

When cutting thick-walled iron, brass, copper, and steel pipes, you will need a pipe cutter. This is a large and quite heavy device. The pipe is fed in between two rollers and the handle is gradually tightened to make the cut. A small version called a *tube cutter* is shown in Fig. 4-12. It is primarily used for cutting copper and plating tubing.

PIPE THREADERS

Although the fittings used to connected pipes are factory-threaded, steel and iron pipes are usually not. You will need to thread them yourself. This can be done manually with a pipe threader stock outfitted with a suitable size of pipe die and guide bushing. The pipe is cut by dies held in a diestock which has long handles to turn it. The stock is able to hold different sizes of dies for pipes of different sizes. Once you have the proper size die in the stock, clamp the pipe in a vise and slip the diestock over the pipe end with the guide on the inside (Fig. 4-13). Push it further over the pipe until the die has caught the pipe surface. With the die pressed firmly against the pipe, turn the stock gradually in a clockwise direction. When you have cut a few threads and the die has taken a firm hold on the pipe, apply a generous quantity of cutting oil to the pipe end and the threads of the die. Proceed with the thread cutting. For every one-half turn forward, back the die up one-quarter turn to clear off the metal chips. Continue this process until the pipe end emerges from the die. To disengage the tool, turn it counterclockwise.

Pipe threading can be a difficult and time-consuming process, and it may not be feasible to spend a great deal of time on this portion of the project. In this case, most plumbing supply houses offer this service. The cost may be offset by the amount of time saved in labor.

PIPE REAMER AND FILE

Once a pipe has been cut, there will be noticeable burrs and ragged edges on both the inside and outside. These will need to be repaired before

101

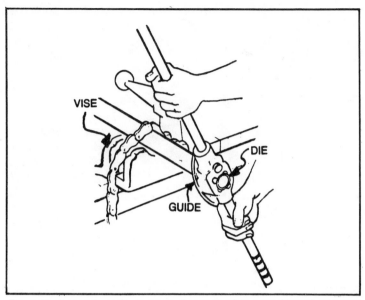

Fig. 4-13. A pipe threader is normally attached to a workbench or some stationary spot in a workshop.

assembly. Those on the outside will make the threading process difficult, if not impossible. Also, any contact with your hands may cause cuts and scratches. If the edges are not smoothed on the inside, they may cause clogs or obstructions at a later date when they are chipped off from the constant flow of water or drainage running through the pipes. This can be prevented by using a reamer such as the one shown in Fig. 4-14. This is a spiral-shaped file which can have a turning handle or may be fitted in a brace. To use this tool, simply insert it into the pipe and rotate it until it does so in a smooth, unobstructed manner. Remove it and run your finger around the inside edges of the pipe to check for smoothness and repeat the process again if necessary. The burrs on the outer edges of the pipe can be removed with a simple flat file.

VISES

One of the difficulties encountered in the cutting of pipe is associated with keeping it immobile during this process. This can make it hard to get a clean, even cut. Most people find that by laying the pipe on a sawhorse and planting a foot solidly on it, the problem is alleviated somewhat. A vise such as the one shown in Fig. 4-15 will hold the pipe firmly in place and leave your hands (and feet) free for other tasks. This vise is mounted in a convenient place on a workbench, and its serrated jaws can be adjusted to fit the size of pipe being cut. When clamping plastic pipe in a vise, be sure to exercise some caution. Plastic can crush or split open very easily if too much pressure is applied to its surface. Remember that the vise is merely used to hold the pipe in place and does not need to be tightened to a point where it can cause damage to the pipe. A vise is a worthwhile investment, as its uses beyond the plumbing field are many and diversified.

MEASURING TOOLS

Two types of measuring devices will come in handy, both in laying out the plumbing system and in measuring pipe lengths. The first is the flexible steel measuring tape shown in Fig. 4-16. For efficiency and convenience, the automatic return type is recommended. These come in almost any length desired, from as short as 6 feet up to as long as 25 feet. It is up to you to determine what your needs will be for the project or projects to be undertaken and purchase a length suited to those needs. For longer lengths of pipe and larger layouts, purchase a winding reel tape, which is made of flexible steel or cloth and comes in 50 or 100-foot lengths. This tape is pictured in Fig. 4-17.

In those situations where a flexible steel tape won't remain in position or extend to its maximum length without drooping, a *carpenter's folding rule* such as the one shown in Fig. 4-18 might be more suitable. These are made of wood and are completely self-supporting in the vertical position

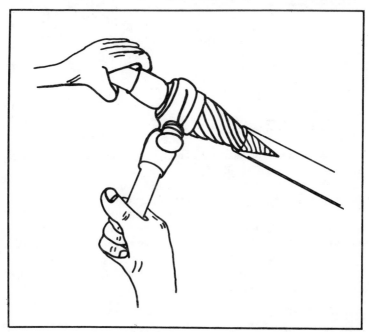

Fig. 4-14. Pipe reamer used to smooth both the inside and outside of a pipe after it has been cut.

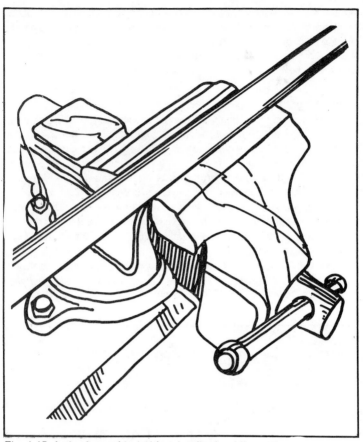

Fig. 4-15. A vise is used to hold a pipe firmly while cutting.

and nearly so in the horizontal position. This device can make measuring easier and usually unfolds to a 6-foot length.

Although a pencil may suffice in some instances to mark measurements on sections of pipe, some types of pipe are hard, glossy, or jet-black, making any pencil marks invisible. In these situations, it will be necessary to use a marking tool other than a pencil. Almost anything will do, as long as its edge is sharp enough to make a small cut at the desired point on the pipe. A chunk of fine-toothed hacksaw blade salvaged from a broken one will do the job nicely. A scratch awl such as the one shown in Fig. 4-19, or even an ice pick, can also be used.

FLARING TOOL

When working with plastic pipe and tubing, connections are commonly made through the use of transition or adapter fittings if the pipe is to

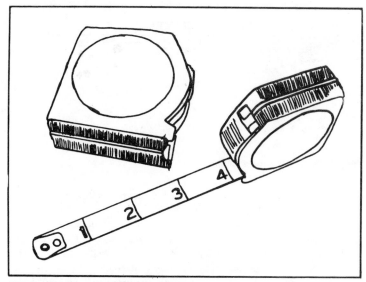

Fig. 4-16. A flexible steel measuring tape.

be attached to an existing metallic section. In some instances, however, the proper size pipe or tubing may be available that will be identical to the copper tubing. Under these conditions, a standard brass flare of the type usually associated with copper tubing can form the attachment, thus doing away with the need for specialized adapter fittings. In order to connect either copper or plastic tubing to valves and fittings of the flare type, its end must be flared, or expanded, and turned outward. This is done with a flaring tool (Fig. 4-20). The tool itself is made up of a clamp which locks the

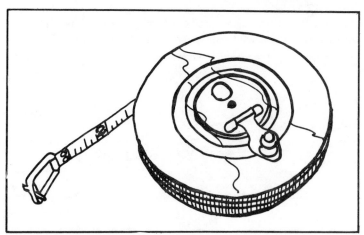

Fig. 4-17. A winding reel tape.

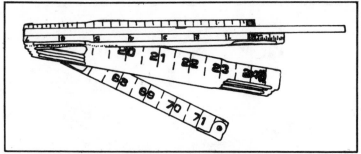

Fig. 4-18. A carpenter's folding rule.

tubing in place after insertion, a vertical yoke with a compressor screw and drive handle, a split die block of hardened metal with holes to accept various sizes of tubing, and a solid metal cone fitted to form a 45-degree flare or bell shape. To make a flare, insert the tubing into the appropriate hole from beneath the device, and clamp it in position after the sleeve nut of the fitting has been slipped over it. Turn the compressor screw which will cause the cone to move down and engage the end of the tubing. As the handle is turned, pressure is applied to the tubing and the flare is formed. A flaring tool is an inexpensive device and can be found in any plumbing supply store.

PROPANE TORCH

When working on a system comprised of copper pipe and brass fittings, the joints will be soldered together. Likewise, disassembly will require melting the connection. This is done with a propane torch (Fig. 4-21). A common torch consists of a bottle of liquid gas connected to a nozzle and valves. The nozzle can be purchased in varying sizes to produce the desired amount of heat and shape of flame. For most plumbing applications, a ½-inch diameter nozzle is recommended. All propane torches

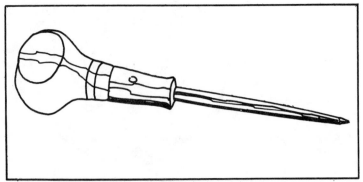

Fig. 4-19. A scratch awl for marking pipe.

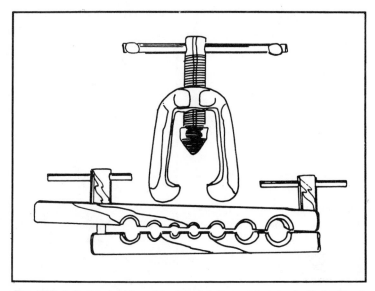

Fig. 4-20. A flaring tool is used when joining copper or plastic to flared fittings.

come with manufacturer's directions for use, but a few precautions are given here as an additional reminder.

■ The burner should be disconnected from the cylinder when the torch is not in use.

■ Do not store the propane tank near heat.

■ Keep the torch out of the reach of children, preferably on a shelf in a work area, *not* in a living area.

To operate a torch, connect the torch burner tightly to the tank or cylinder by screwing it on. Open the control valve slightly until a hissing sound is heard. Ignite the torch immediately with a match or cigarette lighter. Allow the flame to burn for a couple of minutes so that the burner will heat up. Open the control valve to obtain the desired size of flame, and the torch is ready for soldering or desoldering. The flame is applied to the fitting and pipe until the solder melts and the pieces can be pulled easily apart. Assembly is accomplished in the same manner, using fresh solder and soldering flux.

DRILLS

There are two basic stages when beginning a whole house plumbing system from scratch. The first stage consists of assembling all the pipes and fittings and valves in place in the open, unfinished structure while everything is still accessible. This initial stage is called *roughing-in*. Once this phase has been completed, the building is then finished off and the

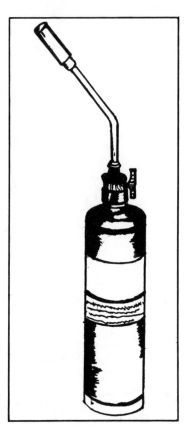

Fig. 4-21. A propane torch is used for systems with copper or brass fittings.

second stage, known as the finish work, can take place. This consists of installing and connecting appliances, basins and sinks, showers and tubs, faucets and sill cocks, trim plates, and so on. To accomplish these operations, an assortment of general carpentry tools will be required. Since a large portion of the piping will travel through the structure of the building, some boring equipment will be necessary.

The most common method used to bore holes is with a bitstock and an assortment of auger bits. Figure 4-22 shows a ratcheting bitstock which can be locked and turned in full revolutions. This will be quite useful in hard-to-reach areas. Simply adjust the ratchet lock in one direction or the other and the bitstock can be turned in quite small arcs. Auger bits come in varying sizes, ranging from ¼ inch to 1 inch in diameter, in increments of 1/16 inch. Although these standard bits are somewhat short and are not useful in boring deep holes, there are several types of extra-long bits available for such applications.

If larger holes are desired, an adjustable expansive bit can be fitted to a bitstock. This bit features an adjustable wing blade on an auger shank

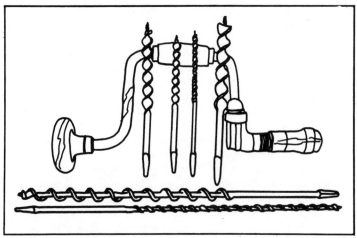

Fig. 4-22. A ratcheting bitstock may be used to bore holes in the structure to allow for the installation of the pipelines.

which can be adjusted to any size hole diameter, as long as it is within the adjustment range of the blade (Fig. 4-23).

The process of boring holes is not an easy one, and a simpler method can be had by using an electric drill of the type used by carpenters. Drills come in many sizes, but the most commonly used are ¼ inch, ⅜ inch and ½ inch. The size refers to the bit-diameter capacity of the chuck on the drill and is also an indication of its power. The larger a drill's capacity, the more powerful its capabilities. A ⅜-inch drill will best suit most plumbing applications. A ¼-inch drill, though quite compact, lightweight, and easy to use, has limited capabilities in plumbing and is only recommended in those situations where it will not be used for a whole house system. Even

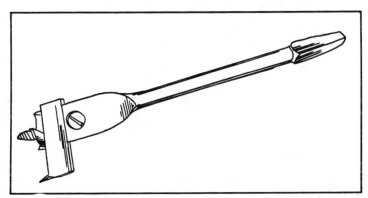

Fig. 4-23. This adjustable expansive bit is attached to the bitstock if larger holes must be drilled.

when drilling only a few holes with a ¼-inch drill, proceed very slowly to avoid overheating.

A ⅜-inch drill is capable of doing quite hard work on a constantly repetitive basis without danger of overheating or burning out a motor. Even though its capacity is only increased by ⅛ inch, this difference makes it much more powerful, and drilling 1-inch holes through wood becomes a much simpler task.

There are many models and sizes of drills available on the market today, and prices vary considerably. As was mentioned previously, when purchasing tools, it is always best to avoid the lowest priced devices. If at all possible, power tools especially should be of the highest quality affordable for the obvious reasons. They will far outlast their low-priced counterparts and save time and money over a long period of time. A typical moderately priced ¾-inch drill is pictured in Fig. 4-24.

Bits for electric drills come in varying sizes and types. They are tailor-made to suit the individual purpose for which they are to be used. Figure 4-25 shows an assortment of twist drill bits, which are used to drill nonabrasive materials. As can be seen, these bits are quite short and are not meant to be used when a deep hole is required.

The power wood-boring bits shown in Fig. 4-26 will be familiar to those readers with some carpentry background. These are most commonly used in general construction work for boring through wood. *Spade bits*, as they are sometimes called, are made of a hardened alloy steel and

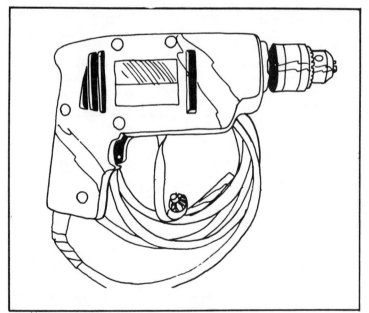

Fig. 4-24. A typical ¾-inch electric drill.

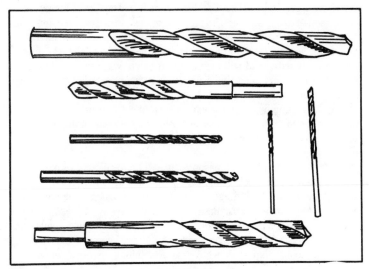

Fig. 4-25. An assortment of twist drill bits.

consist of a flat blade with a centering and starting point integral with a ¼-inch shank. They are very inexpensive, can be used in any size drill, and are available in a variety of diameter sizes. Another advantage of this type of bit is that it can be reused a number of times before it becomes necessary to replace it. Simply remove it and resharpen it with a file and it can be installed for further use. The maximum depth hole that can be bored with a speed bit is approximately 4½ inches, which is sufficient for most

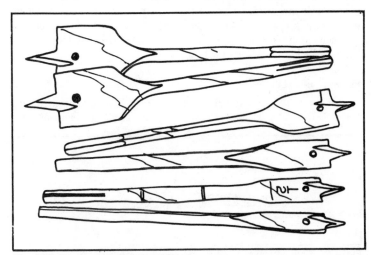

Fig. 4-26. Power wood-boring bits.

plumbing applications. When a hole must be bored deeper, a bit extension can be added to a speed bit. The bit is set in the end of the extension shank and secured with a setscrew. The other end of the shank is chucked in the drill in the normal fashion, thus extending the length of the bit to over 16 inches. This extension will also come in handy when cramped quarters necessitate holding the drill a distance away from the hole, as long as the hole to be bored does not have to be too deep. Alternately, an electrician's bit, which is simply a long carbon steel twist drill about 18 inches long, may be used in a similar manner. A lesser amount of flexibility is found in this type of bit, though, because it is not available in as many varying sizes of diameter.

When drilling through masonry, it will be necessary to purchase a special masonry bit such as the one shown in Fig. 4-27. These bits are specially equipped with tungsten carbide tips designed to remove chips associated with masonry. Limited size variations are available, but they do come in all three shank diameters common to the drills discussed here.

Again, for a few simple repairs and/or alterations, a ¼-inch drill will suffice. For construction of a whole house plumbing system, I recommend the purchase of a ¾-inch drill.

POWER SAWS

Although handsaws were previously discussed in this chapter, no reference was made to their powered counterparts for plumbing applications because their use in actual pipe assembly and cutting will not be necessary. However, in the installation process where either rafters or studs may need to be notched, a power saw will do the job in half the time. It is a simple task to set the blade to the proper depth, make a few quick passes across the stud, tap it with a hammer, and the notch is completed. It is not recommended that you go out and purchase a power saw for its limited applications in plumbing work, but if one is already part of your complement of tools, its use will save some time when used instead of a handsaw. A typical power saw is shown in Fig. 4-28.

PLUMBS AND LEVELS

In most instances it will be desirable to have pipelines running perfectly vertical or plumb. You can attain such an alignment using refer-

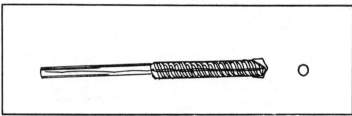

Fig. 4-27. Special masonry bit for drilling through masonry.

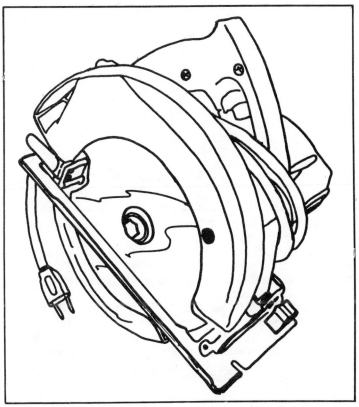

Fig. 4-28. A typical electric saw.

ence points, but this is not always possible. The results of approximate measurements of this sort may not become apparent until the plumbing system is completely installed. Any repairs at this point will be both difficult and time-consuming. To avoid such circumstances, a plumb line and bob, shown in Fig. 4-29, can be used to align the pipes properly. This simply consists of a length of heavy string, most often on a reel, with a pointed steel or brass weight attached to its end. The bob is dangled to achieve a perfectly vertical line, and the measurements can be transferred from one point to another directly below, thus assuring that the two points are aligned with each other.

When installing horizontal pipelines, a slight pitch is needed for ease of flow through the pipes. This is obtained through the use of another common carpentry tool, the *spirit level* shown in Fig. 4-30. There is a small bubble inside the center of the level which is used to indicate when a line is dead level. Whereas a carpenter uses this tool to obtain an exact horizontal or vertical line, the plumber determines the pitch desired by the degree to

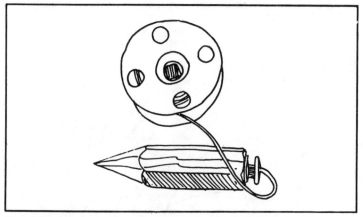

Fig. 4-29. A plumb line and bob is used to insure that lines are perfectly vertical or plumb.

which the bubble inside the level is off-center. A smaller version of the spirit level, known as the *torpedo level*, is shown at the bottom of Fig. 4-30 and is ideal for shorter runs.

Another carpenter's tool that can be useful when laying out a level line or a series of points along which a pipeline must travel is the carpenter's *chalk line* shown in Fig. 4-31. Once you have established the desired path, simply stretch the chalk line along the route and snap it. The result will be a visible chalked line which will provide a perfectly straight guide.

HAMMERS

Practically every household has at least one hammer, and it will be almost impossible to get through a whole house plumbing system without one, especially if there will be any structural changes required prior to installation. Hammers come in a wide variety of sizes and types, and you can pretty much choose whatever will suit your individual requirements for your particular situation. A few of the choices available are pictured in Fig. 4-32.

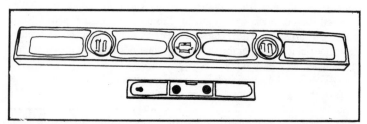

Fig. 4-30. A spirit level is used to provide the proper pitch on horizontal lines.

114

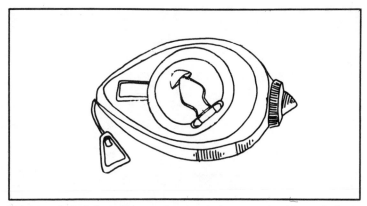

Fig. 4-31. A chalk line will be useful when laying out a level line to indicate where a pipeline will travel.

CHISELS

For hard-to-reach places in cramped quarters where the use of a hand or power saw is not possible, a wood chisel may be substituted. The chisel can be used to notch away a portion of a stud or rafter to allow for the passage of a pipeline (Fig. 4-33). Chisels come in several sizes, ranging from ¼ inch to 2-inch widths, and for plumbing applications, those toward the lower range will be adequate. Make sure the blade is constructed of hardened and tempered high-carbon steel, as this is the most durable and long-lasting, and that it has two precision ground cutting edges.

MISCELLANEOUS SUPPLIES

There are a number of supplies that the plumber should always have on hand in order to carry out any type of plumbing work. These should be kept nearby, preferably in the toolbox, for easy access at all times.

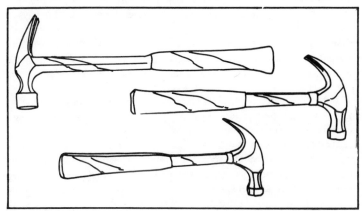

Fig. 4-32. Some of the more common hammers in different sizes.

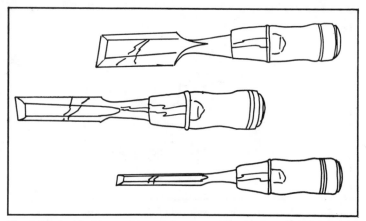

Fig. 4-33. A chisel is used to notch away portions of studs or joists.

When either installing or disassembling copper and brass pipe fittings, a torch such as the one discussed previously in this chapter will be used in conjunction with solder and flux. Solder comes in a number of types and forms. Resin-core solder is primarily used for electrical applications and should never be used in pipework, as is also the case with the acid-core solder. Uncored solder is normally used in plumbing work, and as its name implies, it contains no flux in its core. This type of solder comes in rolls and its quantity is measured in pounds. Solder is made of a mixture of tin and lead and can be purchased in varying degrees of either of the two, such as 50 percent tin and 50 percent lead, or 40 percent tin and 60 percent lead. The higher the percentage of lead, the softer the solder, and an equal 50-50 mixture will probably give the best results.

Because this type of solder has no core, the flux must be applied as a separate operation. Flex normally comes in small tins and the type purchased should be the noncorrosive paste, resin-base type. This paste is applied with a small metal-handled brush called an acid brush. Alternately, a small piece of cloth can be used. All three of these items are quite common and can be found in any plumbing or hardware store for a reasonable price. A roll of solder, a tin of flux, and a small brush are pictured in Fig. 4-34.

Another material which will be necessary when working with copper, iron, and steel pipe is pipe-joint compound, or *pipe dope*, as it is more commonly referred to. Pipes are never connected dry. Instead, the dope is applied to the threads to provide lubrication and a seal. If this is not done, difficulty will be encountered in any attempt to join the pipes, and even if this part of the procedure is accomplished, the system will most definitely leak. Not only does the application of this compound ease assembly and prevent leakage, it is a great deterrent to rust. A plumber's toolbox would not be complete without a can of this material.

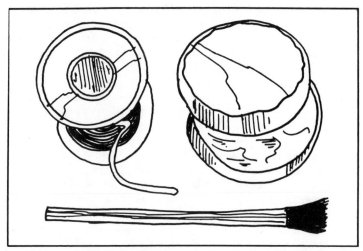

Fig. 4-34. A roll of solder, a tin of flux, and a small brush are used to join or disassemble copper or brass fittings.

With the introduction of Teflon to the consumer market, a Teflon plumber's tape is now available. Although this material costs a little more, it has some advantages over its time-honored counterpart. The finished project has a much neater appearance. Because only a few strands of the tape around the threads complete the job, it can also be faster and a lot less messy to work with. Both of these materials are pictured in Fig. 4-35.

There are quite common items that have not be discussed previously, the first of which is a must for any household. Almost everyone has at one

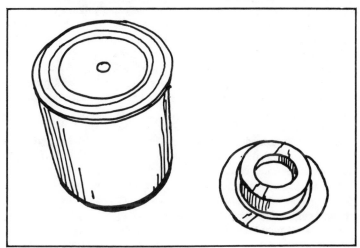

Fig. 4-35. Teflon tape.

time or another had a clogged sink or toilet, and this is most frequently corrected by the use of a *plunger*, or more aptly put, a plumber's friend. This tool is simply a rubber suction cup mounted on a wooden handle. The rubber section is placed over the drain tightly and pushed, applying a degree of pressure which serves to unclog the drain.

In situations where a simple plunger will not satisfactorily clear the drain or toilet, a snake can be used. Also known as the hand auger, this tool is a long coil of metal usually equipped with a handle whose end is inserted into the clogged drain up to the predetermined area of blockage. The handle is then turned, always in the same direction, pushing it in and out until the obstruction is broken and the auger is moving freely. This is a quite simple procedure and will most often succeed on the first or second try. One point to remember: *always* make sure that the auger is cleaned thoroughly before use, as its insertion into the pipes when dirty may cause contamination or spread bacteria.

PLUMBING SAFETY GEAR

As with any undertaking of this sort, there are a certain amount of precautions that will make for a safe completion of a project. It's always a good idea to equip yourself with the necessary safety equipment to avoid any type of injury, large or small. A good pair of safety goggles such as the ones shown in Fig. 4-36 are mandatory and will shield the eyes from dust, small pieces of metal, splinters, and the like. Not having some sort of protection for your eyes is not only dangerous; it is downright foolish.

When working with power tools, it is always good to give them the respect they deserve and exercise a certain amount of caution in operating them. All power saws are equipped with a metal shield to protect the operator's hands from injury when cutting, but remember to always make sure you have a firm grip on the handle before applying the blade. Saws have been known to "jump" out of the user's hands. Should the blade strike an obstruction, the entire device may kick back toward the operator. Due to these factors, firmness of grip and strict attention to the work is a must. Should you possess power tools which are missing the necessary safety attachments, do not attempt to operate them until repairs have been effected. These pieces of equipment can be extremely hazardous when not in completed form.

To protect the hands from cuts and scratches which can occur from contact with rough edges of pipe, a good pair of heavy leather gloves may be worn. Although they may, in some situations, make for a certain degree of clumsiness when working with smaller sections of pipe and in cramped areas, the protection they offer far outweighs this slight inconvenience. In situations where finer handwork is required and there is no danger from ragged edges, they may be removed temporarily and left within reach for easy access when their use is once again dictated.

Another item which deserves mention is a good set of earplugs. Studies have indicated that damage to the ears can occur when sound

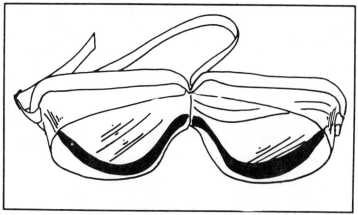

Fig. 4-36. Safety goggles are a must to shield eyes from metal, splinters, etc.

levels above 70 dBa (adjusted decibels) are generated. These studies also state that most power tools run at levels over 90 dBa, thus making close proximity to them for extended periods of time a possible danger to the ears. A small set of earplugs are very inexpensive and well worth the price in protecting this most valuable sense organ. Although it may seem or look foolish to a neighbor or family members, this simple safety precaution can save you from a hearing loss that may not be apparent immediately and which is not usually repairable at a later date. Anyone who has been employed in a noisy factory can attest to the fact that strict safety regulations require the use of some sort of protective device covering the ears at all times. Do yourself a favor and purchase these plugs if the use of power tools will be included in any of your projects.

SUMMARY

I have attempted to cover a large assortment of tools in this chapter and to give a brief description of their use in plumbing applications. Depending upon the size and scope of each individual's project, some of the tools discussed may not be necessary. It may be that you will have to purchase some additional tools not discussed here. Alternately, some of the more specialized tools can be rented, or the work may be done at a plumbing shop if time and/or cost is a factor. In the beginning stages of a project, take the time to sit down and consider all the options available and choose the route which will provide for the least expensive method of completing a project smoothly and efficiently. Due to the great variety in prices and models of tools available today, plan to spend some time going from store to store and locating the best buys in regard to both price and quality. Individual preference will also play a part in tool selection because of the many styles and designs available. Once you have the complement of tools that you will require, they should be stored in an organized manner where they will be readily accessible when needed.

Up to this point, a basic introduction to a home plumbing system, its components and tools, and the regulations relating to its installation have been provided. With this as the groundwork, you are now ready to begin to learn the actual procedures that are provided in the remainder of this text.

Wells and Pumps  5

Wells are the major source of water for those living in rural areas. Those living in cities or towns and their surrounding metropolitan area generally rely upon a municipal water supply and need only be concerned with running pipeline to an existing water main. Some cities supply this water, which is chemically treated to remove impurities, by means of controlled lakes or reservoirs; others obtain it through the use of deep underground wells. A municipal water supply system will be discussed in detail in another chapter.

One of the initial steps before planning a residence in a rural area where an individual water supply must be established should be to check with local authorities as to accessibility and availability. Charts and maps of the area may be available to assist in this process. These maps may indicate where the known water pockets are in the area. Another possible source of information may be neighbors. They can provide information relative to their experience in locating a sufficient water supply.

Based on the information obtained, you should be able to determine what will be involved in supplying water to the proposed residence. Depending upon the situation, there may be several choices; then again, the only option available may turn out to be the most costly. It may be necessary to dig 300 feet below the surface to obtain an adequate water supply, and those trained in this type of work usually charge by the foot.

DETERMINING WATER REQUIREMENTS

Before proceeding to establish a water supply, it will be necessary to do a bit of arithmetic to come up with an estimated usage. This is normally figured in gallons per minute. The method used to determine usage is

121

called the *unit method*. A unit is equal to 1 cubic foot of water flowing each minute, which comes to approximately 7½ gallons per minute. An organization known as the *Uniform Plumbing Code Committee* has performed tests on the various fixtures normally installed in a residential plumbing system and established unit values for each of these fixtures. The results of these tests, which are considered to be standard, are shown in Table 5-1. Using these figures, an estimated usage can be determined. First, add up the unit values for the fixtures that are applicable to the proposed residence. Multiply this by 7 ½ to arrive at a gallons per minute usage. Under normal living conditions, all of these fixtures will rarely be running simultaneously. Table 5-2 provides multiplying factors which are used to determine an approximate flow based on normal usage.

POINT WELLS

A *point well* is a good source of easily tapped water found just below the ground surface. In areas where there is a high and fairly constant water table and where water permeation occurs regularly, sizeable water pockets or flow layers can be found in gravel strata only a few feet deep. The procedure for obtaining a supply such as this is quite simple and can be accomplished by one or two persons with a minimum of time and materials. A special device called a well point is attached to a length of heavy steel pipe and driven into the ground like a stake (Fig. 5-1). As the point is driven deeper and deeper into the ground, short lengths of pipe are

Table 5-1. Standard Unit Values for Common Household Fixtures.

NAME OF FIXTURE	UNIT VALUE
Toilet	6
Bathtub	2
Shower	2
Lavatory or washbasin	1
Slop sink	3
Kitchen sink	2
Laundry tray set	2
Clothes washer	3
Dishwasher	3
Bathroom group with toilet, lavatory, tub, and shower	8

Total Number Of Fixtures	Multiplying Factor to Give Simultaneous Flow
1 to 5	1.0 to 0.5
6 to 50	0.5 to 0.25
More than 50	0.25 to 0.10

attached until a sufficient water flow is established. It is not unusual for two or more points to be driven in this manner. These are then interconnected to a single supply line to a building. This method would possibly be used in situations where the desired gallons per minute flow cannot be established through the use of one point well alone.

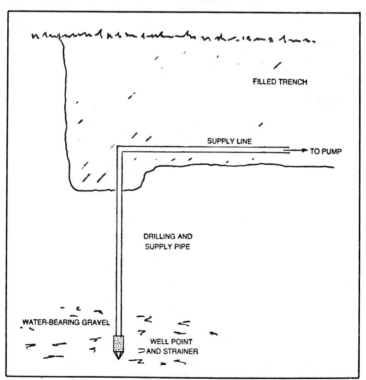

Fig. 5-1. A point well may be dug with a minimum of tools and supplies.

Once there is sufficient water flow, the water service line is then extended to the building and attached to a pump. Shutoff valves are normally installed in the service line at a point close to the pump. This allows for disconnection of the system for repair work, so that the system will not have to be primed when it is reconnected. Figure 5-2 shows a typical arrangement for a shutoff valve in conjunction with a water pump.

GRAVEL WELLS

A *gravel well* is basically the same as a point well in that it taps a water supply located near ground level. This type of well is normally a bit deeper and will provide a better and more constant supply of water. The water supply is obtained through the use of a heavy steel pipe which is driven into the ground until a water-bearing gravel layer is reached. Water is then forced down the pipe under pressure and pumped back up again. The water being pumped out will carry with it a certain amount of sand and gravel, which is normal. This process is repeated until a large cavity is hollowed out at the end of the well pipe. This cavity becomes a sort of a reservoir. After the water has had an opportunity to clear itself out by a settling process, a continual supply of clear water can be had. The water service line is then connected to the well pipe and can be underground to prevent frost damage, or if this is not necessary, it can be installed at the surface or in a well pit. The line is then extended to the building and connected in the appropriate manner to the pump. A shutoff valve is suggested with this type of installation for ease in making repairs when necessary.

DUG WELLS

A *dug well,* or a shallow well, is one which collects water that has seeped through the subsoil or the uppermost layer of soil to form what is commonly called the *ground water table.* A shallow well is rarely, if ever, dug through an impervious stratum or rock layer. Because this type of well is anywhere from 10 to 30 feet deep, there is limited percolation or filtering of the water through the soil. Therefore, this water does not undergo very much natural purification. As with the two types of wells discussed previously, there may be some quantities of gravel or soil present in the water supply. This will vary from area to area, with the more

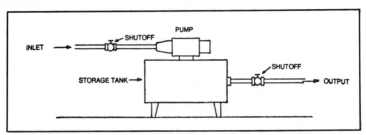

Fig. 5-2. Arrangement for a shutoff valve in conjunction with a water pump.

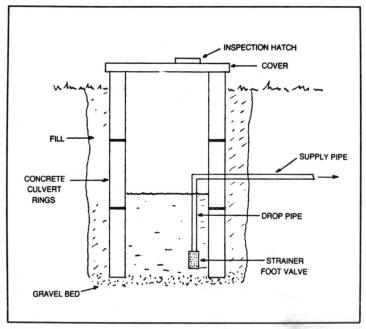

Fig. 5-3. A dug well can provide an adequate supply of water for hundreds of years.

rural areas providing the cleanest supply. A typical dug well is shown in Fig. 5-3.

To obtain water in this manner, a hole is dug deep enough so that the bottom several feet will be filled with water continuously after a short period of time. A dug well is basically a hole in the ground, generally from 4 to 6 feet across, although some may be as much as 20 feet across. There are literally thousands of dug wells all over this country, some of which are more than 100 years old and are still providing ample water. These old wells were normally lined with stone and capped with a small wooden structure to house a windlass or pulley arrangement for a bucket. Today, the walls are constructed of either concrete blocks or large precast poured concrete rings. A large cast iron cover is placed over the hole to prevent anyone from falling in.

The water service line is installed in much the same manner as with the previously discussed wells. A strainer foot valve is attached to a length of pipe and suspended in the well a foot or so above the bottom. The pipeline is trenched in below the frost line and connected to a pump and shutoff valve in the building. Pumps available for use with a shallow well can lift water the required amount of feet, which includes the static lift, the head loss in the pipes and fittings, and also the maximum drawn down in the well during pumping, plus seasonal variation in the water level.

DEEP WELLS

In areas where there may be sources of contamination present in the upper stratum of the soil, requirements may dictate the drilling of a deep well. Even if this is not the case in a given locality, lack of water at distances of 25 to 30 feet will necessitate boring to lower levels. A deep well is drilled down to a water bearing stratum, called an *aquifer*, which will furnish a sufficient supply of water when pumped. The aquifer is overlain by at least one stratum of impervious material that prevents seepage of the contaminated water in the water table of the top layer of soil.

Deep wells can range anywhere from 40 to 300 feet or more, depending upon the locality. For example, in the Southwest, it may be necessary to drill down several thousand feet into more than one aquifer to obtain a sufficient supply of water. The normal procedure is to drill to a depth that provides a satisfactory flow of gallons per minute for the proposed residence.

The drilling of a deep well is a highly technical procedure that is normally performed by a professional with a drilling rig. A wrought iron casing, 2 to 8 inches in diameter, is driven into the ground. This casing prevents contaminated water from entering the soil. When more than one aquifer is being tapped, perforated pipe surrounded with mesh screen is substituted for the blind pipe over the runs through each aquifer. If necessary, a screen is installed at the end of the casing. Steel can also be used in this procedure.

The water service line may be installed in any one of several different ways. One method is to attach a submersible deep well pump to the end of a section of pipe and lower it, along with its electrical power lines, almost to the bottom of the borehole. The pipeline is then routed out through the casing and placed in a trench normally below the frost line. The line is then attached to an appropriate shutoff valve near the building.

Another method is to lower a strainer foot valve into the well on a length of pipe and connect it directly to a deep well pump housed in a small wellhouse built directly over the well. Valves can be placed directly in the line, one immediately on either side of the pump, so that if and when the pump must be removed, the water can be cut off. The remainder of the water service line is then extended in the normal manner from the pump to the building to be serviced. Another shutoff valve can be located immediately within the building wall for additional convenience when repairs are done inside the building.

A third alternative is to place a strainer foot valve deep in the well pipe, with an unbroken water service line extending directly into the building being served. This line is attached at this point to a shutoff valve and deep well pump. Figures 5-4 through 5-6 show the installation of each of these methods, respectively.

The three types of installation discussed here will all provide satisfactory operation, with the major difference being the placement and type of

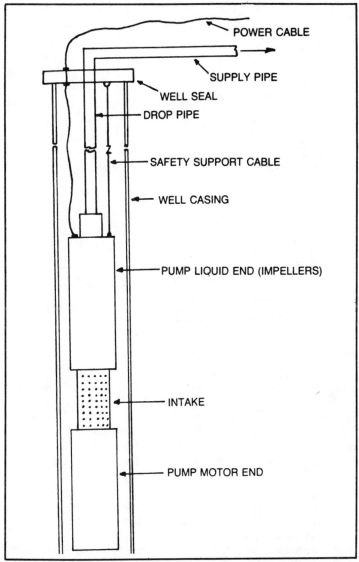

Fig. 5-4. In this installation, a submersible deep well pump is lowered into the well itself.

pump being used. Other factors to be considered in the selection of a particular method are job conditions, availability of equipment, the distance between the well and the building, and estimated costs. A variety of pumps are available for installation with both shallow and deep wells, a discussion of which follows.

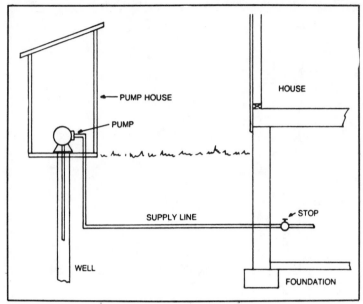

Fig. 5-5. An alternative is to lower a strainer foot value which is connected to a deep well pump located in a small wellhouse.

STRAIGHT CENTRIFUGAL PUMP FOR SHALLOW WELLS

A pump is necessary to raise the water from the well to either a storage tank or the actual distribution system. A straight centrifugal pump is composed of a single moving part which is called an *impeller*. The impeller is a fan-type blade, the size of which is determined by the required flow and the total head. The speed at which the impeller rotates will also vary with the size and revolutions per minute speed of the motor. As the impeller turns, water is sucked up from the well and is delivered to the outlet pipe.

A centrifugal pump is not able to suck air out of the pipes because its fittings are not sealed tightly. It will be necessary to prime this type of pump. Priming is the process of filling the suction pipe with water before starting the motor. Specially designed self-priming pumps are available, but they are more expensive. These pumps consist of an impeller which is constantly submersed in water contained in a special chamber. As long as this water is not evaporated or released in some other way, this pump will prime itself when the motor is turned on. A self-priming pump is shown in Fig. 5-7. Air inside the pipe bubbles through the water-filled chamber and is driven out of the system gradually until the pump is working at full speed.

Priming a pump simply consists of filling the suction pipe with water before the motor is started. Failure to do so will cause damage to the

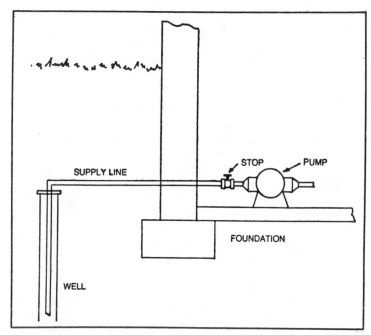

Fig. 5-6. The strainer foot valve may be placed in the well pipe, with an unbroken service line running directly into the building.

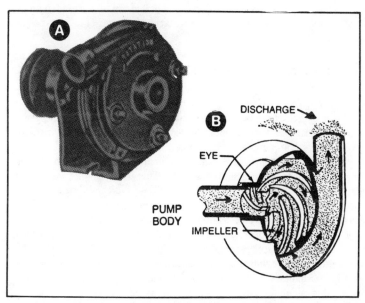

Fig. 5-7. A self-priming pump is a bit more expensive but convenient.

internal parts of the pump itself, as they need this liquid lubrication to run properly. The method used to prime a pump will vary with the type of pump, and manufacturers normally provide instructions which should be strictly adhered to. If for some reason the pump is not primed the first time, the process can be repeated.

RECIPROCATING OR PISTON PUMP FOR SHALLOW WELLS

A piston pump is very simple in its operation and consists of a handle which is pushed down and pulled up manually. As this motion occurs, a piston inside the cylinder moves. There are normally two valves in this pump, one at the bottom and one at the top. As the piston moves up, a vacuum is created. This causes the lower valve to open which, in turn, creates suction and forces the water to be sucked up into the cylinder. As the motion of the piston is pushed downward, the upper valve opens, causing the water to flow out. This cycle is repeated a number of times until a consistent flow is obtained. Although this type of system is somewhat old-fashioned, there are many thousands of pumps such as this still in operation today which provide a sufficient and satisfactory water supply. The operation of a piston pump is shown in Fig. 5-8.

An improved version of the piston-type pump is now available which provides for an automatic draw. This pump has a double-acting piston and four valves instead of two (Fig. 5-9). The principle remains the same as with the previously discussed piston pump, except that each stroke causes the water to be both drawn up and pushed out in a single motion. Thus, a more steady and consistent flow can be obtained in a shorter length of time. If desired, a storage tank can be attached to either of these types of piston pumps.

CENTRIFUGAL JET PUMP FOR SHALLOW WELLS

A centrifugal jet pump is normally used in place of a straight centrifugal pump in situations where it is necessary to obtain higher flows and a greater degree of pressure. It can be used alone, or even in conjunction with a standard centrifugal pump, if water must be drawn from medium levels where one pump will not suffice.

A jet pump is basically a standard centrifugal pump to which a water jet assembly has been attached (Fig. 5-10). This arrangement consists of a number of components designed specifically to increase flow and pressure. First, the suction pipe is narrower at one end than the other. This causes some of the water from the discharge end of the pump to be captured and carried back towards the inlet side. A small diameter nozzle is situated in the water passage which helps to increase the speed of the water flowing through this portion of the assembly. A small length of pipe with an expanding diameter is located inside the pump. Called a *venturi flume*, this pipeline creates a suction pressure which, in turn, draws additional water from the well pipe.

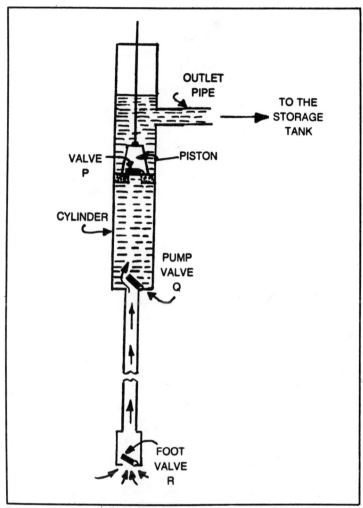

Fig. 5-8. Operation of a piston pump.

With this type of arrangement, an appreciable increase in flow and pressure can be had. A jet pump will provide the needed gallons per hour for a proposed residence in those situations where a normal centrifugal pump may fall short.

CENTRIFUGAL JET PUMP FOR DEEP WELLS

The basic principle of a deep well jet pump remains the same as those used in shallow wells. The major difference lies in the location of the nozzle and the venturi. As seen in Fig. 5-11, these components are located

131

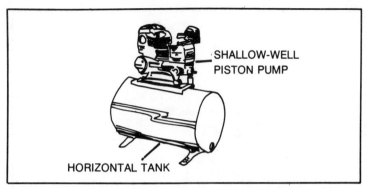

Fig. 5-9. An improved version of the piston pump which provides an automatic draw.

lower down in the well pipe. Another addition is a foot valve provided at the bottom of the venturi. In a system such as this, the flow back which runs from the outlet of the impeller enters the nozzle at the lower end of the venturi. From this point, the high velocity jet travels upward. Because of this high velocity, the pressures are much lower than the existing atmospheric pressures present. This disparity causes the water from the well to be pushed up. This additional water flow adds to the overall flow thrown out by the impeller, thus raising pressure.

The above system will work efficiently and satisfactorily in wells to a depth of about 110 feet. In installations of a greater depth, another type of

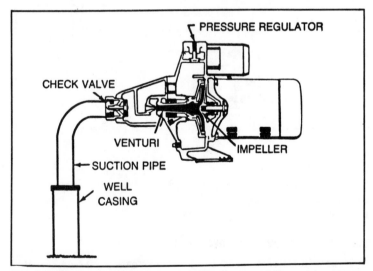

Fig. 5-10. A centrifugal jet pump provides increased flow and pressure through jet action.

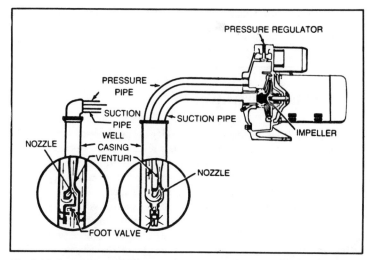

Fig. 5-11. A centrifugal jet pump for use in deep wells operates in the same manner, except additional flow and pressure are obtained due to the location of the components.

assembly, known as a multi-stage jet pump, is suggested for maximum efficiency. In this type of system, a single shaft carries a number of impellers instead of just one. The result of additional impellers being installed is the creation of pressures as high as 40 to 60 pounds per square inch. This will bring about the increased flow and pressures necessary in attempting to draw water up from well below the surface. It is not recommended that this type of pump be used in the shallower or medium depth wells solely for the purpose of increasing flow. Instead of producing this desired effect, they will merely increase pressures rather than flow.

SUBMERSIBLE PUMPS

Submersible pumps can be used with either shallow or deep wells and are installed inside the well casing itself at a level below the ground. Little space above ground is required for the installation of this type of pump, as only the electrical control box and a tank must be provided for. These can quite easily be placed in a corner of the kitchen or other room inside the proposed residence.

The submersible pump has some advantages, as well as disadvantages, when compared with a standard centrifugal pump. It is more expensive, and because of its installation below ground, any repairs will be that much more difficult. A pump located above ground makes a certain amount of noise that will be absent if the pump is submerged, so this is a favorable characteristic.

A typical submersible pump installation is shown in Fig. 5-12. This system basically consists of a long, thin centrifugal pump with a motor connected to the lower end of the supply pipe. Since this pump is immersed in water, it will need no priming. Submersible pumps can be used for wells as shallow as 20 feet and will work equally well for pumping water from wells as deep as 500 feet. They are quite versatile, durable, and popular on today's market.

PUMP SELECTION AND INSTALLATION

The proper selection of a pump and its correct installation cannot be overstressed. This is the final step in bringing the predetermined required water supply to its final destination—your home. After determining the amount of flow in gallons per hour, a pump that will provide for this capacity is installed. A good minimum recommended pump capacity to use as a guideline for an individual home is 540 gallons per hour. When selecting a pump, the water supply to be tapped must be taken into consideration. If the well has a limited water supply, the installation of a high-powered pump would have the adverse effect of drying the well up in a matter of minutes. A professional well driller can be of assistance here. He can provide information regarding the normal yield of a well. When a pump is first started up, there is always a drawdown, which is a dropping down of the water level inside the well. A pump runs intermittently, since once the storage tank is filled to capacity, it automatically shuts itself down. When this occurs, the water level in the well itself has an opportunity to return to its normal level gradually as more water seeps into it from the aquifer. If during the drawdown process any portion of the intake foot-valve is left exposed, the pump will lose its prime. If this goes unnoticed for any length of time, serious damage to the system will result. Therefore, pump capacity must be determined in such a way that this drawdown remains within reasonable limits.

Each individual system will have different requirements and possibly special features that will make the use of one type of pump more desirable than another. If space is a consideration and construction of a housing for the pump is not possible, the obvious choice would be a submersible pump. Conversely, if there is ample space for this type of outbuilding, or if the pump can be located in a basement where noise would not be an inconvenience, a centrifugal pump would make for a better installation and save the builder the cost of the more expensive submersible pump. Another important factor to be considered is the depth of the water supply being tapped. Always remember to stick with a deep well pump for a well any deeper than about 25 feet.

The installation of a pump is a relatively simple procedure and is made even simpler if plastic pipe is used. Most submersible and centrifugal pumps will allow for its use, but be sure to check with the necessary local authorities with regard to plastic pipe for this type of installation. The manufacturer's instructions that normally come with pumps will specify the size and type of pipe recommended for use with the particular pump. All the joints should be completely watertight, and the clamps and fittings

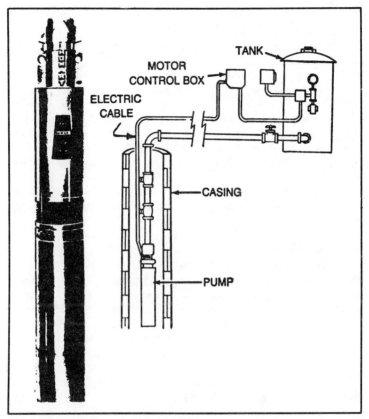

Fig. 5-12. Installation of a submersible pump.

should match the type of piping being used. In a centrifugal jet system, the diameter of one pipe leading from the pump to the jet is larger than that of the other. These connections are made to the correct size fittings at both ends. Make sure the length of tailpipe between the jet and the foot valve are not more than what is suggested by the manufacturer's instructions. If plastic piping is being used, care should be taken not to damage the pipe when it is being lowered into either the submersible pump cartridge or the jet of the centrifugal pump in the well. This can easily be prevented by using a large bending radius and taking care not to let the pipe strike against the sharp edges of the casing.

Once the pump is installed properly, the pump can be primed and the motor then started. There will be a short delay after the motor is started, during which time all the entrapped air is driven out. Once the system is functioning satisfactorily, it should take only a very short amount of time to fill the storage tank and build up pressure. When the storage tank is filled to capacity, the pump will automatically shut off and remain off until it becomes necessary for more water to be pumped into the tank. This supply is carried to the house distribution system for use.

A number of attachments to an above-ground pump can be installed to cut down on noise. The pump base can be bolted to the floor with rubber washers both above and below their holes, as well as rubber ferrules placed inside these holes. This will necessitate the use of slightly smaller bolts than those provided with the unit, but these bolts are not very expensive and can be purchased at any hardware store. Metal washers are then installed over the upper set of rubber washers to provide a tight connection. The use of plastic piping between the pump and the supply lines will make for a generally quieter system, since plastic is an insulator of sound.

Table 5-3 is a troubleshooting chart that will assist the builder in pinpointing some of the more common causes of a malfunctioning pumping system and how to correct these problems.

AUXILIARY SYSTEMS

A *booster pump* is normally used in situations where it becomes necessary to boost the pressure of an existing system to meet its demands. This type of pump can be installed in a home water supply system, in a sprinkler system, in a line leading to an annex or barn, or anywhere else it may be needed. If the lack of sufficient water pressure occurs where the home is connected to a municipal system, the pump would be installed at the head end of the cold water system, immediately on the outside side of the main shutoff valve in the water service line and water main. If the problem lies in the water volume in the mains being inadequate, an installation of this type will be fruitless.

Booster Pumps

The actual location of a booster pump in a system is flexible in that it can be installed on the main supply line or the particular branch where pressures are low. The pump itself consists of a motor and an automatic flow switch. As water demand increases and pressures decrease, this switch will automatically turn the pump on. A typical booster pump is shown in Fig. 5-13.

Sump Pumps

Although not used in the same manner as those pumps already discussed in this chapter, a sump pump works along the same principles in that it carries water from one place to another. Rather than being used to supply water, a sump pump is used to remove water from a basement or other sublevel area where seasonal water accumulations may be a problem. These types of pumps are quite expensive and difficult to install and repair, but more and more homes today are equipped with sump pumps as they prove to be the easiest and most effective way of removing unwanted water.

Table 5-3. Troubleshooting Chart for Pumping Systems.

Malfunction	Probable Cause	Suggested Cure
No water pressure in tank after a very cold night	1. Slim tube to pressure switch has frozen and switch is not working at low pressure. 2. Pipe between pump and tank has frozen.	1. Heat the copper tube by flame from a candle or cigarette lighter. 2. Thaw pipe as explained in Chapter 5.
Pump starting and stopping very frequently.	Tank is waterlogged with no air cushion at top due to: 1. Loose and leaky fitting at the top (test by applying soap suds solution) 2. Defective air regulator or stuck part.	 1. Open fitting, apply Teflon tape and retighten. 2. Tap air regulator lightly with a small wrench to free the stuck part. If air regulator is damaged, repair or replace it. Re-establish air space at top by draining off some water from tank.
Though no water is being used, the pump keeps starting and stopping at long intervals during the night.	1. Leak in plastic pipe. 2. Foot-valve leaking	1. Remove sanitary well cap and inspect the pipe. If found defective, replace. 2. Watch for a day or two. If problem persists, take it out and have it re-paired.

The first step in making an installation of this type is to determine where the major problem area is and what its flow paths are. Once this has been ascertained, it may be necessary to make some adjustments in the concrete flooring and arrange recessed gutters to rearrange the flow and direct it to the lowest point in the floor of the basement. This is where the pump should be installed. To accomplish this, a portion of the concrete floor will have to be removed and a hole dug in the ground beneath it big

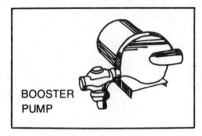

BOOSTER
PUMP

Fig. 5-13. A booster pump is used to boost water pressure to meet demands.

enough to house the sump crock. The crock is the housing for the actual machinery of the pump itself and can be made of a number of different materials, the most common being either plastic or some type of clay (Fig. 5-14).

The placement of the crock in the hole is especially critical. It must be set in place in such a manner as to allow for holes to be cut in its side for connection to the drainage trenches in the cellar floor. If the drainage lines are underground, the crock must be positioned so that it can receive these lines at the proper locations. Once the crock is set in place and properly aligned, the drainpipes are routed into the crock as required (Fig. 5-15). Cement mortar or concrete patching compound of a somewhat soupy consistency is then poured in around the outside of the crock to secure it in place and provide a certain degree of strength to the overall structure. Concrete patching compound is troweled around the top to fill in the ragged edges in the concrete floor and smoothed around the floor gutters where applicable. Figure 5-16 shows how this is done.

The cement should be given enough time to dry before installing the pump into its housing. After the pump is in place, cutouts are made in the crock cover for the pump standpipe and the discharge pipe. The cover is then slit and slipped into place. The electrical cords will exit through the

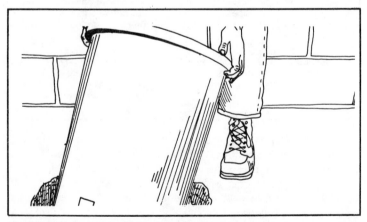

Fig. 5-14. The crock is the housing for the sump pump.

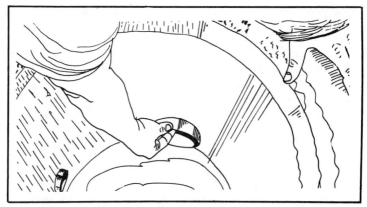

Fig. 5-15. The crock is set in place and aligned so that the drainpipes are routed as required.

same holes. A check valve is installed on the discharge outlet. A suitable pipeline is installed on this valve to carry drain water off and away. Electrical connections are made by plugging the cord into a *grounded* wall receptacle which is positioned a short distance away *on a wall*.

The method of handling the discharge from a sump pump will vary in relation to the type of disposal system in operation in the residence. In situations where the house main drain connects to a municipal sewer line, it is a simple task to tap the discharge pipe directly into the main house drain or to any other nearby drainpipe of suitable size in the DWV system.

If the main house drain is connected to a septic tank, it is not recommended that this discharge be disposed of through this system. A sump pump is capable of discharging a good deal of water in a relatively

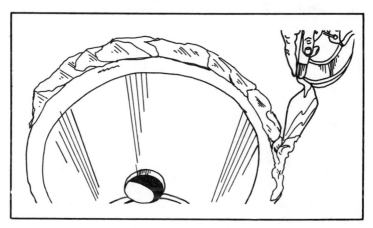

Fig. 5-16. Cement or concrete patching compound is poured around the outside to secure it in place.

short period of time and will cause the normal bacteriological balance of a septic tank to be upset. In short, a sump pump will put too much of a load on the tank and cause the whole system to malfunction.

An alternative to connecting a sump pump to a septic tank would be to route the discharge pipeline through the foundation of the house at a point which is approximately 2 feet or more below ground level. This pipeline should then be connected to another length of pipe which is solid and of a larger diameter, preferably about 4 inches and continued in a trench with a ¼-inch pitch per running foot down and away from the house. Once the pipe has been extended a reasonable distance from the house, it is then changed to a section of perforated pipe, under which a bed of gravel has been installed. The length of the section of perforated pipe will be dependent upon the volume of water that the sump pump can be expected to discharge. If the volume is low and the discharge will be somewhat periodic, the pipeline can be relatively short. However, if it is expected that a large volume on a more or less continuous basis will need to be disposed of, this pipeline should be lengthened considerably. Another factor to be taken into consideration in determining pipeline length will be the absorption capabilities of the soil.

In situations where space is not available for a pipeline of this sort, it may be necessary to build a dry well or a small seepage bed. The simplest way to do this is to dig a hole 2 or 3 feet in diameter and 3 or 4 feet deep. The hole should be positioned so that the incoming drainpipe will lie in the middle of the hole. Gravel is then placed in the hole up to about 4 inches above the pipeline. Install an interface of either paper or straw and then backfill the hole. A typical dry well is shown in Fig. 5-17.

Another method would be to put a couple of feet of gravel into the same size hole and placed an inverted old oil drum on top of the gravel. A hole is cut in the barrel to admit the drainpipe, and a series of holes are drilled in the sides of the barrel. Push the barrel down into the gravel about

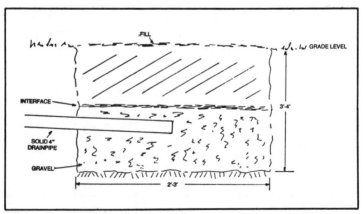

Fig. 5-17. A sump pump may be drained into a small dry well.

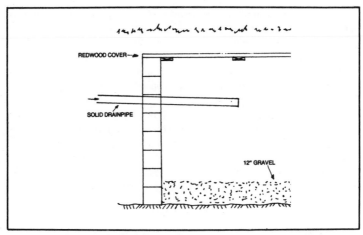

Fig. 5-18. Another arrangement might be a small seepage bed which can be cleaned quite easily when necessary.

a foot, and run the pipe into its hole. The area around the outside of the barrel should be filled with gravel and then backfilled.

Yet another alternative would be to dig a hole about 4 to 5 feet in diameter and about 4 feet deep. A foot of gravel is placed in the bottom of the hole and the sides are lined with ordinary concrete blocks laid dry in a ½-inch overlap. A lid can be constructed of 2-inch thick redwood planks made to rest on the top layer of blocks. The cover is positioned over the hole and the area backfilled. This type of arrangement is pictured in Fig. 5-18. A certain amount of convenience can be had with an installation such as this, because the pit can be dug up periodically and cleaned without causing any damage to the pit arrangement.

SUMMARY

You will find when reading this text that a great deal of emphasis is placed upon the planning time involved in each stage of a plumbing system. This holds true insofar as wells and pumps are concerned. Each decision carefully thought out and planned in the early stages will most certainly be found to be well worth it in later years.

Future needs must be taken into consideration when determining what amount of water supply is sufficient for a proposed residence. Any additions or modifications will place a greater demand on the water supply. This can be provided for during the initial planning stages quite easily. Make sure that the source to be tapped will supply an amount of water that is somewhat more than present needs to allow for any possible future expansions that might take place at a later date.

You are cautioned here to take certain safety measures at the well site. A well should never be open in such a way as to pose a hazard. It should be covered at all times, especially if little children are present anywhere near the installation. This is simply common sense, but it is stressed here nevertheless, since it is not unusual to read of accidents involving injuries resulting from persons falling into wells. It should also be stressed here that a pump is a piece of machinery that can be damaged if not protected from the elements, and proper housing should be provided for this device.

If proper planning and installation procedures are followed, there is no reason why a water supply, once functioning satisfactorily, should not provide an efficient, clean, and continuous supply for many years to come. Thousands of water supply systems installed even 100 years ago are still providing their owners with an adequate source of water today and, if maintained properly, may even last another 100 years.

Sewage Disposal Systems

Modern advances over the years have greatly improved methods for the disposal of waste, as is evident by the amount of money that is spend in cities across the United States to design and construct large waste treatment facilities. One of the major forces behind this movement was and still is the protection of persons from such diseases as typhoid, dysentery, and hepatitis, which can result if precautions are not taken to prevent widespread contamination. Typhoid is quite uncommon in this country today due to the modernization of waste disposal methods, but in countries where such improvements have not taken place, this disease is a common killer and will continue to be until the necessary practices are enacted. It is not unusual to note when reading in the newspaper about a catastrophic event such as an earthquake that many people are stricken with dysentery or another infectious disease. This is due to the breakdown of mechanical systems designed to dispose of human waste. It can be seen that these systems are not only efficient and protect the environment, but they have also saved countless numbers of lives over the years.

All large cities and the majority of smaller towns with more than a specified number of people are required to have some sort of sewage disposal system. However, as more and more people opt for the rural life, there has arisen a need for an individual system for each residence, as the cost of running lines would have a detrimental effect on the financial status of the smaller municipalities. Hence, the individual septic system was designed to fill the need for those "off the beaten path." Not only can a septic system provide a safe, effective method of disposal, but it has been proven to be environmentally safe if properly designed, installed, and maintained.

Any sewage disposal system must be constructed in such a manner that the following guidelines are provided for:

■ The system must not allow any access by rodents, insects, and other disease carriers which, after picking up the disease bacteria, can contaminate water supplies.

■ The water is protected from pollution.

■ The wastes from the system cannot pollute the waters of streams, beaches, lakes, or the breeding grounds of fish. If any of these clean waters are used for drinking or recreational purposes, the waste disposal system should be located far away from them.

■ The disposal area should not be accessible to children and pets.

■ The system must conform to all local and state health codes.

The requirements for a residental sewage disposal system provide for proper investigations of the site, detailed design of the system, and actual installation.

PRELIMINARY INVESTIGATION AND DESIGN PARAMETERS

The first and most obvious step is to determine the condition of the soil where the proposed system will be located. If a purchase is to be made for construction of a new home, this matter should be looked into *beforehand*. A septic system operates by converting all of the waste water and solids from a home into liquid, gases, and a small residue of solids. The gases, which account for only a very small portion of sewage, escape into the atmosphere. The solid residue, or *sludge*, remains in the septic tank itself, where it builds up slowly over a period of time and must be removed periodically. The liquid which remains, called *effluent*, constitutes by far the greatest portion of the volume of material in the tank and is released slowly into the drainpipes of the system on a continual basis. This effluent is absorbed into the soil primarily, although some does evaporate into the air. The soil must have certain qualities and/or characteristics which will allow for the successful absorption of this liquid. There are a number of preliminary steps that can be taken to get a general idea of the condition of the soil at the planned site.

If the area near the proposed installation is already somewhat populated by other homeowners, this will be an indication that there are some working systems in operation. A check with these neighbors can provide some direct input as to any problems they may have encountered during installation that were directly related to general soil conditions. On the other hand, if the area is somewhat remote and there are no existing homes nearby, check with the local authorities to see if there are any soil maps available. These types of maps are designed to give a complete description of the soil and its general characteristics and, if not available locally, can be obtained from the Soil Conservation Service in Washington, D.C. A map such as this will note the different kinds of soil and outline some information about the soil qualities and potential uses. Many areas in the country have organizations devoted to the study of experimental agriculture which may prove to be of help.

Most localities have a building inspector and/or health inspector who, if requested, will come out and look over the premises and give a preliminary opinion as to whether or not a system can be installed. Another option that may be available is a soil engineer, who can make a complete soil survey. He will provide a written analysis of his findings as they relate to soil absorption and the potential efficacy of a septic installation.

If none of the above steps produce satisfactory findings and there is still some degree of uncertainty, percolation tests will be necessary to prove there is adequate soil absorption for a septic system installation. These tests are sometimes done before a piece of property is zoned as residential. The tests will not only test the soil's absorption capabilities, but can pinpoint the best possible location for the system's various components. Even if it is known that percolation tests were done previously, local authorities will normally require that these be done again shortly before actual construction begins. These officials may state that an employee of theirs be present during the actual test and that it must be done by someone skilled in making such a test. It may be possible to perform the test yourself, depending upon each area's requirements.

It is not a difficult procedure to perform this test. The first step is to pick a spot within the area of the proposed leaching field area and dig a hole to the depth at which the drainage pipes will lie. This depth is variable depending upon local conditions and regulations. Most fields are kept as shallow as possible to aid in evaporation, sometimes only 12 to 18 inches below the surface. The pipes should lie below the level of the topsoil. If the winter weather in the proposed area tends to be severe, these pipes will probably be installed deeper than 18 inches and the hole should be dug to the proposed depth of the pipe installation. Dig the hole large enough in diameter to allow room to work and flatten the bottom of the pit out. In the center of this hole, dig another hole 12 inches square and 12 inches deep, square-sided and flat-bottomed. Roughen the sides of the hole with a flat piece of rock or some implement to remove any tightly compressed earth slick that might have been formed with the shovel. A 1-inch layer of sand or fine gravel is spread in the bottom of the hole. Fill the 1 cubic foot hole almost to the top, taking care not to crumble the top edges inward. Let the water sit for half an hour, at which time a measurement is taken to determine how far the water level has dropped. After taking this measurement, fill the hole again and repeat the procedure regularly for at least four hours, noting the water level drop each time. The individual measurements are then added together and divided by four to get the average rate of drop per hour. This figure is the *percolation rate*.

There are a number variations of the method given here. One such method is to simply keep the hole full for at least four hours, with the total amount of drop noted after another two hours have passed. Another method would be to use anywhere from 6 to 12 holes with a diameter of 4 to

6 inches and scatter them about the area of the proposed leaching field installation. The measurements from each hole are added together to arrive to an average percolation rate for the entire field. This method would be useful in situations where the soil consistency changes markedly from spot to spot and to determine the area that has the best percolation rate. After this preliminary test to pinpoint the most suitable site, another series of test holes would then be dug to define the high-rate area so the leaching field can be placed directly upon it.

Once the method has been chosen and the tests performed, the results will need to be interpreted. If the percolation rate is less than 1 inch per hour, this indicates that the soil is almost certainly not capable of absorbing a substantial effluent discharge with any effectiveness, and the proposed field would be a health hazard. A rate between 1 and 2 inches per hour is marginal but workable if the field is large enough and properly and carefully installed. However, it should be noted here that a field installed on such a site would probably have to be replaced or relocated after a few years, so this should be taken into consideration before construction is begun. Obviously, if this is not feasible, the site should be considered unsatisfactory for the system. When the percolation rate falls within 2 to 4 inches, this is deemed average and acceptable. A leaching field installed here will work well if installed properly and will last for many years. A percolation rate 6 inches or higher an hour is considered to be excellent. A rate of 5 inches an hour will make for ideal conditions for a leaching field that will provide excellent drainage and will last for an indefinite period of time.

The figures given here are the common interpretations and will give you a guideline to work with. However, each locale will place its own interpretation upon the percolation test results, and these may or may not equate with the information in this text. Some locales give more flexibility; some give less. It will always come down to abiding by the requirements in the proposed area for the installation. The percolation rate can also be used to determine the amount of space that will be required for the leaching field. For example, the basic requirements for a locale might be based upon 200 square feet of leaching field area for every bedroom at a 6-inch rate, 250 square feet at a 4-inch rate, 300 square feet at a 2-inch rate, etc. This requirement may also vary from area to area, and the basis for making the determination may be a bit different from the example given.

It may be that there are no requirements in the proposed installation area, in which case some basic guidelines will be needed to give you a general idea of the size of the leaching field. Although these guidelines are expanded somewhat, no harm will ever be done by making a field a bit larger than the size called for. Although this may increase the costs of construction, it may provide an extra safety factor in the long run.

Perc rate of 6 inches per hour: 150 square feet per person
Perc rate of 4 inches per hour: 175 square feet per person

Perc rate of 3 inches per hour:	200 square feet per person
Perc rate of 2 inches per hour:	250 square feet per person
Perc rate of 1 inch per hour:	325 square feet per person

Once it has been determined that the necessary requirements have been met for approval of an installation on the proposed site, it will be necessary to ascertain what the design parameters will be. These requirements are also largely set down by the governing bodies in each area and local officials will be able to help in these decisions. In order to make the proper determinations, these officials will need to know family size, the number of bedrooms in the proposed residence, and the number of fixtures in the layout. With this information, they will be able to specify the minimum size septic tank that will be necessary, along with the distance this tank must be located from the house, property lines, water sources. At this point, final decisions should also be made regarding tank material, system dimensions, piping methods, and the piping materials based on the information obtained thus far.

In those areas where such requirements and assistance is not available, it will be the responsibility of the builder to determine these specifications, and a brief discussion is given here to assist you if this is the case. The four basic elements that make up a disposal system are the house sewer line which attaches to the main house drain, the septic tank, the sewage outflow line, and the absorption or leaching pipelines. The tank is usually positioned approximately 10 feet from the wall of the building, but this can vary either way by 1 or 2 feet with no danger of malfunction. The sewage outflow line is the line that extends from the opposite side of the tank and although no specific requirements are stated here, it should be kept as short as possible. The outflow line runs to the absorption field, which can go straight or be cocked off to one angle or another. The next step is to determine where the house will be located and where the main house drain will be. This will give the builder an approximate area as to where the septic tank will have to be located and a site for the leaching field. Remember that the septic field should be as far away as possible from a well and at least 100 feet away and on the downhill side of any water source. Also, take into consideration the fact that the sewer line and the sewage outflow line will require trenches, and the septic tank will rest in a large hole. This means easy access for heavy equipment may be necessary.

The leaching field itself can be built on flat ground or on slightly sloping ground, as long as the slope is 15 degrees or less. Keep in mind that once the field is installed, it should remain open and should not be heavily shaded by thick shrubbery or tall trees, if possible.

Once the basic layout for all the major components has been established using the preliminary data provided here, it will be necessary to begin to compile more detailed specifications and measurements in order to determine what materials will be required for the installation. The following section will deal with each of the components of a sewage disposal system individually to give you an understanding of each one's function in the overall project.

SEPTIC TANK

Septic tanks come in a wide assortment of sizes, capacities, and materials. The size will be determined by the expected amount of sewage that will be flowing through the system. There are a number of methods that can be used to arrive at the proper capacity that will be required, but the most commonly used method is calculated according to the number of bedrooms in the proposed residence. Using a reasonable figure of 50 gallons of waste per day per person, a family of four would need a tank capacity of 200 to 250 gallons to provide satisfactory septic action. The U.S. Public Service recommends a minimum tank capacity of 750 gallons for a two-bedroom home, a 900-gallon tank for a three-bedroom home, and a 1000-gallon tank for a four-bedroom home. Local codes will usually provide for the minimum capacity requirements in your area. Some local codes prohibit the installation of anything smaller than a 1000-gallon tank for a single family residence, and some requirements are even greater. In the absence of any local regulations for guidelines, a 1000-gallon size is suggested here, as this will provide for any future additions and will guarantee a satisfactorily operating system. A standard septic tank is shown in Fig. 6-1 which has a 1000-gallon capacity.

In determining the best material for the septic tank, local regulations may again influence this choice. It may be that one or two specific materials are either required or disallowed, and these guidelines will have to be adhered to. Tanks are available either assembled and ready to install, or they can be built using specified materials. The least expensive and short-lived tanks are made of heavy asphaltum-covered steel in a cylindrical shape with a matching steel cover. Another choice might be a precast concrete tank. These are becoming more and more popular and will provide satisfactory operation. A number of shapes can be had with the concrete tanks, among them cylindrical and rectangular. These are equipped with reinforced concrete covers fitted with one or two cleanout and inspection hatches. Fiber glass tanks are also available in a number of configurations. A home built septic tank can be constructed using precast concrete rings set on a poured concrete base, which is then capped with a precast concrete cover. Smaller tanks can be constructed of large rings of vitrified clay tile. If you want to start from scratch, forms can be built and the whole tank made from poured concrete. The only drawback to this type of construction is that the concrete must cure for a month or two before being put into service. Whatever option is chosen, strict adherence to the local codes must be maintained.

A septic tank is simply a large container with a specified capacity, an inlet at one side, and an outlet at the other side. The inlet is normally composed of either downpointing sanitary tees or openings with baffles hanging down to prevent sewage from flowing at the surface from the inlet

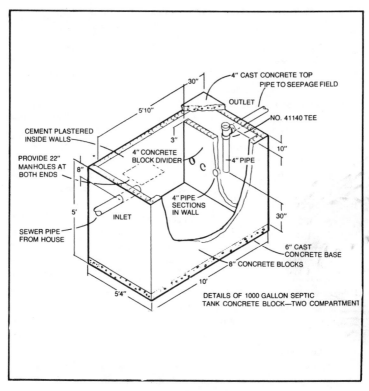

Fig. 6-1. Cutaway diagram of a standard 1000-gallon septic tank.

to the outlet until it is properly decomposed. The incoming sewage flows downward and is decomposed, through bacteriological action, into liquids, gases, and mineral sludge. The gases travel back through the house sewer to the vent stack and escape into the atmosphere. The mineral sludge gradually settles down at the bottom of the tank and is removed at periodic intervals, usually every few years. The bacteria in the tank, which thrive due to the absence of oxygen, are able to decompose the incoming sewage in approximately 24 hours. The liquid effluent exiting the septic tank, therefore, is full of bacteria and quite foul-smelling. It should *never* be allowed to run above ground or come in contact with wells, springs, or other water supply sources, as stated previously. Rather, the effluent is carried underground through pipes with watertight joints leading to the leaching field. The decomposition process may be slowed down somewhat by strong detergents and bleaches such as those used in washing machines or other appliances. When large volumes of water saturated with these chemicals enter the tank, the bacteriological action may be disrupted completely. This will also churn up both the sludge and the solids in the tank which are in various stages of decomposition.

149

Once these materials exit the tank and arrive at the disposal field, they enter the soil and slow down its rate of absorption rate and may, in extreme cases, cause flooding. Precautions can be taken to avoid this. Strong bleaches and detergents should be avoided, as should a continuous washing of several loads in a short period of time. Several appliances should not be operated simultaneously. Washing clothes at times when either the bathrooms or kitchens are at their peak usage should also be avoided to prevent the dumping of large volumes of water into the septic tank at one time.

If flooding does occur, bring in a professional sanitary engineer with some experience in this field for consultation purposes. It may be necessary to enlarge the septic tank or add more trenches to the leaching field. The soil's inability to absorb the wastes effectively can sometimes be traced to a high amount of grease or oil. This can be alleviated by installing grease traps, particularly on the kitchen lines, which will probably be the culprit. Grease tends to clog the pores of the soil. If it is apparent that some of the appliances in operation in the home are using unusually high quantities of water, these should be replaced, if feasible. If the problem still persists after these precautionary measures have been taken, it may be necessary to repipe the system and send only the toilet waste line to the septic tank. Run the rest of the flow to either dry wells or drainage pits for disposal. This may be prohibited, and a careful check should be made with the local authorities as to the codes applicable to a change such as this before beginning.

The septic tank is set in to a depth which will provide a downward pitch of about ¼ inch per foot run on the sewer pipe running from the house to the tank inlet. The inlet is normally 1 to 3 inches higher than the outlet, depending upon the length of the tank. The tank should be set level in the hole with the top buried at least 1 foot deep, but not more than 3 feet. Anything deeper than this will make the removal of sludge a difficult operation. The opening at the top should be large enough to allow for a shovel to be inserted to loosen the sludge and turn the contents into a slurry which can be pumped out easily.

The tank discussed thus far consisted of a single container to receive the waste. Another type of tank is available with two main sections that can be housed in one two-compartment unit or is built as two separate, interconnected tanks. An example of this type of tank, called the siphon-septic tank, is shown in Fig. 6-2. The first section is the tank itself, which is called the settling tank, and its design and purpose is basically the same as a single-unit tank. The second section is called the siphon tank. In the compartmented single-unit design, there is an integral submerged outlet which leads from the settling tank to the siphon tank. If the units are separate, they are attached by means of a short length of pipe. The siphon tank consists of a large space with a trap at the bottom and an overflow drainpipe attached to the outflow side of the trap. The inlet side of this trap is covered with a screened hood.

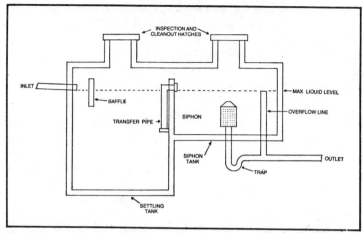

Fig. 6-2. A siphon septic tank works in much the same manner as a single-tank septic tank, except that this tank will empty itself periodically.

The actual operation of the settling tank is much the same as the ordinary septic tank already discussed. As waste is dumped into the tank, an equal amount of effluent flows into the siphon tank. Once the level in the siphon tank reaches a certain height, the resulting pressure forces a large discharge of effluent through the trap and out to the leaching field until the tank is nearly exhausted by siphonage. This tank will empty itself periodically, which is not the case with an ordinary septic tank. The effluent itself is under sufficient pressure to insure that an equal flow is discharged to all parts of the leaching system on a regular but gradual basis. This allows the field enough time to absorb the current discharge and recover itself sufficiently in order to be ready for the next one. The amount of time between discharges will be determined by the size of the tank and the amount of waste discharged into the system daily.

LEACHING FIELD

The technical term for a leaching field is a subsurface-tile absorption system, and this portion of the system consists of a predetermined number of trenches in which drainpipes are installed. The actual size of the field will have already been determined using the results of the percolation test and local specifications already discussed in this chapter. Trench width and depth may also be specified, but the width will most commonly be between 18 and 24 inches and the depth will be 2½ to 4 feet. A backhoe is normally utilized for the digging of these trenches. The trenches should run parallel to each other, with center-to-center spacing not less than four times the width of the trenches. For example, for 3-foot wide trenches, the spacing would be at least 12 feet from one center to the next center, or 9 feet between adjacent trench sides. Unless automatic siphons are included

151

in the system to enable pumping excess water accumulations periodically, the trenches should not exceed a length of 75 feet.

Figure 6-3 shows the layout of a typical leach field with a 1000 square foot disposal area. In order to determine the square footage of a field, select the desired trench width and work from there. For example, if a 1000-foot field is required and the trenches are to be 2 feet wide, a number of combinations can be utilized. The example shown in Fig. 6-3 shows one such configuration with the system divided up into five 100-foot lines. A better solution would be 10 50-foot lines, which would probably prove to be more practical and effective. Another method would be to split the field into two sections, with half the leaching field running off in one direction and half in another. Any combination that will make for an efficient system that meets the necessary requirements can be used here, as long as the total amount of actual drainage trench area equals the square footage determined earlier in this chapter.

For satisfactory operation of a leaching field, the area selected for its installation should meet the following requirements:

■ The area should have 4- to 6-foot depth of good, permeable soil.

■ The water table should be at least 4½ feet below ground.

■ There should not be any ledge rock to a depth of about 7 feet.

■ There should be a sufficient area of land available meeting these characteristics, with a reserve for any future expansions that may be necessary, where applicable.

■ The area should be graded, if necessary, so that no water stands in any natural depressions or hollows.

PIPE FOR THE LEACHING FIELD

The selection of the material to be used for piping will, again, be determined by local plumbing regulations and individual needs and preferences. Some of the choices commonly used for trenching purposes are perforated plastic pipe, perforated fiber pipe, drain tile with ¼ inch to ½ inch open joints, or perforated tile. Solid pipe is used at points where drainage is not required or desired, while perforated pipe is used in the absorption sections of the field. Although the actual sewage disposal field installation will be discussed later in this chapter, it should be pointed out here that when laying perforated pipe, the perforations are always facing the bottom of the trench for the obvious reasons. The different lengths of pipe are connected with standard fittings designed to be compatible with the type of pipe being used. The pipe used in leaching fields in normally rigid, but some of the plastic piping available for this application is a bit flexible and will work here. The piping running from the septic tank connects to a distribution box which, in turn, sends the effluent out through any reasonable number of pipes to the field itself. These boxes are normally made of steel and are readily available with a number of knock-outs through which connections can be made to the various lines (Fig. 6-3).

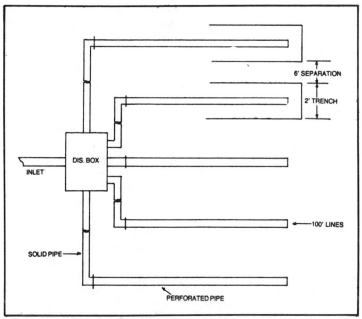

Fig. 6-3. This leaching field is designed for a 1000-foot disposal area. With five 100-foot lines. Any combination may be used, as long as the field meets the necessary square footage requirements.

SEWER PIPELINE

The sewer line is the length of pipe which connects the house to the septic tank. Again, a number of different types of materials which adhere to local regulations, can be used here at the option of the builder. The piping is most commonly 3 to 4 inches in diameter, depending on the diameter of the main house drain to which it is connected. If a cleanout has not already been installed in the main house drain inside the building, this is a good place to put in an above-ground cleanout plug, especially if the septic tank is far from the building.

The sewer outlet, or effluent outflow line, is the piping that joins the outlet of the septic tank to the leaching field drainpipe. The pipe used here should be of the same kind and diameter as that used in the leaching field, except that this piping is solid. If the field being installed consists of a single drainpipe line, all that is required is to switch from solid pipe to perforated pipe at an appropriate distance from the septic tank. This pipeline does not necessarily have to be straight. If a distribution box is a part of the system, this line will run from the tank to the box. The main purpose of the distribution box is to allow for the effluent to enter the box and exit in a more or less equal manner to the drainpipes.

A bypass line may be installed in the system for the purpose of carrying waste water directly from a washing machine or other appliance directly to the leaching field, thus bypassing the septic tank. As was stated previously, the chemicals in detergents can sometimes stop or slow the bacteriological action in the tank, and this is an alternative available to the builder. If too much waste water runs into the septic system, this can not only flood the field, but may flood the field with raw sewage which will create an extremely unpleasant situation.

PRECAUTIONARY MEASURES

A number of other measures are given here to provide for an efficiently operating septic system. To avoid overtaxing the septic tank, runoff water systems should not be piped into it. These systems would include roof gutters, downspouts, storm sewers, or any other outside drainage lines. These drain waters often carry quantities of silt and other materials which the septic tank can not readily digest. They will merely settle to the bottom of the tank and join with the sludge, increasing the frequency with which the tank will have to be pumped out. If this type of drain water is routed directly to the leaching field, it will quickly clog the drainlines. The best method of disposal for runoff water is to drain it into a ditch, dispersing it over a wide area to allow it to be soaked into the ground gradually. It can also be run to a cleanable dry well built exclusively for this purpose as an alternative solution.

Once an understanding of the basic components which make up a sewage disposal system is obtained, you should take the time to sit down and make a sketch of the system as it will look when completed. This need not be drawn to scale, as it is meant to simply give an idea of what materials will be needed and whether the system will perform its function satisfactorily. From this drawing, it should be easy to prepare a complete list of all the parts, their sizes and quantities, that will be needed for the job. With this list in hand, it will be quite simple for a plumbing supplier to quote a price for the materials and order those which are not normally stocked.

SEPTIC SYSTEM INSTALLATION

The actual installation of a septic system involves a great deal of physical labor and some heavy equipment to make the necessary trenches and pits. The first step is to lay out the lines of the excavations exactly where they will be, using a drawing of the system as a guideline. Line up the house drain with the house sewer line, measure it out, and drive stakes to indicate the location of the trench. Outline the pit for the septic tank, allowing an extra few feet all around it to provide ample working room. Mark off the sewage outflow line and then do the same with the leaching field. If the leaching field will be a subsurface tile system, it may be necessary to set out stakes indicating the outlines of several trenches. All of this preliminary work will greatly assist the excavator in knowing exactly where to dig. This digging can be done by hand, but it's much easier and quicker, although more expensive, to hire someone with a backhoe to do the whole job in a few hours. If possible, hire someone who has some knowledge of this type of installation and has a good reputation. It may even be a good idea to solicit a number of bids for the job.

At this point, check over the site of the proposed installation. Trees growing near the pipeline may cause a problem. Their roots can possibly displace or even break up sewer pipes or drainpipes in the field and may need to be removed. Some shrubs with extensive root systems may cause similar problems. For example, a lilac bush's roots will grow right into a pipeline through perforations or even tiny cracks in the joints and will literally fill the pipe and clog it completely. The trees and their roots will prove simpler to remove than those of some shrubs and bushes. If any of their roots are left in the ground, they will continue to regenerate. Some form of herbicide may be necessary in these situations.

Another critical area in regard to installation lies in the proper relationship of the house sewer trench bottom to the septic tank pit bottom. The house line comes through the foundation wall or extends from under the house at a downward slope. This pitch must be continued all the way to the inlet of the septic tank, necessitating the correct positioning of the tank in the pit to allow for ease of connection. Similarly, the sewage outflow line must also pitch downward and away from the tank slightly before it enters the drainpipe in the leaching field. Thus, that connection must be made at a point somewhat lower than the septic tank outlet. Each of the drainpipes in the field itself will have a slight downward pitch. In some cases, it may be necessary to do some recontouring in order to maintain proper depth and pitch.

The house sewer line trench should be dug anywhere from 6 to 8 inches deeper than the level at which the pipe will lie to allow for gravel. This will be discussed later. The septic tank pit should be dug as closely to the exact depth required as possible. This is done so that its weight will rest upon a minimum of loose fill and a maximum of undisturbed native subsoil. The trenches for the drainpipe in the leaching field and the sewage outflow line are also dug 6 to 8 inches deeper than the pipe depth.

All of the required excavation will result in a certain amount of dirt that will need to be disposed of in some way. Arrangement should be made at this time with regard to either having it hauled away or possibly using it on the property for landscaping of some sort. Only a small portion of the dirt will be used to pack in the septic tank and cover it to a depth of 1 or 2 feet. Some will also be used for backfill. The majority of dirt removed from the leaching field will be replaced with gravel. At any rate, a decision will have to be made as to what to do with the dirt.

SEPTIC TANK INSTALLATION

Once excavation is completed, the tank can then be lowered into the pit. This can be done in a number of ways, depending on the size of the tank. If it is somewhat lightweight, it can be positioned by a few persons.

The tank is most often lowered either with a backhoe or a special truck designed for this purpose. The piping is run after the tank has been properly positioned. The bottom of the pit should be leveled before installation, with the inlets and outlets properly aligned, as explained previously. Once the tank is installed, the lid should be set in place to prevent dirt and debris from getting into the tank and for obvious safety reasons. The covers for concrete tanks are normally set in place with a machine, but steel tank covers can be dropped on by hand.

A great deal of tanks are equipped with a small removable lid positioned above the inlet port and outlet port, but if this is not the case, some sort of access should be provided for. There are a number of ways to do this. The easiest method is to make a box column of construction grade redwood 2 × 10s or 2 × 12s about 2 feet tall, or whatever height will bring the top of the column about 6 inches above ground level. Cut another piece of wood or heavy sheet metal to fit on top of the column. This will serve as the lid and can be secured with four short lag screws. A hole is then cut in the tank cover to fit the inside dimensions of the column. If the cover is steel, the column can be attached by means of some screweyes, steel angle brackets, or corner braces screwed to the column and then bolted to the cover. Roofing cement or some other suitable compound is then applied to the bottom of the column, which is then sealed to the cover. A diagram showing a completed inspection hatch is shown in Fig. 6-4. This same method can be used with a concrete cover by lagging the brackets down with shields and bolts. The column is then set in an extra-thick bed of heavy mastic and held securely in place by hand as the backfill is placed. Once the fill has settled and compacted, the hatch will remain secure indefinitely.

An alternative to the above method is to use a length of vitreous clay drainpipe or concrete tile of an appropriate diameter. These would be set in mastic atop a steel cover or mortared in place atop a concrete one. Similarly large lengths of large diameter heavy steel pipe or even sections of culvert could be used, making sure that whatever is chosen is securely sealed to the cover opening. A lid for a hatch such as this could be made of precast concrete with a larger diameter than the outside diameter of the pipe. The weight of a lid made of this material will be sufficient to hold it firmly in place unassisted.

If the tank being used is of the steel type, the lid should always be set in place before any backfilling is done, not only to prevent dirt and debris from entering the tank, but also because of the metal's flexibility. The uneven pressure of the backfill may throw the tank walls out of skew, which will make it difficult to fit the cover in place. If such a problem does result, it will be necessary to at least partially dig up the tank to relieve some of the pressure on the sides before the cover can be fitted properly. In any case, the pit should always be backfilled as soon as possible after installation is completed, mainly to avoid any safety hazards presented by the open pit.

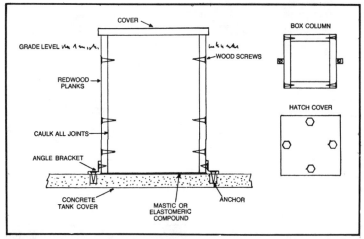

Fig. 6-4. An access hatch may be constructed to allow for periodical inspections.

HOUSE SEWER LINE INSTALLATION

As was stated previously, the sewer line must pitch downward to the inlet on the septic tank. This degree is slightly variable, but ranges from ⅛ inch per running foot to ½ inch per running foot. If the pitch is less than ⅛ inch, liquid drainage will tend to be slow and insufficient to move solid material down the pipe and into the tank. Improper pitch will result in clogging in the line. If the pitch is greater than ½ inch, the flow will be so rapid that the liquid will tend to rush past solid waste and leave it behind, again with clogging being the result. It may be difficult or impossible to run a consistent slight pitch in the house sewer line from one end to the other. This would be the case when the septic tank must be positioned several feet below the house foundation in close proximity due to topography. An alternative method of sewer line installation can be used. If a soil pipe or sewer line is set at an angle of 45 degrees or more, both liquids and solids can flow down the pipe unobstructed. The 45-degree angle is the minimum, and any other practical angle up to 90 degrees can also be used in this situation. If such an arrangement becomes necessary, start laying the house sewer line by joining a section of pipe to the house drain. Lay the pipeline to the tank in the normal manner with a ¼-inch downward pitch. At the proper point, which is determined by measurements, attach a 45-degree fitting and continue the sewer line downward at a 45-degree angle. At the next appropriate point, attach another 45-degree fitting in the reverse position, along with a short stub of pipe to allow for connection to the septic tank. This arrangement is shown in Fig. 6-5.

Before actually proceeding to lay the pipe in place, the trench bottom must be cleared out of any loose dirt or anything that would obstruct the pipe. A piece of board, called a grade board, is then placed in the trench

157

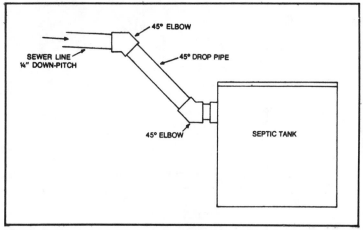

Fig. 6-5. If it is not possible to maintain a constant pitch from the house to the septic tank, this method will work nicely.

which extends the full length of the trench and is angled to a continuous downward pitch. This board can be an inexpensive grade of ordinary 1 × 6 or 2×6 wood stock. Any kind of wood will do, although redwood is recommended because of its longevity. It will also cost more than other choices available, such as pine, spruce, or fir. The board is placed on edge and nailed to stakes which are driven into the bottom of the trench (Fig. 6-6). The top edge of the grade board should lie at exactly the same height as the pipeline bottom. The pitch must be continous and even, so it's best to use planks which are not bowed. The purpose of this board is to prevent the pipeline from setting, sagging, and cracking open as the trench backfill settles and compacts. This board will not be required under sewer lines running at a 45-degree pitch.

After the grade board has been placed in the trench, the slope should be checked once again to make sure it is correct. This is done by measuring the drop from one end of the pipeline to the other. Attach a string to the top of the pipe where it joins the house drain, and stretch the line taut and level to the other end of the pipeline. The level is then checked by attaching a line spirit level to the string and measuring the distance from the string to the top of the outlet end of the pipe. For example, in a 20-foot run of ¼-inch pipe, the total drop from one end to the other would be 20/4 inches, or 5 inches. If the drop is correct, lower the taut line until it is touching the top of the grade board. Check to make sure that the top of the grade board is flat along its entire length.

Once it has been ascertained using this method that the pitch is correct, the next step is to pack backfill in around the grade board. The dirt used for backfill should be free from any debris or rocks larger than a golf ball. If the original dirt excavated to make the trenches does not meet these qualifications, it may be necessary to import some clean fill for this

158

portion of the job. If this is the case, fine gravel makes for a good substitution and will do the job better than soil. The fill should be tamped down quite firmly, but not so hard that it interferes with the predetermined slope of the grade board.

The next step in this installation is to place the pipeline in place in the trench. This can be done using one of two methods, and both will work equally well. If working space is not a problem, the bulk of the pipeline can be assembled out of the trench on a relatively flat surface. Then the complete pipeline is lowered into the trench and is ready for connection to the house drain and the inlet provided for this purpose on the septic tank. The second method is to connect one section to the house drain and continue from this point, section by section, to the septic tank. Remember that this section of pipeline must be of the solid type and that all connections must be completely watertight. No sewage should be allowed to escape at this point in the system. Some types of piping are connected with fittings which are solvent-welded to make the joint. If this kind of line is assembled outside the trench, it should be allowed to cure overnight before the pipeline is placed into the trench. This allows the joints a sufficient amount of time to gain full strength, thus preventing any possibility of weak or leaky connections. If the pipeline is assembled in the trench where it will not be moved around once in place, this will not be necessary and the backfill can be tamped in place immediately following the connecting of the pipes. Some types of pipe use a push-coupling system and are connected dry. These are simply pushed together to make the necessary joints. No waiting time will be needed if this type of piping is used.

The pipe should be lined up carefully to insure that it is positioned exactly on top of the grade board. The fill is shoveled in around the sides of the pipeline and packed down firmly, being careful not to disturb the position of the pipe. More fill is then shoveled in to fill the trench to a depth of 3 or 4 inches over the pipeline. This portion of the fill should not be packed at all, as the line may be crushed. The remaining space in the

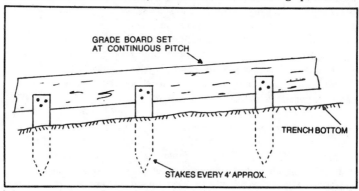

Fig. 6-6. Grade boards are used to maintain pitch and prevent the pipeline from settling.

trench can then be backfilled either by hand or machine, being careful not to apply any large amount of weight to the trench if a machine is used. Now, pack the backfill down somewhat firmly, leaving the surface humped up a few inches above the surrounding land level. The filled area of the trench will eventually settle and compact until it is level with the area's terrain.

Connecting the pipeline to the house sewer line and the inlet on the septic tank is not a very complicated procedure, and no difficulties should be encountered here if measurements and slope are correct. It may be that the house drain is a different diameter than the pipeline, but this should pose no problem at all. An adapter installed at this connection will make this transition quite easy. If different types of pipe are involved in this connection, they can be joined with a solvent-welding process in the fashion described earlier.

The connection of the pipeline to the septic tank inlet is a relatively easy procedure, but this may be complicated a bit by the lack of any connection fittings or pipes provided for this purpose with the tank itself. The holes in the tank are normally a little larger than what is required, and the method chosen to make this connection will be determined by the type of tank being used.

If a steel tank is used, an adapter which completely or at least nearly fills the hole is joined to the end of the sewer pipeline so that it extends about a half inch on the inside of the tank. If there is a lipped flange around the hole, the space is filled with plastic cement, oakum and lead wool, or some form of sealant. If this is not the case, the pipe end should be sealed by whatever means necessary to make for a tight connection. This will probably involve the heavy application of a nonhardening compound that is built up in a tapered ring both around the pipe and against the wall of the septic tank.

If the tank is of concrete construction, it is a simple procedure to run the end of the house sewer pipe through the inlet provided to a depth of about half an inch. The pipe is then sealed in place with mortar which is packed into the excess space and built up in a smooth ridge around the pipe on both the inside and the outside of the tank.

If pipe stubs and/or connection fittings have been provided with the septic tank, the whole procedure will be quite simple. These fittings are usually made either of clay tile or cast iron, both of which will be easier to fit with the proper adapters. These connections are normally made by placing the pipe in place and caulking with some form of sealing compound.

Regardless of the type of connection required by the materials involved, liquid leakage will not be a serious problem at this point. The reason is that the pipeline itself is installed with a portion of it extending into the wall of the septic tank at a downward pitch. This means that any waste traveling through the line will enter the tank without making any contact with the seal itself. No pressure is involved here, either, so if the tank operates as it should, no backup of effluent will flood the inlet area. The primary purpose of this seal is to prevent the admittance of any outside dirt or ground moisture and keep any gases from escaping from the system at this point. Assuming that the house sewer line is installed properly to insure that it does not settle, no mechanical strain will be present here to endanger the connection. It can be seen here that if proper installation procedures are maintained, this seal will hold indefinitely.

SEWER OUTFLOW LINE INSTALLATION

This section of pipe is the line which forms the connection between the septic tank and the leaching field drainpipe system. The connection at the outlet provided on the tank is accomplished in much the same manner as the house sewer line connection to the inlet opening. Again, the purpose for a tight connection here is to prevent any effluent from escaping into the surrounding soil at this point in the system. This line is installed at a slight downward pitch to provide for efficient flow.

If the pipeline being used is reasonably rigid or 45-degree elbows have been installed to maintain pitch, the grade board can be eliminated here. The pipe should be laid upon solid, undisturbed soil, carefully smoothed to the proper pitch and cleaned free of rock or debris. If it is necessary to add loose fill to maintain proper pitch before placing the pipeline in the trench, make sure it is tamped down firmly. This line is normally of the same type as that used for the house sewer line.

It may be possible to join the leaching field and the septic tank with just a single length of pipe. In areas where winters are severe and freezing may be a danger, it is recommended that this pipeline be as short as is practical. There is always a danger when using a long sewer outflow line that small discharges of effluent in the pipe may freeze before reaching the field. This will have the effect of slowly building up an accumulation of ice in the line which will eventually create a clog. The outflow line is more prone to freezing than the house sewer line for a number of reasons. The discharges will be of a higher temperature at the outset of entry into the system, which lowers the potential danger of freezing somewhat. Also, the outflow line is normally not buried quite as deep. If freezing may pose a problem to the efficient operation of a system, a check with the local authorities may provide some assistance in learning what safeguards are normally provided for in your particular area of the country.

LEACHING FIELD INSTALLATION

The final step in the installation of a disposal system can be accomplished in a number of ways, depending upon the layout design selected as most suitable. The first is a straight-line continuation of the sewer outflow line as the drainpipe which, as discussed previously, is converted to perforated pipeline at some reasonable distance from the septic tank. Another possibility would be to run two branches from the outflow line in the shape of a Y where ground contours would make this method a better choice.

The first thing to do when using a single-line arrangement is to drive stakes and set a grade board in place running the full length of the trench, as was done with the house sewer line, with a few slight differences. Local code may dictate the downward pitch required here. Normally, and for the purpose of discussion, the pipes are sloped away from the septic tank at a rate of 2 to 4 inches per 100 feet of run. This will amount to slightly less than ¼ to ½ inch for each 10 feet of pipe. A steeper angle here is recommended. On situations where a single length of pipe will be running 75 to 100 feet, the first 50 feet of pipeline is pitched about 4 inches per 100 feet, with this pitch decreasing slightly after this point, and sometimes the last 15 to 20 feet pitched slightly upward. This is done to prevent any of the effluent from pooling up at the end of the line. Local regulations should be checked when using this method, as there are a number of different schools of thought with regard to the running of single-length drainpipes, with some of the methods being almost the opposite of others.

In situations where the contours of the land are sloped, the pipelines can be laid out with the branches going with the contour at a slight downward pitch. The interconnecting lines between branches are positioned at a relatively steep angle, dropping down sharply to each of the successively lower levels. An installation using this method is pictured in Fig. 6-7.

Once the grade boards have been installed and checked for proper pitch, the trench is filled with gravel. The type and grade of gravel used will most often be regulated by local guidelines. If this is not the case, the type most often used is washed gravel of ¾-inch size. Regulations specifying gravel can range anywhere from the ½-inch to 2½-inch size. Gravel of the washed variety is the best choice. The washing process eliminates the possibility of quantities of silt being sluiced off the gravel by the effluent, thus reducing the porosity of the surrounding soil by plugging it up with these tiny mineral particles. Regardless of the gravel selected, it is worked carefully around the grade boards and stakes, being careful not to dislodge them from their position in the trench. The top of the gravel should be level with the top of the grade board and should cover the bottom of the trench completely. The gravel should be laid to a minimum depth of at least 6 inches, but again, local codes may dictate a depth greater than this.

The next step is to lay the pipeline in place in the trench. As was stated previously, perforated piping is used in the leaching field itself, normally of the same diameter as that used for the sewer outflow line. The perforations are positioned facing down in the trench and connected at the sewer outflow line with the appropriate fitting. Subsequent sections of pipe are then placed in the trench, making sure they are positioned directly on top of the grade board. The connections in this portion of the system will vary according to the pipe material being used.

Once the pipe is in place for the entire length of the trench and properly connected, more gravel is laid in the trenches, the amount of

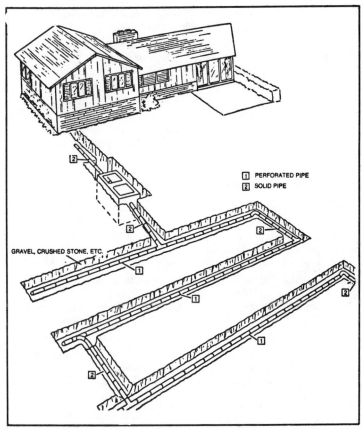

Fig. 6-7. If the site of the leaching field is sloped, the lines may be run in this manner, running at a slight downward pitch with the contours of the land.

which may or may not be determined by local codes. A good minimum to be used as a guideline in the absence of local requirements is a 2-inch depth covering the pipeline. More gravel, up to a depth of 3 or 4 inches, can be added at the discretion of the builder, but anything exceeding 4 inches will serve no real purpose. It is not necessary here to pack this gravel in any way, and doing so may cause the pipes to be broken, crushed, or moved from their proper position. The surface is merely leveled out with a rake so that the gravel is evenly and uniformly distributed in the trench.

Some sort of interface should be placed over the gravel to prevent any backfill dirt used to cover the gravel from filtering into the leaching area. Ordinary building paper, which is available at most building supply houses, is a good choice here. However, anything can be used, as long as it will allow for evaporation. This will disallow the use of plastics, tarpaper, and the like. Straw is another possibility, and a 2-inch layer is recommended.

Regardless of the material selected, make sure it completely covers the pipeline and extends the entire width of the trench.

As was mentioned previously, the backfill soil to be used to cover the remaining portion of the trench should be free of debris and large rocks. The first part of the job should be done by hand, shoveling the fine dirt carefully onto the protective covering so as not to disturb or puncture it. The remainder of the backfilling can then be done with a machine, if desired, being careful that no damage is done to the actual trenches. This soil is not packed down, as it will eventually settle to the level of the surrounding area naturally. A completed installation in a trench is shown in Fig. 6-8.

Leaching Field Installation Using a Distribution Box

The second installation method involves the use of a distribution box, in which the drainpipes for the leaching field are connected by means of this box to the sewage outflow pipe. The procedures remain basically the same, with only a few exceptions.

The first step will be to set the distribution box in place, preferably upon a level pad of undisturbed subsoil to prevent further settling into the ground. The sewer outflow line is then connected to the box, using the appropriate fittings and method of sealing these in place. The grade boards are then placed in position in the trenches which will house the outgoing pipelines. These boards are adjusted to match the level of the distribution box outlets and are pitched downward at the proper angle. The pipes are now attached to the outlets and laid in succession from this point to their

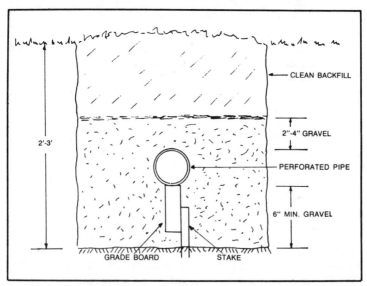

Fig. 6-8. A pipeline installed and buried in a trench in the proper manner.

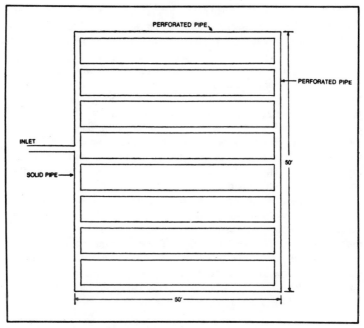

Fig. 6-9. Leaching field installation using a grid network.

completion. The remainder of this type of installation will follow the method discussed previously for a single-line system.

Leaching Field Installation Using a Grid Network

This type of installation is shown in Fig. 6-9 and is somewhat of a different situation than either of the two methods already discussed. To begin, attach a tee fitting to the end of the sewer outflow line at the point where the leaching field will begin. This tee will need to be sealed in such a manner as to make it completely leakproof, and the crossarm of the tee set must be exactly level. Connect a short stub of pipe to each side of the tee crossarm. This pipe should extend the required distance to the next tee. This process is continued until there is one tee in place for all but two of the required number of drainpipes. The remaining two are the ones located at each end of the grid, and a 90-degree elbow is installed at each of these two locations. Make sure all the fittings are tight, completely leakproof, and lie on a level line with each other. Gravel will not be necessary in this portion of the system, as this really is not a part of the actual leaching field, but rather an effluent distribution line. Grade board is not specified here, as this line can simply be placed on the undisturbed subsoil if this is preferred. The grade board can be used if desired.

Grade boards placed in the trenches for the drainpipes need only be used if the pipelines are lengthy. A good rule of thumb would be if the lines wil be longer than one length of pipe, a grade board should be used. Set up the grade boards, if necessary, and spread a layer of gravel in the bottom of the trench. Double-check the pitch at this point to make sure it is correct. If all seems well at this point, install the perforated drainpipes in the trenches and connect them into the tees and elbows. These joints do not have to be leakproof and can simply be pushed together if the type of pipe being used will allow for this. The final procedure of adding gravel, installing an interface, and backfilling can now take place. However, in some installations, a tailpipe is added to join all of the ends of the drainpipes together. This can improve both the effluent flow and distribution and will also provide for a degree of circulated air in the lines. If this method is used, the tailpipe is assembled in the same manner as the connection to the sewer outflow line, using tees and elbows. Although the connections of the drainpipes to the tailpipe do not have to be leakproof, those along the actual tailpipe should be. This section of pipe is not normally perforated, as it is not considered to be a part of the leaching field. Gravel will not be necessary here. There are some situations where the tailpipe will be included as part of the field, in which case it may be of the perforated type and would need to be surrounded by gravel. This would normally be determined by local regulations and possibly the preference of the builder and the amount of space in the system.

INSTALLATION OF A VENTING SYSTEM

The installation of a venting system may be required by local authorities. This may involve only one or two vents in some cases or several in others. The purpose of these vents is twofold. Most importantly, they serve to admit fresh air into the system which provides a freer circulation. This aids both in the evaporation of the effluent and oxidation of the substances carried into the drainpipes. The second purpose is to partly exhaust any slight amount of gases that might be present in the drainpipes. The methane gas produced by the system can present a possible safety hazard, as it is highly flammable.

The vents are normally made of the same material as the pipelines and are joined at the appropriate points to the drainpipes with tee fittings. These connections should be quite solid and completely leakproof. The stubs of the pipe are set vertically and should protrude above the level of the ground by about 12 to 18 inches. The top of the pipe is then capped with a hood specially designed for this purpose. This is done to prevent any debris from entering into the system. The exact location of these vents is most frequently determined by local regulations. In the absence of any specified locations, the vents can be placed at the ends of branch lines or grid legs, or they may be located in a tailpipe line. A typical vent installation is shown in Fig. 6-10. The exact location of vents in the system is not critical in any case, as long as they are installed where they will provide a reasonable amount of air circulation.

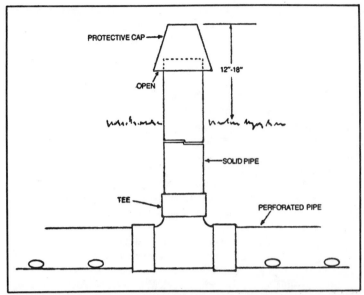

PROTECTIVE CAP

12"-18"

OPEN

SOLID PIPE

TEE

PERFORATED PIPE

Fig. 6-10. Vents are usually required by local codes, but even if they are not, they serve a very useful function and should be included.

AN ALTERNATIVE TO AN ABSORPTION FIELD

Another method of sewage disposal is the seepage pit or seepage bed shown in Fig. 6-11. This type of installation is much different than an ordinary leaching field in several respects. First of all, very little reliance is placed upon evaporation, as the pipes are buried much deeper. Trenches are not used, but rather the pipe is laid in a single, extensive bed of gravel. It is important to note here that many areas prohibit the use of seepage beds, so be sure to inquire as to their usage before any definite arrangements are made for this type of installation. There may also be specifications with regard to the soil condition and characteristics which may preclude this type of system. Because the pipes are installed at a greater depth, one of the advantages of a seepage pit is that there is less danger from freeze-up.

Figure 6-12 is a variation of the seepage bed. Called a deep bed, the construction is done in much the same manner except that the bed is set as much as 15 feet into the ground. This arrangement is most commonly used in situations where the topsoil or subsoil layer runs exceptionally deep and does not have good percolation capabilities. By installing the system at a much deeper level, the field is placed upon a stratum of gravel or some other geological layer that is sufficient to provide the required absorption. The presence of these layers, as well as their depth and thickness, is normally determined by taking core-drilling samples. If it is found that these layers will provide the necessary absorption and they are within a reasonable distance below the surface of the ground, the system can be installed. Due to the amount of excavation work that will be required using a deep bed, it may not be suitable for use with larger fields.

Installation of either of these two arrangements is done in much the same manner as an ordinary leaching field. All connecting sections are solid, while the drainage lines will be of the perforated type. Seepage pits are normally arranged in a grid fashion, with the grids constructed the same way as previously discussed. Instead of being in trenches, the piping is laid out on a large, continuous bed gravel that extends outward about 3 feet beyond the outermost pipelines. Grade boards are usually installed to maintain the proper pitch. In a deep bed system, the drainpipes will usually be only a few feet apart. The depth of gravel beneath the pipe is far greater than in an ordinary leaching system. A depth of 2 feet is about average. In a system such as this, pitch is not especially critical, and the grid may even lie level or pitch slightly downward anywhere up to about ⅛ inch per running foot.

A large quantity of gravel is then placed over the drainage lines, normally about 3 feet or so. The grid may be open-ended or may be closed by means of a tailpipe. An interface is laid over the gravel to prevent the backfill from filtering down into the gravel. The sewer outflow line leading to this portion of the system is normally placed at an angle of 45 degrees or more. Generally speaking, a distribution box is not usually installed with a seepage bed.

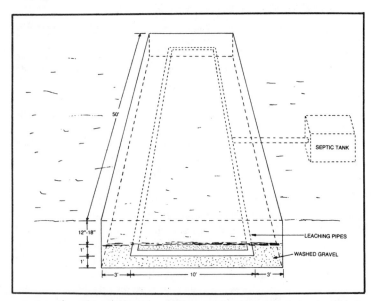

Fig. 6-11. A seepage bed may sometimes be allowed if it is not feasible to install a leaching field due to inadequate absorption characteristics of the soil.

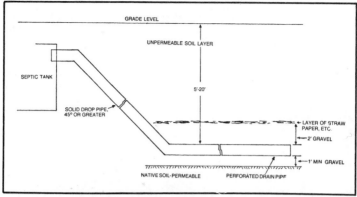

Fig. 6-12. A deep bed will provide a method of sewage disposal in areas where the topsoil or subsoil layer runs exceptionally deep and does not have good percolation capabilities.

SUMMARY

In planning and installing a septic system, great care should be exercised to insure that everything is done in strict accordance with all rules and regulations regarding such an installation. Certain practices governing these systems have been proven to be the most effective methods and will provide the homeowner with a *safe* process of disposing of human wastes and grey water.

The installation of a septic system will almost always require some permits and tests in the early planning stages. If you are planning to purchase a piece of land for a proposed residence, check with local authorities to find out whether the soil is capable of absorbing these wastes in an adequate manner. Some municipalities require percolation tests on any land before it is zoned residential, so the results of such tests will usually be available. A second series of tests may also be required at the time permits are requested for the actual installation. In any case, these tests are very important, since they have been proven to be an accurate method of determining soil capabilities.

Be sure to allow for future additions or improvements to the residence to avoid having to increase the capacity of your septic system at a later date. This forethought and the extra cost will prove to be invaluable, as anyone who has had the experience of having to provide additional trenches or a larger sized septic tank can attest to.

Design and Layout of a Plumbing System

The actual design of a home plumbing system will be less complicated if both design and installation are to be done by the same person. The drawings themselves will not have to be quite as specific and technical, since less interpretation will be necessary. The main purpose of these drawings will be to give the person doing the installation a basic layout of the system, the different sizes and types of piping to be used in different parts of the system, and the location of the fixtures themselves.

Alternately, the home plumber may opt to have complete drawings and specifications made up by either an architect or engineer. This may be quite expensive and may also require you to have some knowledge of reading these detailed design layouts. For the purposes of this text, all the factors which will be involved in making up a simple layout of a plumbing system will be discussed. There are a number of essential points that should be considered in both design and installation, as well as different methods that may be used in system planning.

The first thing that must be considered is the local and/or national codes that must be adhered to during the design and installation of the plumbing system. Even if there is no code in effect in a given area, it is suggested that the major precepts of the National Plumbing Code be followed. Make it a habit in the early planning stages to check through the applicable codes to see which articles in particular might affect the design of the system. A copy of these codes can usually be obtained at any town or city hall, usually at no charge. Any features specifically recommended or possibly mandatory should be included in the plans. Those procedures, techniques, and materials that are code-specified must also be made a part of the installation. One important point to remember when reading the codes is that the words "shall be" or "must" always mean that particular item is mandatory. The word recommended, or its equivalent, means that this is what the authorities would like to see, but it is not mandatory. The interpretation of any particular article can sometimes be variable, depending upon the judgment of the local authorities. In any case, remember to stick with the code to avoid future difficulties involving redoing the system due to noncompliance.

DESIGNING THE SYSTEM

This is the first and most important step in a plumbing installation. The main purpose at this point is to define exactly what will be done, how to best suit the intended purposes, and the materials to be used. This involves determining the location of the water supply line, its length, lie, and diameter, the point of entrance into the structure, the type, size and location of a water pump, and other similar items. It will also be necessary to decide what plumbing fixtures will be used and their approximate location within the building, the number of water distribution outlets that will be needed both now and in the future, and the type, size, and approximate length of major runs of piping that will be required. Water pressure and flow details will also have to be worked out. The layout of the DWV system is quite important. It must be lined out with approximate sizes and lengths of pipeline determined, as well as the type of pipe to be used. The locations for the waste stacks, vent stacks, major branch drainage lines, the house main drain, and its point of exit from the building will have to be worked out. Likewise, if a septic system is a part of the installation, these details will need to be taken care of.

Keeping in mind all that must be considered in designing the plumbing system, the first decision involves choosing the method to be used ir laying out the actual design. There are two basic ways to do this.

Making Drawings

Drawings of the entire plumbing system or subsystems can be as complex or as simple as is desirable. These plans can be drawn up in accordance with current acceptable drafting practices, using all of the proper symbols. Alternately, if the same person will be doing the installation, simple schematic sketches will be sufficient and are the easiest to make. This type of drawing shows, with the aid of standard symbols or with symbols made up by the designer, the principle outline of the piping system, including valves and fittings. An example of this type of layout is shown in Fig. 7-1. As can be seen, many of the minor fittings such as slip nuts on traps are not included, while all the major fittings and valves are. The main purpose is to line out all of the major pieces of the system in order to make it a simple task to come up with a materials list. A more complicated drawing would detail every length of pipe, fitting, plumbing fixture, and accessory part in the entire system. The drawing may actually be scaled, or the necessary dimensions such as those for pipe lengths may be written in. Very complex drawings may also include exact locations of the pipes in the walls, show the points where they make directional changes, and similar details. This type of drawing is complete in every

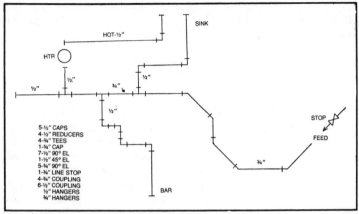

Fig. 7-1. A sample drawing of a design layout indicating fixtures, lines, and pipeline sizes.

way, and their advantage is that nothing whatsoever is left to chance. All the plumber need do is follow the directions explicitly and complete the job as directed. However, to the beginner in the plumbing field, these drawings will probably be quite confusing, not to mention quite time-consuming to make. If this method is chosen, it is suggested that you stick to the more simple drawings.

The Empirical Approach

This method is probably the easiest approach for the neophyte builder. It is practical, time-saving, and only requires that the plumber proceed with a certain amount of caution and common sense. The first step will be to decide what kind of pipe will be used, the sizes, and where the fixtures will be located. The next step is to walk around the structure and spot all plumbing fixtures in their appropriate locations in as accurate a manner as possible. Then figure out, simply by looking around, just what the best way is to get a pipeline from one place to another. At the same time, figure out what fittings will be required in the installation of each particular pipeline. Make up a list of parts and measure out the approximate pipeline lengths. At this point, it is a good idea to make up a quick sketch of the installation and possibly some marks and points of reference directly upon the framing members, floors, walls, etc. Once this is completed, move on to the next section of the system and repeat the process.

Once this procedure is completed, you will have in hand a number of parts and supplies lists that can be compiled into a complete materials list. The next step is to purchase all the materials and start installing the system.

COLD WATER SYSTEM DESIGN

It has already been explained in a previous chapter how to determine the amount of water that will be needed to supply a residence. Another factor which will have to be considered is the pressure of the water being delivered in order to provide ample pressure throughout the house. What most people do not realize is that the pressure where the water leaves the main and the pressure where it enters the house are not the same. The pressure in the latter case is somewhat lower. As the water travels upward from the basement to the first floor, the pressure is still lower; by the time it reaches the end of the line on the second floor, it is even lower. Thus, it can be seen that operating water pressure is not equal throughout all the parts of a plumbing system. The reason is that the inside wall of the pipeline offers a certain amount of resistance to the flow of water. Every fitting, valve, or meter in the pipeline provides additional resistance. The weight of the water itself in the vertical pipes (the horizontal ones make little difference) must be overcome, which results in further pressure loss. As can be seen, enough pressure at the main to provide a modest flow of water at the first fixture on the line will not be sufficient to produce more than a drip at the far end of the system.

The amount of draw on the system at any given time will also produce a definite effect. For example, assume that the water pressure in a system remains constant, even though this is rarely true. If one faucet is opened all the way, a good water flow is obtained. Another faucet is opened, but this time the flow is somewhat less than at the first faucet. This may also cause the flow at the first faucet to be reduced somewhat. A third faucet is opened and each flow is lessened again. In theory, if enough faucets were opened, the flow at each one could be reduced to a trickle. The reason is that when the flow rate is increased by opening more taps, the pressure in the pipeline is decreased. The pressure available at the main is fixed, and there isn't anything that can be done about this.

In designing a distribution system, therefore, all of this must be taken into consideration to avoid the problems of low pressure at all the taps and a sufficient flow or volume of water so that a reasonable number of fixtures can be used simultaneously. This can be accomplished by providing sufficient pressure at the main and, more importantly, installing the correct sizes of pipes throughout the system. At the same time, the system should combine maximum practical efficiency and the lowest possible costs for both materials and installation with the least amount of time and effort.

Water Pressure Requirements

Water pressure is measured in *pounds per square inch* (psi). For example, a bathtub faucet will function fine with a minimum pressure of 5 psi. A shower, on the other hand, requires 10 psi. Practically any standard household fixture will operate satisfactorily at pressures within this 5 to 10 psi range, with one exception. A sill cock with 50 feet of garden hose attached to it will require a minimum pressure of 30 psi to achieve its desirable flow rate of 5 gallons per minute. Aside from this exception, it is safe to say that a pressure of 10 psi will be adequate and, in some cases, more than enough to satisfactorily operate the household plumbing fixtures. Although 10 psi is adequate, a much more satisfactory operation will be achieved with higher pressures. Most household plumbing fixtures do operate at higher pressures.

Now that it is known that a minimum of 10 psi is required, it is necessary to determine what the pressure must be at the main to overcome the various pressure losses along the line to achieve the minimum at the farthest point in the system. The easiest way to do this is in steps, starting with the top floor. It will be necessary to know the approximate location, within a foot, of each plumbing fixture. Start at the point where the supply riser, the vertical feedpipe coming up from the first floor, joins the lateral, or horizontal, pipeline that will feed the top floor fixtures. Measure the whole length of the pipeline from one end to the other, following all of the angles and turns that the line might make. This process should involve only the main supply line that will pass close by the plumbing fixtures. It is not necessary to include any branch lines or the short lengths of pipe or tube needed to connect the fixtures to the supply line. Be sure to add up the number of fittings that will be required in the line. As explained previously, each foot of piping represents a certain amount of friction loss or resistance to the flow of water, which results in a drop in pressure. Likewise, each fitting creates an added pressure loss.

Tables are available which list the exact amount of pressure loss for any conceivable type of fitting and every different type of pipe that might be used. Using such a table will provide exact amounts of pressure in any given pipeline. However, for the purpose of this text, it is not necessary to use these tables. Tables list the fitting pressure loss in terms of pipe diameter's equivalency, or equivalent length of pipe in feet. The latter method will be used here, since it is simpler. Consider that each fitting in the line is the approximate equivalent of a 90-degree elbow, whether it actually is an ell or not. The exception to this rule involves the inclusions of valves in the supply line. A gate valve causes considerably less friction loss than a 90-degree elbow, while a globe valve causes about eight times as much. Assume there are no main line valves, since there almost never are any in the distribution side of a residential water supply system.

Table 7-1 provides the figures that will be necessary for these calculations. Assuming a top floor piping length of 30 feet and nine fittings and the use of ½-inch pipe for the main feed line, it is a simple matter to determine if that size will be satisfactory. From this figure, it can be seen that each fitting is equal to about an extra foot of pipeline. Therefore, the nine fittings are worth an extra 9 feet of pipe, which added to the original 30 feet equals 39½ feet of effective pipe length. This is now rounded off to an even 40 feet for convenience.

Arrangement of Plumbing Fixtures

It is now necessary to consider the plumbing fixture arrangement along the pipeline. If all the fixtures are at the far end of the line, all of the needed water will flow through the entire pipeline and the fittings to reach

174

its destination, and the whole 40 feet should serve as a basis for calculation (Fig. 7-2). If four of the fixtures are in the middle and the fifth is at the far end, the situation is a bit different. If the end of the line fixture is a low-volume one, most of the water will be used at the midpoint of the pipeline (Fig. 7-3). A conservative estimate might be an 80 percent flow in the first half, or 20 feet of the line. With this in mind, it is safe to assume that most of the water will only flow through the first half of the line. Thus, instead of using the full 40 feet of the pipeline in the necessary calculations, it will only be necessary to use just half, or 20 feet. Similarly, if the first four fixtures were spread along 30 feet of the line, that figure would be used. In other words, the effective pipe length is adjusted accordingly with regard to the locations of the fixtures.

Pressure Loss Calculations

To continue in these calculations, assume that a minimum flow of 12 gallons per minute (gpm) is needed in order to adequately feed the fixtures. Referring to the information in Table 7-2, it can be seen that 10 gpm is the highest flow listed for the ½-inch pipe size. This does not necessarily mean that ½-inch pipe cannot be used for flows of over 10 gpm. At that level, a practical limit is rapidly being approached. This brings up two more points worth mentioning at this stage.

First, the information provided in this chart is compiled of approximate figures only. They may or may not be exactly correct, or nearly so, for any one particular kind of pipe. Some types and brands of pipe may carry lower figures, while other may be higher. The best way to make accurate or at least more accurate, calculations is to refer to pressure loss tables applicable to the particular type of pipe being used in the installations.

Secondly, no chart providing this information will allow for the combining of flow rates listed in order to arrive at an accurate pressure loss. For example, if a 2 gpm flow is added to a 10 gpm flow to arrive at the needed 12 gpm flow in a ½-inch line, the total pressure loss would appear to be 50 psi. This figure is really not accurate. Note that there is a great jump in pressure loss between a 5 gpm and a 10 gpm flow. As water flow increases, pressure loss does so at a much greater rate. Adding a 1 gpm and a 2 gpm pressure loss together in a ½-inch pipe totals 4 psi but, as

Table 7-1. Pipeline Equivalents for Fittings.

FITTING SIZE, INCHES	PIPE EQUIV., FEET
⅜	½
½	1
¾	1¼
1	1 ½
1 ¼	2

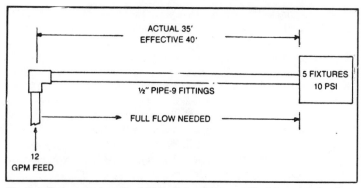

Fig. 7-2. The whole length of 40 feet should be used in this example, since all the fixtures are at the far end of the line.

indicated in Table 7-3, a 3 gpm flow results in a 6 psi pressure loss. Similarly, an odd flow rate lying between the listed figures must be extrapolated toward the high side. A 7½ gpm flow in ½-inch pipe will not lie halfway between the figures for 5 to 10 gpm, or 30.5 psi, but rather closer to 40 psi.

Keeping this in mind, refer back to the example being used in this calculation. You must decide what to do with the nearly overloaded ½-inch pipeline. It is now obvious that a 12 gpm flow in the line might result in unacceptable pressure loss which, at a guess, may be 60 psi or more. One solution to this problem might be to switch to ¾-inch pipe. That is more expensive and also a bit more difficult to work with. Another possibility is to make a few more assumptions and estimates, sticking with the ½-inch size just to see what the results will be. It can be assumed that the required

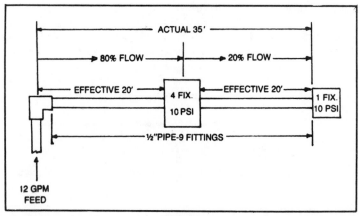

Fig. 7-3. In this example, the effective pipeline is just 20 feet, since four of the fixtures are located in the center of the line.

176

12 gpm flow may actually be closer to 10 gpm, or even less most of the time. Further, even if the flow does reach 12 gpm, the pressure loss will not be bothersome to the occasional users. This further means that the pressure loss perhaps will seldom reach the listed 10 gpm pressure loss figure of 47.0 psi. Arbitrarily, a somewhat higher figure of 50 psi can be used for calculation purposes.

If the pressure loss in 100 feet of ½-inch pipe is 50 psi, as shown in Table 7-2 (47.0 rounded off to 50), and the full 40 feet of pipe is the effective length, the pressure loss will be 40/100 of 50, which amounts to 20 psi. Similarly, if the pipeline were only 20 feet long, the pressure loss would be 20/100 of 50, or 10 psi.

Head Pressure

Now that the amount of pressure loss is known, and it has already been determined that 10 psi is a minimum requirement to operate the fixtures on this top floor properly, the pressure required at the inlet end of the top floor supply line is 30 psi. All the final results of the calculations thus far are shown in Fig. 7-4. The next step is to determine whether this amount of pressure can be delivered to the top floor. This will depend upon three principal factors: the available pressure at the main, how far the water must travel to reach the top floor, and how high the water column must rise in the process. To work this out, only a rough idea is needed at this point, so the thing to be considered will be just the head pressure, or the pressure required to lift the weight of the water.

The first step in this process is to determine the depth of the water main below the street. Determine the elevation of the shutoff valve within

Table 7-2. Pressure Loss for Pipeline Sizes.

FLOW GPM	PRESSURE LOSS PER 100 FT. PIPE SIZE—TRADE				
	⅜	½	¾	1	1 ¼
1	2.5	1.0	0.25		
2	8.5	3.0	0.5	0.25	
3	17.5	6.0	1.0	0.35	0.1
4	29.0	9.5	2.0	0.5	0.25
5	42.0	14.0	3.0	0.75	0.35
10		47.0	8.5	2.5	1.0
15			18.0	5.0	2.0
20			29.0	8.5	3.5
25			48.0	12.0	5.0
30				17.0	6.5
35				26.0	8.5
40					11.0

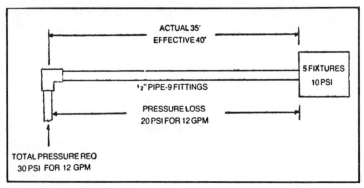

Fig. 7-4. Final results of calculations.

the house about or below street level, and add to or subtract from the main pipeline depth, whichever is applicable to the particular installation. To this figure, add the vertical distance from this shutoff valve to the point where the riser supply line feeds the top floor horizontal supply line. For example, assume that the main is 8 feet underground, the shutoff valve is 5 feet above street level, and the distance from the shutoff valve to the top floor joint is 18 feet. This totals 31 feet. By referring to the water pressure conversion chart in Table 7-3, it can be seen that a 30-foot head will require a pressure of 12.99, and a 1-foot head will require 0.43, for a total of almost 13½ psi to get the water up to the top floor (Fig. 7-5). To this must be added the 20 psi loss in the top floor pipeline and the 10 psi required for the fixtures, which is a total of 43½ psi required at the main. Also, up to this point, frictional losses in the supply line, the riser line, the various fittings, and the water meter have not been taken into consideration at all.

Table 7-3. Head Pressure Loss.

HEAD (FEET)	PRESSURE (PSI)
1	0.43
2	0.87
3	1.30
4	1.73
5	2.17
6	2.60
7	3.03
8	3.46
9	3.90
10	4.33
20	8.66
30	12.99

Unless the pressure in the water main is exceptionally high, it is apparent that the ½-inch pipeline size for the top floor supply line may be inadequate or perhaps even unworkable.

A possible solution may be to either move some of the plumbing fixtures themselves closer to the supply riser outlet, or even decrease the number of fixtures on the top floor. This will work, but for reasons of design and preference, it may not be desirable. Nothing can be done to decrease the amount of pressure required for the fixtures, but the amount of pressure loss in the supply line can be changed. As was stated previously, the bigger the pipe diameter, the less the pressure loss for a given flow rate and input pressure. With this in mind, let's work the calculations again, using instead the ¾-inch size pipe.

To make this calculation a bit different than the previous example, assume that most of the water use will be at the midpoint of the run, and that there are six fittings needed to arrive at this point. The length of the pipeline is 20 feet, and by referring to Table 7-1, it can be seen that each fitting in ¾-inch size is the equivalent of 1¼ extra feet of pipe. Thus, the total effective pipe length is 27½ feet, which is rounded off to 28 feet. By estimating between the lines, a pressure loss per 100 feet of ¾-inch pipe for a 12 gpm flow is about 13 psi. Thus, 28/100 of 13 is just over 3.6 psi. Add this to the necessary fixture pressure of 10 psi to arrive at a total of somewhat less than 14 psi as a requirement at the top floor distribution line. It can be seen that this is considerably less than that required by the ½-inch pipe. This does not mean that the entire run must be ¾-inch pipe. The second half of the line might possibly be ½ inch or even smaller, depending upon the fixtures, with no ill effects and a savings in cost (Fig. 7-6).

Top Floor Riser and First Floor Pressure Losses

In order to determine what size riser pipe will be required, it is necessary to arrive at a pressure loss for that distance. Obviously, if the

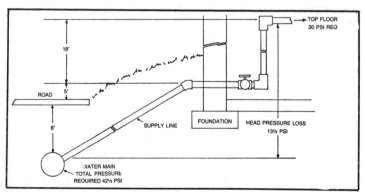

Fig. 7-5. In this example, the distance from the street main to the top floor is 31 feet.

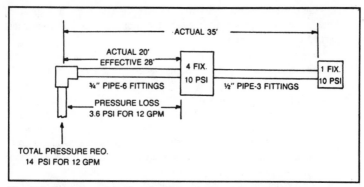

Fig. 7-6. The second half of the line may be ½-inch pipeline.

first section of the top floor supply line is to be ¾-inch diameter pipe, a smaller size cannot be used for a riser, although a larger one could be. The ¾-inch riser will run from a tee at the first floor level to a point approximately at the second floor level. Assume this distance is 8 feet, and refer to the example in Fig. 7-7. It has already been determined that a 12 gpm flow is needed, and that the pressure loss per 100 feet of ¾-inch pipe is 13 psi. Thus, if the riser is installed using ¾-inch pipe, the pressure loss in 8 feet of pipe would be 8/100 of 13, or a little over a pound.

It will also be necessary at this point to consider the head pressure loss, since this is a vertical riser. By referring back to Table 7-3, it can be seen that this amounts to 3.46 psi. If both figures are rounded off and added together, a total pressure loss in the riser of about 5 psi is the final result. By adding this to the pressure requirements at the second floor of 14 psi, it can be seen that a total of 19 psi will be needed at the first floor tee to provide an adequate upward flow.

These same procedures are followed in much the same manner to arrive at the requirements for the first floor supply line. Using the example shown in Fig. 7-8, it can be seen that there are five plumbing fixtures requiring a 12 gpm flow. All five fixtures are closely grouped and can be treated as a unit. The pipeline distance from the first floor tee to the fixture grouping is 13 feet, with seven fittings in the line. By making the first calculation for ½-inch pipe, an equivalent of an extra 7 feet of pipeline is arrived at, for a total effective length of 20 feet. The pressure loss for 100 feet of ½-inch pipe is roughly 50 feet. Since this line is only 20 feet long, the pressure loss is the equivalent of 20/100 of 50, or 10 psi. This is not an unreasonable figure and, if necessary, ¾-inch pipe can be substituted for a substantial reduction in pressure loss. For the moment, let's continue using the 10 psi figure already arrived at using ½-inch pipe to see if the final net result will be adequate. The required fixture pressure of 10 psi is added to the 10 psi for a total required first floor pressure of 20 psi. There is not much difference between the required pressures at the first floor tee for the two floors. The first floor is 1 pound higher. The only effect of this is that the pressure available for the second floor will be a bit greater than originally calculated. As long as the first floor pressure is greater than the top floor, no ill effects will be experienced. In fact, even if the first floor pressure requirement was less by anywhere from 2 to 5 pounds, no harm would be done and the system would most likely be satisfactory. In a situation where the requirement at the first floor were substantially lower, it would be necessary to meet the higher pressure as nearly as possible to provide an adequate functioning installation.

To continue, since it is already known that the riser from the first to the top floor will be ¾ inch minimum, it will be necessary to maintain this size for the entire length of the riser from the basement all the way through the system. Making calculations in the same manner as for the first to second floor riser, if the head is 8 feet as before, the head pressure loss will be approximately 4 pounds. Assume that the riser must take a slight jog and is 10 feet long, including two fittings equal to 2½ feet of pipe (Fig. 7-9). A flow of 12 gpm for the top floor, plus the same requirement for the first floor, results in a total of 24 gpm. A 24 gpm flow through 100 feet of ¾-inch pipe results in a pressure loss of approximately 42 psi. Since the riser is only one-tenth of this length, the friction pressure loss will amount to about 4 pounds rounded off. The total loss in the riser is 8 pounds from the main shutoff valve to the first floor tee. The total pressure required to feed the first floor is equal to the required first floor pressure of 20 psi, plus the riser pressure loss of 8 psi, for a total of 28 psi. This is a higher pressure than required by either the first or the second floor branches in

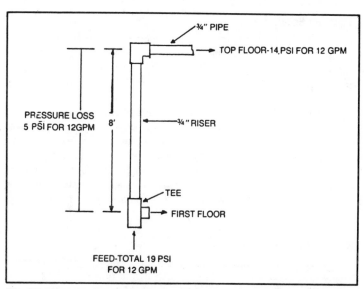

Fig. 7-7. The riser must be at least ¾ inch, since it cannot be smaller than the pipeline on the top floor.

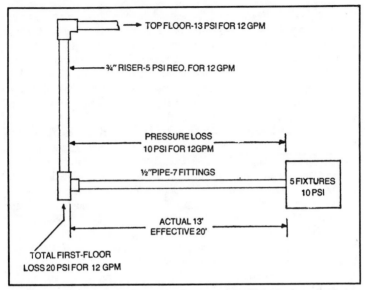

Fig. 7-8. In this example, the effective length is 20 feet, since there are seven fittings in the line.

themselves, so this will be adequate as long as the higher pressure requirement is met.

Basement Requirements

The next step is to determine the requirements for the basement of the structure. Referring to the example shown in Fig. 7-10, assume again that a minimum of 10 psi and a rate of 12 gpm are required. All five fixtures are grouped within 10 feet of the main shutoff valve, with only two fittings. Using ½-inch pipe size, this makes for an effective pipeline length of 12 feet. Using an approximate pressure loss figure of 50 psi per 100 feet of ½-inch piping, the result is that 12/100 of 50 equals a pressure loss from friction of 6 psi. This is added to the required fixture pressure of 10 psi for a total of 16 psi needed for the basement piping. Since it has already been determined that 28 pounds are needed to supply the first floor and the basement pipeline connection will be made at approximately the same point, there is more pressure than necessary once again. The result is a satisfactory arrangement.

Water Supply Line

The final step in this process is to work out the details for the water supply line. As has already been established, a flow rate of 12 gpm is required for each floor. For ease of figuring, a 35 gpm flow will be used. It is also known that a minimum pressure of 28 pounds is required to properly

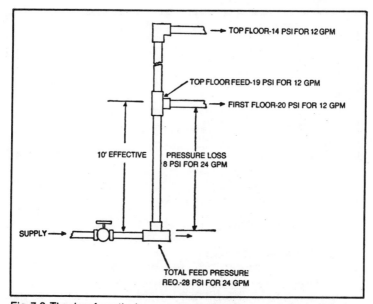

Fig. 7-9. The riser from the basement to the first floor must also be ¾ inch.

supply the plumbing fixtures themselves. Assume for the purposes of this example that the distance the pipeline must travel from the water main to the main shutoff valve in the structure is 50 feet, and that the water must rise vertically from the main to the valve a total distance of 6 feet. Also

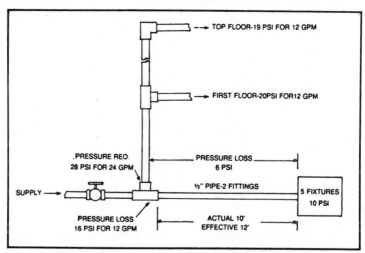

Fig. 7-10. This example shows the method for determining the requirements for the basement of the structure.

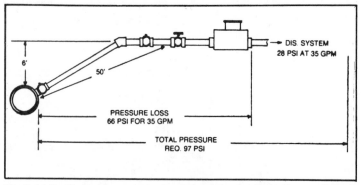

Fig. 7-11. It will also be necessary to work out the details for the supply line.

included in this example are a stop valve at the main, a curb stop valve at some point outside the house, and a main shutoff valve. The water valve in this system should always be a gate valve which has low restriction, rather than a globe valve that greatly impedes water flow. Also included in the system is a water meter. Since ¾-inch pipe size is often specified for water tap and house supply lines, it will be used in the calculations. This arrangement is shown in Fig. 7-11.

The first thing that must be done is to determine the effective pipe length. The presence of a gate valve in a ¾-inch line is the equivalent of an extra ¼ foot of pipe. Since there are three such valves in this installation, this can be rounded off to an extra 1 foot of pipeline. Refer to Table 7-2. Note that for the required 35 gpm, the pressure loss in 100 feet of ¾-inch pipe, is not listed; this size is not adequate for normal uses. To prove this further, assume that the pressure loss may be as high as 100 psi or more, so 51/100 of 100 equals 51 psi. To this is added the head pressure loss which, according to Table 7-3, is 2.6. This figure is rounded off to 3 and and added to the 51 psi for a total of 54 psi. It is now necessary to determine the pressure loss through the water meter.

Water meter pressure loss will vary depending upon flow and also the particular construction and characteristics of the meter itself. Specific information on the particular meter used should be obtained. Assume a pressure drop of 15 psi. This results in a total pressure required at the main of 66 psi, and a grand total of 97 psi for the entire house when added to the requirements for the first and top floors.

It should be obvious at this point that a ¾-inch supply line would be impractical in this example. It is possible to provide a water main pressure of 97 psi, and many will run a good deal higher. But 80-90 psi is considered a bit high for a residential system, even though some do operate at 100 psi or more. The system would be functional at a reduced water main pressure of 80-90 psi, but at the expense of a total flow of less than 35 gpm. For example, if the system were recalculated on a basis of a total flow of 25 gpm, the pressure requirement at the water main would be 56 psi, which would result in a workable installation. As can be seen, a functional system can be arrived at through a process of juggling estimates and figures, as well as pressure/gallonage tradeoffs.

Options

Once the calculations are completed, it may be that the available water main pressure is less than the final figure arrived at as a requirement. There are a number of routes that can be taken. If the difference is somewhat marginal, you may simply decide to accept the situation as it is and live with it. Remember that these figures are simply reasonable estimates, and a difference of a few pounds will very rarely be noticeable during the actual operation of the system. If the available pressure is significantly less than what is required, it will be necessary to do some further calculations and possibly some redesigning of the system itself. The first thing to consider would be a large supply line. For example, the pressure loss at a 35 gpm flow in a 1-inch 50-foot line would be only 26 psi. Once the head pressure and water meter loss are added to this, the figure will still be considerably less than the 97 psi arrived at in the previous example.

Another solution might possibly be to increase the size of the riser pipe from the cellar to the first floor to, say, 1 inch. Alternately, the calculations can be repeated to reassess flow demand, refine the numbers to lower stages, and through a process of less rounding off, to come up with a tighter and more specific pressure requirement for the supply at the entrance to the distribution system.

All of the information given thus far applies to those structures connecting to a street main. For those in rural areas with a private well or other water source as a supply, the situation will be a bit different. If the well is already in a particular location with respect to the house, you will obviously have to deal with whatever pressure is available and make whatever adjustments are necessary in the structure. If the well is still in the planning stages, it should be located as close to the dwelling as possible and in line with the most convenient position for a main shutoff valve or a pump location. Not only will this cut down piping costs, but it will also have a beneficial effect upon pressure requirements. Also, in selecting a pump, it should be sized accordingly to deliver whatever water flow rate is desired and whatever water pressure is required for the house distribution system (within reason).

Although the method given here for determining pressure requirements may seem like a complicated and time-consuming process, it is really quite simple and straightforward. Each plumbing system will have its own individual features, but this approach will give you a good idea of what sizes and types of piping and materials will best suit your individual needs. If a system is installed with no regard for pressure drop and other factors, there is a good chance that the system will not perform adequately. This is not to say that no system installed in this manner will work; in fact, it may function perfectly well, simply by sheer luck and the chance that the available water pressure happens to be adequately high.

Another very important lesson is learned in using this method of calculation. By condensing and compressing the entire plumbing system as much as possible, better flow and pressure will be attained. Similarly, using the smallest practical pipe sizes and a minimum number of fittings will save a good deal of money at the same time. An example of this is the stacked plumbing system, where plumbing fixtures are located back to back, directly adjacent to one another and/or directly above one another. The reasons for this should be obvious at this point.

Tried and true methods can be substituted for these calculations with much the same satisfactory results. What has worked well previously will undoubtedly work well again, provided that the installation conditions are approximately the same as well. Thus, drawing on either your experience or that of a professional plumber may preclude using the calculations described in this text. It is up to each person to choose a method that will provide a workable arrangement.

SYSTEM LAYOUT AND OTHER CONSIDERATIONS

Regarding the actual layout of the system, common sense will be a good guide. Pick the shortest and most direct routes from point to point, using a minimum number of fittings. Keep the pipelines out of exterior walls wherever possible to avoid the danger of freezing. If pipes must be run on exterior walls, always place them to the inside of the thermal insulation. If plastic piping is being used, be sure to keep them well away from any sources of heat such as furnaces, chimneys, smoke pipes, and refrigerator and freezer coils. In cases where the connections will be covered over, such as with some types of bathtubs that are built-in, arrange the pipelines so that an access hatch can be provided for. Make sure that all pipelines are adequately supported, with space allowances for any possible expansion or contraction. The lines should be installed in such a way that there will be no danger of accidentally puncturing them during final construction of the walls and/or floors. It is also wise to install the necessary drains in the system for any possible future additions, such as a washing machine, and to provide at least one outdoor sill cock or hose bib for a garden hose or sprinkler system.

The selection of the type of pipe to be used in the system is left up to you. Make sure to check local plumbing codes as to guidelines regarding allowable piping for each particular application. A number of parts will be needed for the water supply line other than the pipeline itself. This will be somewhat variable under different circumstances. In a water main tap-in, there will probably be a saddle and stop cock located at the main, which may or may not be provided by the utility, a curb cock with a curb box and cover located somewhere along the line, and a main shutoff valve located just inside the house. Depending upon requirements, a water meter may be a part of this line, and a meter stop cock may also be installed.

Along with the necessary lengths of pipe for the actual cold water distribution system, an assortment of fittings will also be needed. A collection of 90-degree elbows, 45-degree elbows, couplings and tees, caps, and reducing bushings will be about all that is required. Likewise, each stub-out for a plumbing fixture connection should be fitted with a stop valve, either straight or angled as the situation demands. It might also be to your advantage to install one or two riser or main feed line stop valves. This will provide a means to cut off full sections of the distribution system while leaving others operational.

HOT WATER SYSTEM DETAILS

The method for determining both pipe sizes and pressure losses in a hot water distribution system is much the same as for the cold water system. The only difference is that there is no main supply line to be concerned with. With a few exceptions, such as the toilet tank, refrigerator ice maker, outdoor hose bib or hydrant, all of the normal household plumbing fixtures that require cold water will also need hot water. Alternately, some fixtures such as automatic dishwashers will only be supplied with hot water. Since these variations will not make any significant difference in most cases, it is really not necessary to go through the whole calculation process again. The hot water lines will be installed parallel to the cold water lines, and the same pipe sizes can be used, as well as the same fittings. The actual pipe lengths may differ somewhat due to the spacing provided between the two sets of lines, so this should be taken into consideration. In some instances, pipe size may differ, but this is the exception rather than the rule. This would be possible in situations where a fixture will require only a modest amount of hot water in comparison with cold water. This can be determined by making the necessary calculations based upon the desired flow rate at the fixture and the pressure losses in the feed line.

One of the most important things to be considered in a hot water distribution system is the demand, which is done in a different manner altogether. In making calculations for the cold water system, the basis was an assumed probable demand. That same demand stands for the hot water system as far as correct pipe size is concerned. By using the same pipe size for hot water as for cold, proper flow will be achieved. But this flow rate is defined in terms of gallons per minute, and no one would be able to heat that much water on a continual basis. Thus, to estimate the probable demand for hot water, usage is defined in a different manner, usually the number of gallons needed over a period of time, such as a 16-hour day.

The first thing to do is to calculate the total number of gallons of hot water that will be needed during a normal day. Obviously, this figure will only be an estimate, and Table 7-4 can be used as a guide. Each fixture has a range for usage. The best way to determine individual usage is to reflect on an average day and try to come up with a reasonable figure of estimated hot water consumption. It is also important to note at what times during

Table 7-4. Standard Usage Figures for Common Residential Fixtures.

Fixture	GAL. REQ.
Tub	20-40
Shower, per min.	3
Lavatory	1½-3
Kit. sink, per meal	2-4
Dishwasher, one cycle	3-8
Clothes washer, one cycle hot	30-40
Clothes washer, one cycle warm	10-20

the day this water will be needed. This will be helpful in determining what kind and size of water heater will be most suitable. The reason is that as hot water is drawn from the heater, it is replaced with cold water. Obviously, it will take a certain amount of time for this water to be heated. Therefore, if a 35-gallon tank is installed, and 30 gallons of hot water are required at one point and another 30 gallons at another point, as long as the two times are separated by a reasonable period of time, the system will be adequate. Figure 7-12 shows daily hot water consumption.

Once requirements for hot water demand have been established, a suitable water heater should be purchased. As far as layout of the hot water distribution lines is concerned, remember to allow a minimum of 6 inches between the hot and cold lines. This will provide for ample working room both during installation and any future repairs or modifications. It will also prevent any heat loss by transfer from the hot line to the cold. The water heater should be positioned so that it is as close as possible to the major points of hot water usage, with the pipelines routed in a direct, efficient manner. If there is more than one area of heavy hot water usage, it might be wise to calculate the benefits of using more than one heater and installing two hot water distribution subsystems.

DRAIN-WASTE-VENT SYSTEM DESIGN

It cannot be stressed enough how important it is to properly arrange and install this portion of the system. An improperly or sloppily installed distribution system may cause the system to function less than efficiently and will require some repairs. If the drainage system is poorly constructed, the consequences may be far greater. An improperly arranged DWV system presents a serious health and/or safety hazard due to the harmful bacteria present in the lines.

The process of determining proper size and length is not a complicated one. Drainage pipes are sized according to their optimum drainage capacity in terms of a measure called a *fixture unit*, which is the equivalent of 1 cubic foot of liquid per minute of flow. Table 7-5 provides a chart of typical plumbing fixtures and their assigned number of units. To use this chart, follow the rule of thumb that an individual toilet is connected to a 3-inch drainpipe, and a shower to a 2-inch drainpipe. Other individual items can be served by 1½-inch drainpipes. Some drainpipes will collect liquid from more than one fixture. Add up the fixture unit total for the line and find the proper size for either horizontal or vertical lines. In most cases, the soil stack must be sized at a minimum of 3-inch diameter, although some areas will require a 4-inch minimum. If no codes apply, choose the 3-inch size, preferably with an outside diameter that will fit neatly within a conventional 2 × 4 stud wall space for ease of system installation. The soil vent portion of the stack should be of the same size, although some codes allow for smaller diameters. In areas where freezing is a danger, it is suggested that the larger 3 or 4-inch size be used, as a portion of the piping will extend above the roof.

Sizing of the vent stacks which serve one or more plumbing fixtures should be done according to the fixture units themselves. The smallest size, 1½ inch, can only be used with an individual lavatory or an individual floor drain, with no other fixtures attached to the vent line. The 1½-inch

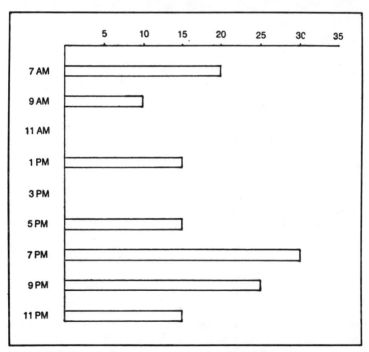

Fig. 7-12. A chart such as this may be used to determine daily hot water consumption.

Table 7-5. Fixture Unit Chart.

Fixture	Units
Kitchen sink	2
Dishwasher	2
Washing machine	2
Laundry tubs	2
Lavatory	1
Bathtub	2
Shower	2
Toilet	3
Bar sink	1
Darkroom sink	2

size is generally considered to be a minimum for vent stacks and will serve eight fixture units where sanitary tees are employed for connections, or 12 when wyes are installed. The 2-inch size will accommodate 16 and 36 fixture units, respectively, and is the largest size that will normally be encountered in most residential structures. As far as the house main drain is concerned, 3-inch pipe will be adequate as long as there are only two toilets in the home. If more than two toilets are present in the structure, the 4-inch size will be required.

Once the proper sizes for the drainpipes have been determined, it will be necessary to establish lengths as well. This will rarely be a problem in the case of the soil and vent stacks because a 1½-inch vent stack would have to be more than 65 feet long before exceeding code requirements. Horizontal drainpipes, however, are regulated as to length not only by the plumbing codes but also by practicality. If the total length and drop in pitch of the given pipe diameter is too great before it reaches a stack or is vented, the rushing liquid can build such a vacuum as it travels down the drainpipe that it will siphon water out of the trap behind it and thus open the trap seal. These horizontal pipes will have their length determined in terms of pipe diameter. Regulations state that a plumbing fixture cannot be connected any closer to a stack or vent than two pipe diameters, or any further away than 48 pipe diameters. The figures given here will be helpful in determining maximum lengths:

1¼ inch drainpipe:	5 feet
1½ inch drainpipe:	6 feet
2 inch drainpipe:	8 feet
3 inch drainpipe:	12 feet

In the case of the 3-inch drainpipe, which is normally used to connect the toilet to a soil stack, the length be kept as short as possible, but no shorter than 6 inches in any case. None of the horizontal drainpipes should be any longer than the figures given above, as measured from the trap outlet connection to the nearest wall or a vent or waste stack. The only exception to this would be in situations where a series of plumbing fixtures are all connected to one branch drainpipe, as long as a toilet is not included in the series. The drainpipe runs to a soil or waste stack, and the fixture at the farthest end of the drainpipe line is vented through a separate stack of substantial size. This provides for adequate venting for the remaining fixtures on the line and is called *wet venting* (Fig. 7-13). It is always a good idea to check with local authorities before making this type of installation, as some codes do not permit it. If wet venting is allowable, the maximum drainpipe lengths are as follows:

1½ inch drainpipe:	2½ feet
2 inch drainpipe:	3½ feet
3 inch drainpipe:	5 feet

As with the previously given measurements, the pipe length is measured from the connecting point of the trap outlet to the nearest wall of the serving waste or vent stack.

If it is found that the proposed layout does not meet these minimum specifications, a number of methods are possible to correct the situation. The best solution would be to shift the location of the plumbing fixtures in order to satisfactorily shorten the drainage pipes, if this can be done

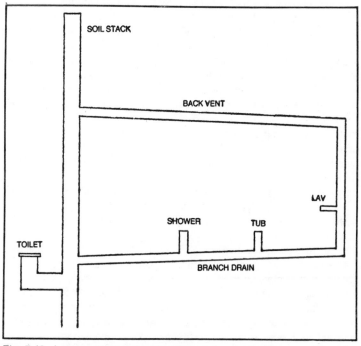

Fig. 7-13. A wet venting arrangement may be used when one fixture is located some distance from other fixtures.

Table 7-6. Specifications for Different Pipe Diameters.

Pipe Size	Drop (inches) Min.	Max.	Length Min. (inches)	Max. (feet)
1 ¼	1 ¼	1 ¼	2 ½	5
1 ½	1 ½	1 ½	3	6
2	2	2	4	8
3	1 ½	3	6	12
4	2	4	8	16

conveniently. Not only will this provide better drainage flow, but it will obviously require less work in the installation and less cost for piping. Also, rearrangement in some cases means a possible reduction in the diameter of branch drainpipes, since the fixture unit load will probably be decreased.

Another solution would be to introduce additional soil or waste stacks at appropriate points or possibly additional vent branches or stacks. this would result in both additional piping and cost, but may be the only option in cases where it is not possible to rearrange the fixtures.

Proper pitch will also have to be maintained in the drainage system. All drainpipes must be continuously pitched down and away from the building to allow for smooth drainage flow. Likewise, all horizontal branches must be sloped at a slight downward pitch from each fixture to the appropriate soil or waste stack. The waste continues downward through the stacks into the main house drain, which also pitches downward into the sewer line. Various pitches are permissible and will operate efficiently. As with drainpipe lengths, the degree of slope will be dependent upon the pipe diameter. For residential purposes, the simplest approach is to remember that both the minimum and maximum drop in inches of the horizontal drainpipe, measured from the trap outlet connection to the stack inlet connection, is exactly the same as the diameter of the pipe being used for the three most common drainpipe sizes used for residential purposes: 1¼ inch, 1½ inch, and 2 inch. For the 3-inch size, the maximum drop is the same as the pipe diameter, and the minimum is just half. Table 7-6 gives both the minimum and maximum specifications for the different pipe diameters.

Another possibility may be to simply use the same pitch through the entire system. The permissible range of pitches runs from ⅛ inch per running foot of pipeline to ½ inch. This range will vary depending upon pipeline size, but if the ½-inch pitch is used throughout, no difficulties will be encountered regardless of pipe size as long as the pitch remains constant. If it is not possible to keep the pitch constant, such as when the connections do not line up, run the first part of the drainpipe at ½ inch, and at the appropriate point, drop into the stack with a 45-degree angle change by means of fittings (Fig. 7-14).

VENTING DESIGN

The proper venting of a system is most important, and most people seem to have difficulty understanding the design principles. Following the guidelines provided in this text, the procedure should be easily understood. In situations where the system consists of one tightly grouped collection of plumbing fixtures, they can simply be drained directly into a soil stack (Fig. 7-15). This grouping can consist of any number of variations, such as one bath, two baths back-to-back, a kitchen and bath back-to-back, etc. The upper portion of the actual soil stack will serve adequately as the vent stack, as long as no other fixtures are connected anywhere above the individual grouping. This method is commonly called *direct venting*.

Figure 7-16 shows the wet venting method. The toilet drains directly into the stack. A common branch drain for the remaining fixtures enters the stack from a position slightly above this connection. As can be seen, the toilet is vented by the soil stack, and the other fixtures are vented by another stack at some other point in the system at the far end of the branch drain line.

In a situation where there is one grouping of fixtures at one point and a single fixture located some distance away, a different method will be required. If the distance is more than the permissible drainpipe length, it may be necessary to provide individual venting for the long fixture. This

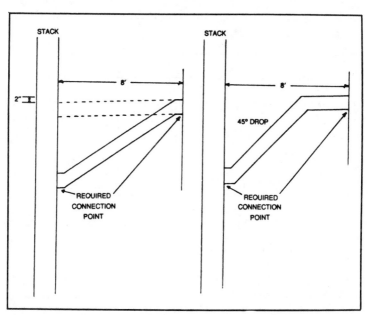

Fig. 7-14. If connections do not line up, 45-degree elbows may be used in this manner.

193

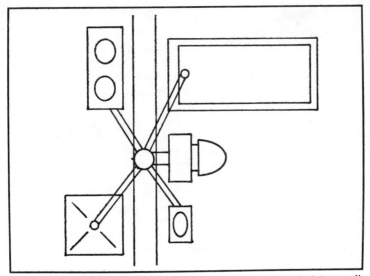

Fig. 7-15. In this system, all the fixtures are drained directly into a soil stack.

will involve continuing the fixture drain line upward past the fixture itself and venting it out through the roof (Fig. 7-17). Alternately, another commonly used method which can be used for either one or several plumbing fixtures is called *back venting*. The drainage line continues up

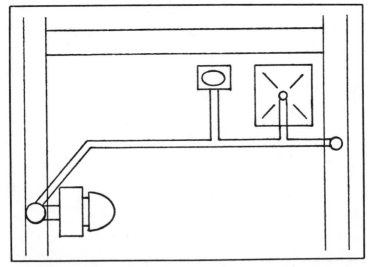

Fig. 7-16. A wet venting arrangement with two fixtures using a common branch drain.

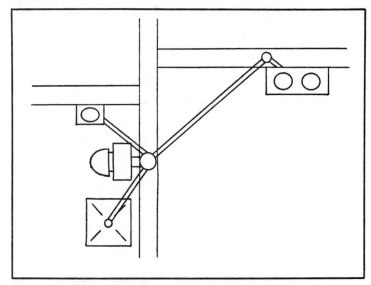

Fig. 7-17. In this example, one fixture is vented individually.

past the fixture for a minimum distance of inches above the drainage overflow level of the fixture or fixtures themselves (Fig. 7-18). At that point, the line turns and extends back to the main soil stack for connection. This will work fine as long as there are no fixtures connected to the stack at a higher level. If this is the case, the vent pipe must rise beyond the lower

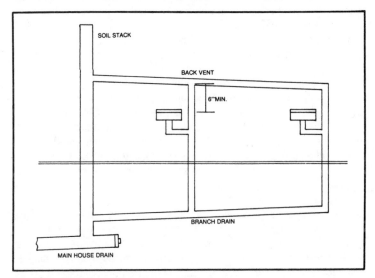

Fig. 7-18. A back venting arrangement.

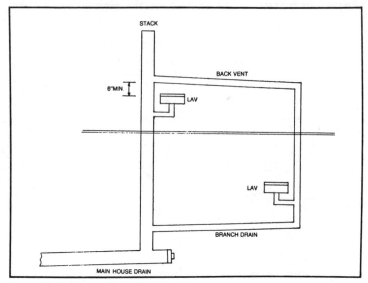

Fig. 7-19. A loop venting arrangement.

fixture to a point of connection to the main stack which is at least 6 inches higher than the overflow drain level of the higher fixture (Fig. 7-19). Several fixtures are often vented through a main stack in this manner, which is called *loop venting*.

The procedures discussed thus far are quite common in residential installations. Similarly, it is not unusual to find that various combinations of them are found within an individual system, or one method repeated more than once at different points. For example, a large sprawling home might have a kitchen-bath grouping of plumbing fixtures at one end and another master bath-guest bath fixture grouping at the other. In this situation, a direct venting system could be used for each grouping, with each having its own main soil stack, and with both soil stacks routed into the main house drain (Fig. 7-20). As can be seen, cross-venting may be provided at the homeowner's option for better air and drainage flow and pressure equalization. Alternately, a direct venting system could be used at one point and a back venting system used to vent a scattering of additional fixtures to another main soil stack at a different point (Fig. 7-21).

In situations where there are several groupings or lone fixtures some distance away from each other, yet another method can be used. This involves the installation of a separate vent stack. The bottom of this stack may be connected to either a horizontal branch drainpipe or to another stack. The upper end will continue upward through the roof. Any number of branch vent lines serving any number of fixtures can be attached to this stack in a back venting or reventing arrangement, as long as the size of

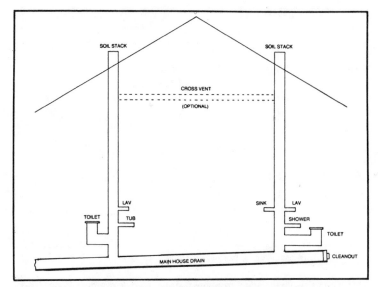

Fig. 7-20. Separate groupings of fixtures have separate soil stacks.

piping used in the stack is adequate to handle the number of fixtures being connected. The stack will generally be 1½-inch, although 2-inch pipe is sometimes used. A system with a separate vent stack is shown in Fig. 7-22.

It has already been stated that drainage pipes slope at a downward pitch into the stack, and that this slope should be constant throughout the system for ease of flow. The same holds true for horizontal vent pipes in the reverse. These lines slope upward into the stack and should be sloped at the same pitch as that used for the drainpipes. This is because the inflow

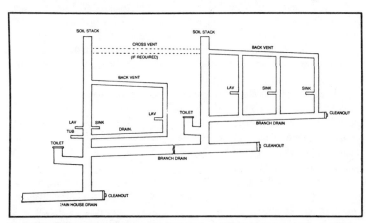

Fig. 7-21. Different methods may be used at different points in the system.

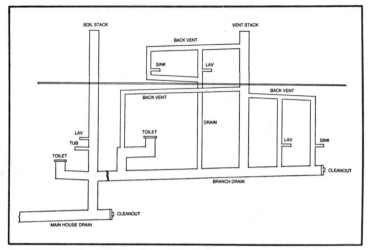

Fig. 7-22. In this example, the system has a separate vent stack.

of air comes down the stack from the top and into the drainage lines. Also, the upsloping vent lines allow the gases present to escape on a natural upward slope, thus preventing them from becoming trapped anywhere in the system. In the interest of efficiency and economy, these vent lines should be kept as short as possible, within reason and in accordance with local plumbing codes.

TRAPS

Traps serve a vital function in the DWV system. Every single drainage fitting on a plumbing fixture must be fitted with one, with the exception of toilets. Traps come in a number of different shapes and configurations, the most common being S, half-S, P, or U. The type used with each fixture will be dependent upon the physical characteristics of each particular installation. For example, a full S-trap, shown in Fig. 7-23, would be the choice in a situation where the kitchen sink drains down through the floor. If the sink drains down into the wall above floor level, a P-trap might be used (Fig. 7-24).

The *drum trap* is sometimes used in shower installations (Fig. 7-25). These traps are known to be debris collectors. Since the traps cannot be cleaned out with an auger, they should only be used in situations where another type would be impractical from a physical standpoint. Because the traps will need to be cleaned out by hand, they should only be installed in an accessible manner.

There is one exception to the rule of one trap for each fixture. In a situation where a double-bowl kitchen sink is a part of the system that has two drain outlets, they can be connected to a single trap by means of a special assembly designed for this purpose. Solid, one-piece traps should

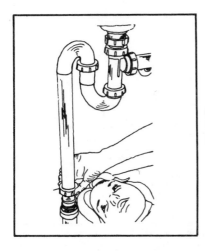

Fig. 7-23. A full S-trap is used when a sink drains through the floor.

contain a cleanout plug at the bottom of the bend. Sectional traps that come apart in two or more pieces by removing the slip nuts will obviously not require such a plug. These can be easily disassembled for cleaning. The traps are sized according to the size of pipe at each particular fixture. For example, a 1½-inch drain is fitted with a trap of the same size.

Figure 7-26 shows the installation of a U-trap, which is generally located in the main house drain close to the point of its exit from the house. This trap serves as a sort of master water seal for the entire DWV system. U-traps are quite large and have a cleanout plug at the top of each arm of the U. It is not unusual for this trap to collect a small amount of solid material which may have to be cleaned out periodically.

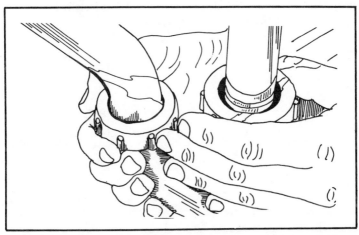

Fig. 7-24. A P-trap is used if the sink drains down into the wall above floor level.

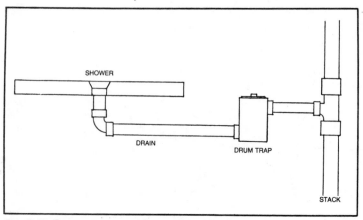

Fig. 7-25. Drum traps should only be used when another type would be impractical from a physical standpoint.

MORE DWV DESIGN CONSIDERATIONS

During the actual layout process, remember to maintain the proper constant pitch, as mentioned previously. Keep the lines as short and as simple as possible. Similarly, keep the number of fittings and directional changes to a minimum, too. Drainpipes from fixture to stack should turn no more than 90 degrees total if at all possible. It is also important to arrange everything for the best possible gravity flow by providing as gentle a path as possible for the liquid. For example, if a right-angle turn is needed at some point, it is better to install two 45-degree fittings with a length of pipe between them than an abrupt 90-degree elbow. Another way to provide for this type of directional change would be to install a long sweep elbow. Figure 7-27 shows these arrangements.

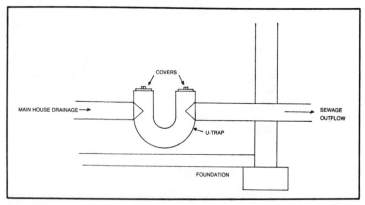

Fig. 7-26. Proper installation of a U-trap, which serves as a master water seal.

In situations where the waste flow makes a change from a horizontal drainpipe to a vertical stack, or simply a vertical drop, it will be all right to use an abrupt 90-degree elbow, as shown at the left of Fig. 7-28. Similarly, a sanitary tee can be installed in the manner shown at the right of Fig. 7-28. A common tee would not be suitable in this situation because it does not have the downward-curved corner that a sanitary tee has.

If the flow is from vertical to horizontal, down a stack and into a drain line, an abrupt transition is not permissible. In this case, the installation of a 45-degree elbow in the vertical line makes the transition more gently

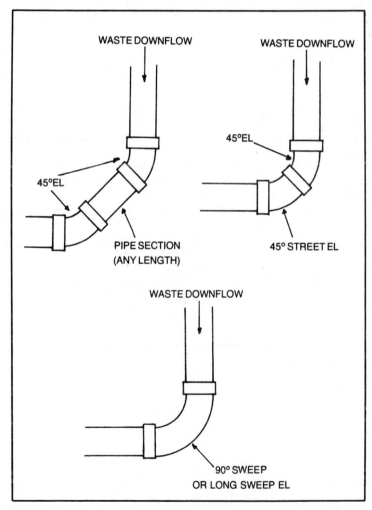

Fig. 7-27. Sharp directional changes are avoided with the use of long sweep elbows.

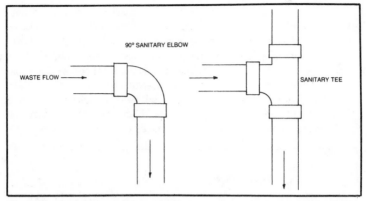

Fig. 7-28. Other types of directional changes may be accomplished with 90-degree elbows or sanitary tees.

into a Y fitting at the end of the horizontal line (Fig. 7-29). Remember that an abrupt change such as this should never be made without the installation of a cleanout plug. Hence, the use of a Y provides the necessary plug by means of its other arm.

One thing to watch for in this portion of the system is the proper installation of fittings. A common mistake is connecting a sanitary tee upside down, which will cause it to be operating against the flow. For vent purposes, this will make no difference, since the air flow is in reverse of the drainage system.

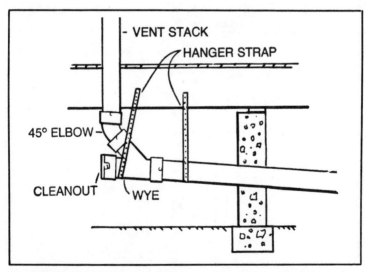

Fig. 7-29. The installation of a 45-degree elbow in the vertical line makes the transition much gentler.

SUMMARY

Proper layout and design of plumbing systems is imperative in order to arrive at a completed project which is both practical and efficient. As has been stated, the time for these concepts to be worked out is *before* any actual construction is started. The middle of a project is no place whatsoever to begin to make design changes. Make sure that you have covered every aspect of your intended system and that you have looked at it from all angles. If there is doubt as to the appropriateness of a certain portion of the installation, delve deeper into your research so that you have a clear picture of what is needed and what is provided to fulfill those needs.

You will most certainly be able to call upon some fairly expert guidance by seeking out professionals in the plumbing field. Your local supplier of plumbing materials will probably be able to answer the great majority of your questions and will often be able to call upon approved designs for similar plumbing installations from memory. If you don't know, ask. The truly ignorant person is one who does not ask questions. A proper layout and design will save many hours of extra labor and will certainly be the most cost effective approach to a personalized plumbing installation.

Installation of a Plumbing System

Once the proposed system has been planned and laid out and all the materials assembled, the actual installation can take place. This will be a time-consuming job, but if everything that has taken place to this point has been done in a thorough manner, the installation will not be as difficult as you might think.

The matter of pipe selection has been covered in another chapter in this book, and the job difficulty will vary somewhat depending upon the materials themselves. Using plastic components will make for a much easier installation due to its flexibility and light weight.

Each portion of the installation should be completed before continuing on to the next subsystem. For the purposes of this discussion, I will begin at the point where the system connects to the street main. Keep in mind one important fact. Since some sections of pipe will be quite rigid and you will be working in cramped quarters at least part of the time, it will be important that accurate measurements be made so that all the pieces will line up correctly and fit easily.

INSTALLING THE SUPPLY LINE

The water supply line is installed in basically the same manner for all water sources. Some requirements may be a bit different, depending largely upon the source being tapped. It makes no difference whether this subsystem is installed before the distribution system. This is left up to the builder. Local plumbing codes may also play a part in the requirements, specifying certain fittings or certain procedures.

Before beginning to lay the water supply line, it will be necessary to locate the water main. This is normally obtained from the municipal or water company authorities. It may be that a tab stub, which is provided to tap into the line, already exists. Some companies install these tab stubs at intervals in the line to allow for future access. If this is the case, the company may require that the line be run to a tab stub already provided, whether this is convenient or not. If no stub exists, the homeowner may be permitted to tap into the main at any reasonable point.

Once it has been determined where the line is to be, a trench must be dug from the main to the point of entry into the structure. This may require the use of heavy equipment, and great care should obviously be taken not to cause any damage to the existing lines. The trench should lie well below the frost line, if applicable, and should meet any local requirements. The interior of the trench should be clear of any debris and quite smooth. Any changes in elevation should be accomplished in a gradual manner rather than abrupt steps.

If a tab stub is in place on the water main, it will include a stop valve which is called a *corporation stop*. In this case, no further adjustments will have to be made to allow for connection. If no tab stop exists, a special saddle is installed on the main with a stop valve attached. The tap is made with a special wet-tapping tool, which bores a hole in the water main without allowing any water to escape. As the tool is removed, the stop valve is closed and installation can begin.

The next step is to attach an appropriate length of pipe to the valve and run the pipeline to the house, providing a number of valves, which were already discussed in Chapter 1. This pipeline will continue to a point inside the house where a meter valve may be installed. This is identical to the valves installed outside the structure and simply serves as another point where the supply may be shut down if necessary. The next thing on the line will be a water meter, if applicable. If a meter stop valve is included in the system, the water supply line will terminate at this point, with the water meter attached to the meter stop valve by means of a short coupling made of either plastic or metal. If there is no meter stop valve, the water supply line will terminate at the meter itself. Another valve, called the *main shutoff valve,* is closely coupled to the water meter. If no meter is required or desired, the water supply line terminates at the main shutoff valve. Connection is made to any or all of these devices with transition fitting assemblies.

At this point, it is wise to take the time to check the installation for any leaks. Before beginning, make sure that all the valves in the line are in the off position. The main shutoff valve handle should be fully turned down, and the handles of the stops (which only require a quarter-turn from full open to full closed) should be at right angles to the pipeline. Begin by slowly opening the corporation valve at the main, which will result in a loud inrush of water being heard immediately. This sound should cease almost immediately, depending upon the length of the pipeline to the next valve. If this does not occur and some slight sound is still heard, check that portion of the system for any possible leaks. If the sound stops right away, open the valve all the way and check all the fittings and the connection to the next stop valve for any leaks. If all is well at this point, open the next valve and continue in this manner until the main shutoff valve in the house is reached. This one is kept closed. Although everything seems satisfactory at this point, you should allow a few hours time and then go through the whole procedure again. Remember that once this line is buried, it will be a most difficult and unpleasant chore to go back and dig it up again. A little time spent here in carefully and thoroughly checking the system will guarantee a working system.

If all seems secure and nothing is leaking, the next step is to bury the supply line. The backfill should be free of rocks and any other debris, and the first portion of dirt should be shoveled in by hand to avoid any damage to the pipes by heavy equipment. The curb valve will be protected by a curb box, as explained in Chapter 1, with the cover being installed once the box has been firmly backfilled almost up to the top. The remainder of the trench is then filled completely, making sure to thoroughly seal the pipe on both sides of the foundation.

INSTALLING THE COLD WATER SYSTEM

Although the cold water distribution system begins at the main shutoff valve, the water meter, or the output line from the water pump, it is not necessary to start at this point. Various segments of the system may be installed, connecting them all together and eventually working back to the supply source. Let's begin at the source. The attachment is made with a transition fitting of appropriate size. The piping is then attached and angled up and away according to design specifications. This can be done as a single riser providing main feed, or a feed line can be run across the basement ceiling, with a series of risers running upward to the rest of the house. In any case, simplicity and a minimum of fittings should be remembered as providing the best pressure and an economical system.

It is recommended that the dry-assembly method be used, which is simply cutting, fitting, and installing the lines dry before permanently connecting them together. Once the risers and main feed lines are installed with branch tees spotted at appropriate points, each branch line can be installed from feeder to fixture stub-out. If the plumbing fixtures are not being installed at this time, it will be necessary to install caps for two reasons. First, this will keep dust and dirt from entering the system; second—it will make it possible for pressure tests to be run. It is wise to allow about 6 inches on the stubs. This will provide an ample length of pipe to work with during fixture installation. An alternative to this method is to go ahead and attach stop valves directly to the stubs as the branch lines are installed, finishing up that segment of the job completely. Stubs for those few fixtures that may not require stop valves can be capped.

Make sure the location of each fixture is carefully calculated so that the stubs protrude from the wall or floor at the appropriate points. Make sure there is enough clearance both behind and under the plumbing fixtures to allow for ease of installation and room for installing the risers.

Once the system is installed and the pipes are connected with the appropriate sealant, check over the entire system for any missing sections or unsealed fittings. Check to make sure that there are caps on all the stubs, or that all stop valves are in the off position. Give the system the required amount of time to dry, and turn the water on at the main shutoff valve, or activate the pump, whichever is applicable. Just as with the test done on the supply line, a rushing noise should be heard as the water flows through the pipes. In a small system, this should cease almost at once. If the system is fairly extensive, the noise should diminish, but at a slower rate. If the sound diminishes somewhat but does not stop completely, there is probably a leak somewhere in the system. If the sound remains at its original level, there is probably a wide-open line somewhere. The appropriate valve should be shut off immediately in this case to prevent any flooding or water damage. If everything checks out and the sound stops immediately, the whole system should be inspected with the valve still open, making sure no joints are seeping or slowly dripping. Again, at this point, it is wise to wait another few hours and repeat this process once again as a form of insurance. If all seems well, this portion of the system is complete.

INSTALLING THE HOT WATER SYSTEM

In most instances, the cold water supply line splits at some point, with one line continuing on for cold water distribution and the other line extending to some type of hot water heater. The hot water feed line running to the remainder of the distribution system is connected to an outlet provided for this purpose on the heater itself. The installation of a water heater has been discussed previously in another chapter, but it should be installed in such a manner as to allow enough area around the tank for the piping connections. These are usually made at the top of the tank, but adequate working space should be provided around the heater regardless. It is also a good idea to allow room to add on an insulating outer jacket once installation is completed. This will prevent a good deal of heat loss, thus reducing operating costs. Ready-made fiber glass jackets are available at most hardware stores, as they are becoming increasingly popular. Likewise, they can even be constructed at home using common fiber glass thermal building insulation.

Plastic pipe is never used for direct connection to a hot water heater. The usual procedure is to provide a metallic interface at both the inlet and outlet. The inlet and outlet of the water tank will be equipped with standard female pipe threads. In some cases, both are at the top; in others, the hot water outlet is at the top and the cold water inlet is at the bottom. In any case, use a galvanized steel nipple about 10 to 12 inches long with galvanized pipe fittings to make quick directional changes if necessary. At this point, the pipe material can then be changed to plastic by means of transition adapters. In metallic piping systems, union fittings are often installed in either the hot or cold water lines. Many transition fittings will serve this same purpose. If the transition adapters being used do not come apart into two separate threaded parts, it might be wise to consider installing conventional union fittings at the tops of the galvanized nipples so the tank can be easily removed if necessary.

If the cold water inlet is at the bottom of the tank, a short nipple can be installed with a tee on the end of the nipple. The cold water inlet is piped in the tee branch, and a drain cock is attached to the open end of the tee. This allows a convenient means of draining the tank (Fig. 8-1). It is also a good practice to install a stop valve in the cold water feed line and possibly one in the hot water line as well.

Once the hot water heater is in place, the lines for water distribution can be installed. This is done in much the same manner as the cold water system, as the lines will probably be running parallel throughout. Remember to keep the two lines about 6 to 8 inches apart to avoid heat transference. Make sure that the pipes will not rub or chafe on framing members as they expand and contract.

Check over the system once it is installed and all connections are sealed to make sure all stop valves or caps are in place. The system is now ready for testing. Adjust the water heater to the desired temperature and turn the cold water feed line stop valve on. The tank should begin to fill. As the tank fills, check the cold water inlet connections for any leaks or seepage. If none appear, wait for the tank to fill and check the outlet connections. Eventually, the lines will fill partly. Check all the fittings on the line, as well as all the stop valves. Air will be present in the lines at this point and can easily be bled out by opening one or more of the stop valves.

It is wise at this point to also check the operation of the water tank itself. Turn on the electricity or fire up the gas burner and let the tank run for awhile. Within a short while, hot water should be present in the system and the galvanized steel outlet nipple should feel warm to the touch. If no hot water is needed at this point, shut the tank off. Once it has been determined that this portion of the system is operational, you can proceed with the next part of the installation.

INSTALLING THE PLUMBING FIXTURES

The actual installation of the fixtures will vary, depending not only upon the particular fixture but also upon the style and manufacturer. The best procedure is always to follow the manufacturer's instructions to the letter, making sure to observe all the dimensional details in particular. Many plumbing fixtures are not installed until surrounding finish work has been completed. Bathtubs must often be installed during the construction process, since many tubs must be supported by the structural framework of the walls. Shower stalls are often built during the construction process. Toilets may or may not be left until all or most of the finish work is done. Water-using appliances are normally left until the last stages of construction.

The actual sequence insofar as making the plumbing connections is really left up to each individual's preference. Some will have to be made early for convenience of installation, while the whole system can even be installed with stubs in place for later connection. This is a matter of choice.

INSTALLING THE DWV SYSTEM

This portion of the system is a bit more complicated than installing the water distribution lines, mainly because the pipes will be larger and the runs may be more extensive. If care is taken, no problems should be encountered. As with the other subsystems, it makes no difference at which point in the system you begin. Assume that the house sewer line has already been brought to the foundation wall and start from there.

For ease of plumbing, the soil stack should be located directly behind the toilet and hidden in the wall, so the first step will be to accurately locate where the toilet will be installed. The exact location of the soil stack is left up to the builder, as long as it is near enough to the toilet to serve its intended purpose.

This portion of the installation will require some carpentry work. Cut a hole in the floor large enough to accommodate the closet flange to be used. Be careful not to locate this hole in such a position that running the waste line from the toilet will be difficult or impossible. A clear space will be needed here. Spot the point where the soil stack will extend upward through the wall. Center a 3-inch (or 4-inch) pipe coupling or a short section of pipe over the point and trace a circle around it. Bore a hole a bit larger than the pipe diameter down through the wall sole plate and through the flooring into the open. The easiest way to accomplish this is with a

Fig. 8-1. This method will provide a convenient means of draining the hot water tank.

speed bit chucked in a hand electric drill. Bore a series of small holes around the perimeter of the circle, and cut the scrap piece out with a keyhole saw. Dangle a plumb bob from the header plate of the wall so that the bob point is exactly centered in the hole just bored. Mark the corresponding point on the top plate and carve out another hole. It may be necessary to bore through a double plate here. Just be sure to check that no joists are being bored into. It is a good idea to wear some sort of protective goggles while doing this type of work.

If this is the top floor of the house, this hole will bring you out into possibly an attic. If there is another floor above, the hole will be merely drilled into a ceiling cavity. It will be necessary to line up yet another hole to get through the next floor. Take a short section of 3-inch pipe with the end cut square and smooth, and apply a liberal dose of carpenter's chalk to the edges. Insert the pipe through the hole in the top plates, but don't touch the flooring above. Holding the pipe in one hand, set a level against the pipe and adjust the pipe until it is lined up vertically. Then push it upward slowly, maintaining the proper vertical position, until it touches the floor. Twist the pipe a few times to allow the chalk to leave a ring on the undersurface of the floor. This will make it a simple process to bore out this hole in the same manner as the others. Actually, the whole thing will be quite awkward. Try a simpler method.

Locate the center of the chalked circle. Using the electric drill, bore a single hole directly upwards, making sure the vertical line is properly maintained. Go upstairs. Using the small bored hole as a reference point, draw a larger circle around it. The hole can now be drilled downward.

The next step is to locate a point on the underside of the roof sheathing that exactly lines up with the center of the hole bored through from below into the attic. This is done with the plumb bob. Again, using this point as a reference, trace a circle around a piece of pipe for a coupling to provide guidance. Unless this roof is flat, the hole will not be large enough unless the drill bit and the saw blade are kept perfectly straight during the cutting process. This is a bit difficult to do. The steeper the roof pitch, the harder it will be to get an exact fit around the pipe. The simplest solution to this is not to bother with a tight fit and settle for a loose one.

In any case, bore a single small hole maintaining as close to a vertical pitch as is possible. Go up on the roof. If the house is a new one and the roof is not completely finished, the job will be a bit easier. Hold a pipe coupling in a vertical position, centered on the hole just drilled through the roof. Trace around the coupling to provide a guide for cutting. This can be done with either a keyhole saw or a power jigsaw. If a keyhole saw is used, saw approximately straight up and down, leaving the cut edges of the roof sheathing roughly vertical. The sheathing will be cut at an angle to its surface. Make the cut about ¼ inch larger than the outline.

If a jigsaw is used, the shoe of the saw will lie flat on the sheathing and the cut edges will be at right angles to the sheathing, not on a vertical plane (Fig. 8-2). More room will be needed to run the pipe through the hole, so the cut should be made about ¾ inch larger than the outline.

If the finish roofing is already in place, it will be necessary to remove a section of the roofing before beginning the measuring. Make sure the section removed is somewhat larger than the dimensions of the bottom plate of the roof jack to allow for working space during the measuring and cutting process. Once this is done, continue in the manner already explained.

ADDITIONAL CARPENTRY WORK

In a house that is still under construction, it will be much simpler to install piping, since there will not be as much breaking through both wall coverings and floors. However, it may be necessary to cut through some of the studs and joists to run the piping in any case, and a brief explanation of some of the carpentry involved will be helpful.

In order to run pipelines inside partitions and under floors which have ceilings underneath them, it will be necessary to measure out the clear spaces or the depths of the joists and studs. Generally, this will only be necessary with regard to the drainage pipes, since the piping used in distribution systems is usually much smaller in diameter. The information provided in Table 8-1 will aid the builder in determining how much clear space is required for the different pipe diameters used in drainage systems. It will only be necessary to measure the actual depth of the studs and/or joists if their dimensions are not known. Figure 8-3 indicates what measurements must be taken. With this information in hand, refer to Table 8-2 to arrive at both the amount of clearance that is available and the maximum size pipe that can be run. At this point, it is wise to plan for fittings at locations where the most space is available, since more working room will be required for this part of the installation.

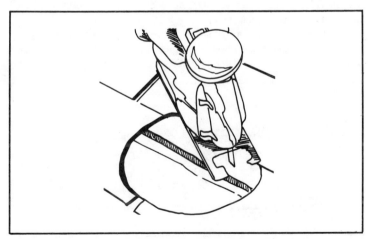

Fig. 8-2. Using a power jigsaw will make the job a bit easier.

Table 8-1. Clearance Space Required for Different Types and Sizes of Pipe.

Material	Diameter in Inches			
	1 1/2	2	3	4
Cast Iron Pipe	-	4	-	6 1/4
Cast Iron Fittings	-	4	-	6 1/4
Plastic and Copper Pipe	1 3/4	-	3 3/4	-
Plastic and Copper Fittings	2 1/8	-	3 5/8	-

If it is found that the partition walls are not thick enough to take the necessary drainage line, it will be necessary to make the walls thicker. The method for doing this is shown in Fig. 8-4. After the facing has been removed and the line installed, add 2 × 4-inch studs. If the wall is an outer one and cannot possibly be broken into, the required clearance can be obtained by adding 2 × 6-inch or even 2 × 8-inch studs.

Notching Studs and Joists

The purpose of both studs and joists is to provide structural support for either the walls or the floor. Thus, notching through studs and joists reduces their strength to a certain degree. In order to restore this strength to its original capacity, each notch made to allow for clearance for the piping should be properly reinforced once the pipes are in place. This is accomplished in a number of ways, depending upon the location of the notch on the stud or joist itself.

A notch at the top of a joist should be reinforced by inserting a tightly fitting block, possibly a section of the piece of wood removed to provide space for the line. If the notch is on the underside of the joist, a firmer

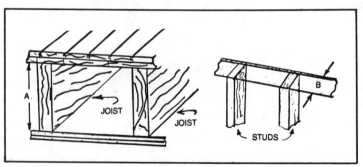

Fig. 8-3. Proper measurements of the studs and joists is essential.

Table 8-2. Clearance Space Available Versus Pipe Size.

Size of Stud/Joist	Approximate Clearance in Inches Available	Maximum Size Pipe It Will Take
2″ × 4″	3 3/4″	3″ copper/plastic
2″ × 6″	5 3/4″	2″ cast iron
2″ × 8″	7 3/4″	4″ cast iron

means of support will be needed since the joist will be required to take the load of the piping. This can be done with either a steel mending plate or a strap with screws. Likewise, a stud that is notched will also require a mending plate. Examples of these methods of support are shown in Fig. 8-5.

Notches at the top of studs in a non-load bearing partition wall can be cut up to half the thickness of the studs if only two studs in a row are notched, and at least two more studs are left unnotched. This rule applies if the notching is done only in the upper half of a stud's height. If the stud's lower half is notched deeper than one-third its thickness, it will have to be reinforced. In any case, no notching should ever be deeper than two-thirds the thickness of the stud, even if reinforcing is provided. The amount of structural support lost cannot be replaced by reinforcement of any kind.

It is wise to avoid notching if at all possible. This can be done by running the pipes parallel to the joists and in between them. This will not apply when working in an attic, since the lines here can be crossed on top of the joists. Similarly, when working on the first floor, they can be crossed below the joists. A layout designed with a minimum of notching is shown in Fig. 8-6.

Another way to reduce notching is to run the lines through floors, subfloors, and stripping. Toilet drains will always run in between the joists

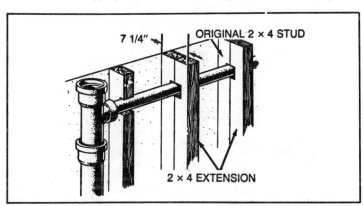

Fig. 8-4. A method which may be used to provide the necessary clearance space.

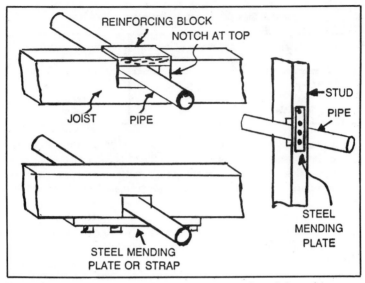

Fig. 8-5. Any joint or stud which is noticed must be reinforced in some manner.

or below them. If notching is necessary, never make one in the middle of a joist. It is better to make the notches toward the ends. No notch should be deeper than one-fourth the height of the joists. See Fig. 8-7.

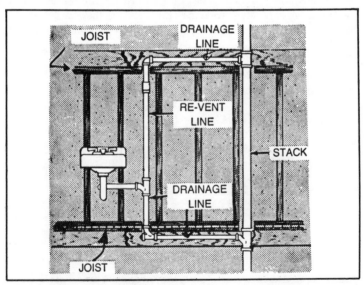

Fig. 8-6. In this system, a minimum of notching is provided by running the lines parallel and in between the joists.

If it is possible, drilling a hole is preferable to allow for the passage of pipes. A lesser amount of strength is lost in this manner, and the pipes are simply threaded through the holes. If this method is used, make sure the holes are centered between the top and bottom of the joist and that their diameter does not exceed one-fourth the height of the joist.

Running the Pipe

Great care must be taken in providing clearance for the drainage lines to avoid any structural damage to the construction of the house. As long as proper support is given and the guidelines regarding notching and/or drilling are adhered to, this procedure should be an easy one. Once the required clear space is provided for, the pipes can be made ready for installation.

The running of the pipe will vary quite a bit from home to home, depending upon the system layout, the specific piping material, and the number of fittings. Assume the system is a simple one and follow the installation step-by-step.

Start at the sewer line stub and run the main house drain pipeline back to the soil stack location. If the sewer line is the same size as the house drain, simply couple on the next section of pipe. If not, make the joint with the proper adapter fitting. Drop a plumb bob down through the soil stack hole in the floor above, making sure it is centered exactly. Take careful measurements, and end the run of the main house drainpipe with a wye fitting. The slanted arm of the wye should be cocked upward with a 45-degree elbow connected to it. The center of the opening in the top of the elbow should line up exactly with the point of the plumb bob. The lower open continuation of the wye can be fitted with a cleanout plug, or the house drain may travel on from there to pick up another stack. This connection is shown in Fig. 8-8. Be very careful when installing this section, as the exact location for this upturn in the pipeline is quite critical. The sections should be assembled dry first, the measurements double-

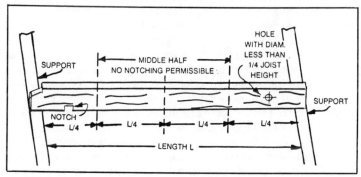

Fig. 8-7. Notches should be made toward the end of the joist and no deeper than one-fourth of the height.

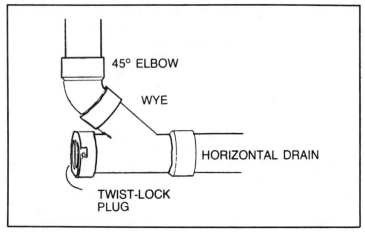

Fig. 8-8. The house main drainpipe is ended with a wye fitting.

checked, and the joints sealed once it is ascertained that everything is as it should be.

Another fitting, or possibly a combination of fittings, will be needed at some point just below the level of the joists of the first floor. Any number of possibilities can be used here, such as a single or double special waste and vent fitting, a waste and vent tee, or whatever suits the required specifications. Calculate the location of the next fitting in line with respect to its proper alignment with the drainpipes that will eventually connect to it. Measure off the necessary length of pipe, insert the section into the upturned 45-degree elbow, and plant the required fitting on top. If the alignment is correct, a permanent connection can be made. Any other fittings necessary at this point should be installed in the same manner.

Once the level of the first floor is reached, a number of variations may also be in order. If the stack is to remain unbroken all the way up, the whole line may be installed at this point. This is not too likely, as even in a single story house more than one length of pipe will probably be required. An easy way to do this is to have another person slide a 6-foot section of pipe down through the hole in the first floor and hold it steady while a permanent connection is made. At this point, the end of the house drain line should be supported from below to prevent any strain from being placed upon it. Make sure to allow time for the connections to dry, if applicable, and be careful not to move the freshly sealed piping around during the drying process. This will cause a loose joint, which will have to be redone, and may throw the whole line off its alignment. After the first section of piping is in place, the whole process is repeated with another length, continuing up through the roof. Once the joints have been allowed to seal, the piece extending above the roof surface is cut off to about 1 foot from the roof, as measured from the highest point at the back or uphill side of the stack. If using metallic pipe, you should allow for this in the measurement of the last section of piping, since cutting will obviously be difficult once the pipe is permanently installed. The final step is to attach the pipeline in an appropriate manner to the roof itself. One method of doing this is with watertight roof flashing (Fig. 8-9). Whatever method is chosen, make sure the pipe is secured in some way. Any open areas around the outside of the piping should be sealed, possibly with roofing compound, to prevent any rain water from leaking into the structure.

The method explained here assumes an unbroken pipeline from basement to roof. Although this is not that unusual, it is more common for there to be a number of plumbing fixtures which will have to be connected at different points. In some areas, it's permissible to add a lavatory drain to a direct-vented system above the input level of the other fixtures on the line. This would require the insertion of a reducing sanitary tee or a reducing wye at some point a couple of feet above first floor level. Other plumbing fixtures may drain into the soil stack from the second floor, or perhaps a vent tee might be installed at some point. Continue the stack under these circumstances by first locating the proper position for the first fitting in the line. Cut a suitable length of pipe and slip it down through the first floor, making a permanent connection in the uppermost soil stack fitting socket. Weld the fitting to the top of the pipe section and position it properly. This is shown in Fig. 8-10, and the process is repeated all the way up to the roof.

Installing Branch Drainage Lines

In order to install the branch drainage lines, start from the stack fittings and work back to the fixtures locations. To start with the toilet waste pipe, join a sanitary elbow to a length of pipe that will fit into the

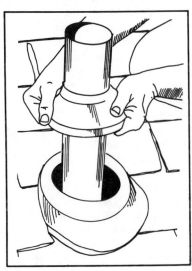

Fig. 8-9. The line is connected to the roof by means of watertight roof flashing.

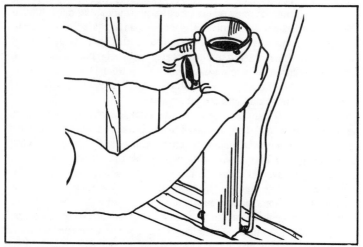

Fig. 8-10. Fittings are attached by means of reducing tees or wyes.

proper 3-inch socket provided for it in the soil stack assembly. The length of this pipe should be calculated so that the upturned opening of the elbow is exactly centered below the hole cut earlier for the closet flange. A short stub of pipe is then connected to the upturned socket of the elbow, the length of which is calculated so the closet flange can be slipped onto the pipe from above at its proper location and height with a full-socket fit. This installation procedure is shown in Fig. 8-11. If the waste pipe lies just below the level of the floor, a sanitary street elbow can be positioned to match exactly with the closet flange, allowing for the exclusion of the extra stub of pipe. Similarly, a street closet flange can be inserted into the socket of a sanitary elbow, as long as the upturned elbow socket can be positioned as required. Once all the connections are permanently made, the flange can be installed and permanently fastened to the floor. This is done by means of screws usually provided with this fixture (Fig. 8-12). Keep in mind that proper pitch must be maintained throughout, even in the case of these normally short lengths of piping. When using sections of pipe such as these which are usually large diameter and quite short, the pipes will be somewhat cocked in the fitting sockets. This will require the use of a bit more sealant than would normally be required, but should present no problems with the system.

The piping for bathtubs, showers, kitchen or bathroom sinks, and the like will be smaller, but the installation is done in much the same manner. These lines will probably be relatively short, and the 1½-inch diameter is most commonly used. The 1¼-inch pipe may be used for a bathroom lavatory with a 1¼-inch diameter drain flange. The stubs of these lines can either protrude from the wall or come up through the floor. Regardless of the location, the first thing to do is determine as accurately as possible at which point the stub will exit the floor or wall. Do whatever boring or drilling is necessary to provide clearance for the piping. Run the drainpipes from the appropriate fitting inlet socket at the stack back to an elbow with as few changes of direction on this route as possible. The elbow, usually a 90-degree fitting, will change the pipeline direction from horizontal to vertical. Insert another length of piping of the desired length into position either through the floor or sole plate, and make up the elbow joints. If the stub exits through the floor, this will be all that is required. In the case of a wall stub, a second 90-degree sanitary elbow will be needed to make the turn back to the horizontal. In addition, a short length of pipe will be used to extend the piping out from the wall. These two methods of installation are shown in Fig. 8-13 and 8-14, respectively.

Drain lines will be required for either bathtubs or shower stalls, and these are installed in much the same manner. Bathtubs or tub/shower combinations are generally served by a 1½-inch diameter drainpipe, while a stall shower uses a 2-inch pipe. In any case, the first step is to determine at what point the trap for the shower or tub will terminate. This will vary a bit, depending upon the type of trap being used. Once the termination point has been established and fittings appropriate to the type of trap being used are on hand, the necessary holes are bored. The actual trap connections may be either above or below floor level. In either case, start at the stack drainage inlet provided for this piping and run the appropriate size pipe back to the hole made in the floor. Make a turn with a 90-degree sanitary elbow and stub upward to the connection point of the trap or make a direct trap connection, as applicable. A typical connection using a P-trap is shown in Fig. 8-15.

With the installation of the drainage lines for the tubs and/or showers, the DWV system is now complete. The system used for explanation

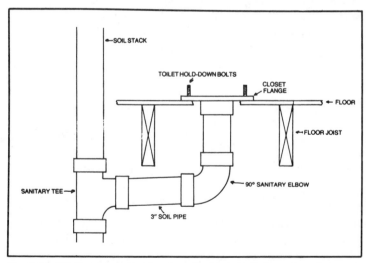

Fig. 8-11. Pipelines installed to allow for the installation of a toilet.

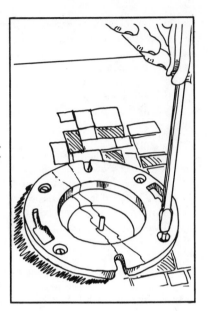

Fig. 8-12. The flange may be installed and fastened to the floor with screws.

purposes in this text is a quite simple one and, depending upon individual needs, a typical residential installation may be made up of many more elements and combinations of this simple system. There might be additional plumbing fixtures, vent or revent lines, a continuation of the main house drain, and the installation of possibly another soil or vent stack. Regardless of individual circumstances, the general guidelines will be pretty much the same. The most important things to remember in any installation are tight, leak-free joints, adequate support for the pipes, and a

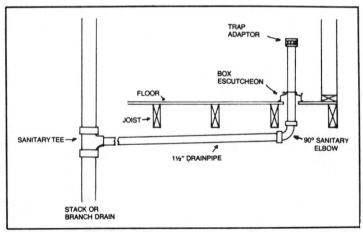

Fig. 8-13. A method by which drainpipes are run up through a floor.

220

constant drainage pitch throughout. If these three things are maintained, the end result will be a properly functioning system.

TESTING THE DWV SYSTEM

Testing is not an absolute necessity in every case, but it is really a wise move at this point, since it will give the builder an indication if any of the fluids or gases present in DWV pipes are leaking into the living quarters. Some areas have quite strict codes regarding this portion of the system and may require some sort of pressure test under the guidance of an inspector. Once the building is finished, it will be a difficult task to repair leaks.

To run a test of the system, allow a couple of extra inches of drainpipe at every point where the pipes are stubbed out for connection to plumbing fixture traps. Close off each of the stubs with a pipe cap sealed in place. Remember that the low end of the system must also be plugged. This can be done by installing a plug at the point of connection to a sewer main. In the case of a septic tank installation, the plug is inserted at the tank inlet pipe. If the sewer line has not yet been installed, the open end of the house main drain is capped or plugged with something that will form a relatively tight water seal.

A commonly used method of testing the system is to insert a garden hose into the top of the highest soil or vent stack in the system. Let the

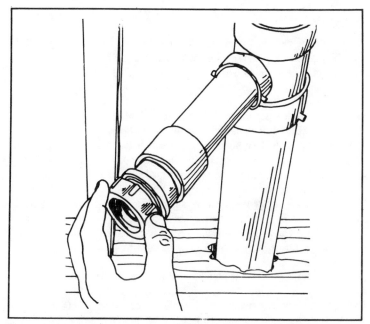

Fig. 8-14. If the stub exits a wall, the method is a bit different.

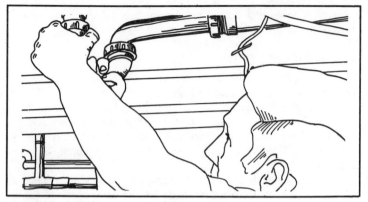

Fig. 8-15. A typical connection using a P-trap.

water run until the entire system fills up with water. Since this portion of the system will not be under pressure under normal operation, this will be sufficient to point up any possible leaks or seepage. Once the system is full, begin at the starting point of the system and go through the entire system, section by section, checking for leaks or seepage. If any are discovered, they should be repaired, given time to dry properly, and the test should then be repeated. If all is well, you can be assured the system will function properly when in actual operation.

SUMMARY

After the proper planning has been completed, the installation of a plumbing system is a step-by-step procedure which should be carried out with strict adherence to detail. By its nature, a plumbing system is a construction project which must be designed to last for many years. Each step in this construction is important to the overall system. The weakest link in the entire plumbing chain can cause the entire system to be completely useless or inadequate for all practical purposes should a malfunction occur. This is a system design, so care must be taken during every step of the construction process to prevent such failures. There is nothing quite so pleasing and rewarding as a functional system which you yourself have designed and installed. By the same token, there is nothing quite so disheartening as a system which fails due to a thinking lapse. If one step or construction act is not properly completed, the savings which are normally associated with a do-it-yourself project can be entirely lost and many additional expenses incurred when compared with a contractor-installed system.

The installation of home plumbing systems is a relatively simple task. Consistency in planning and installation techniques are bound to assure the builder of a properly functioning plumbing system which meets the individual needs for which it was designed.

Drains, Sinks, and Lavatories: Installation and Repair

The plumbing industry is governed by rules and regulations which were established years ago. Certain guidelines must be followed which have been proven to provide the best operation of the overall system. The same holds true with regard to the actual fixtures used to supply the home with both water and a drainage route. Certain standards must be met by the manufacturers of plumbing fixtures, both in regard to the type and quality of materials. Some of the commonly used and approved materials are stainless steel, cast iron, enameled pressed steel, and vitreous china. There are also requirements as to the number of fixtures that must be installed, depending upon the size of the home and the number of people expected to live in it. This information is readily available from local authorities and must be strictly adhered to. Plumbing codes may also specify such things as overflows on sinks and/or lavatories, and even the necessity for a window in the proposed bathroom, although most people provide for this anyway. The reason for a window is the need for ventilation.

INSTALLATION OF A LAVATORY OR SINK

The only difference between a *lavatory* and a sink is their location in the home; lavatory is simply the technical name for a bathroom sink. The installation of either sink is basically the same and quite simple. As has been mentioned previously, each fixture must be fitted with an individual trap unit that attaches to the fixture drain flange at one end and the drainpipe stub at the other. If the pipe installation has been measured exactly, a one-piece trap can be used. On the other hand, a two-piece trap makes for a much easier installation, since the parts can be swiveled in opposing directions to correct minor mismatches or produce deliberate offsets. The type of trap used will depend upon the location of the stub. A

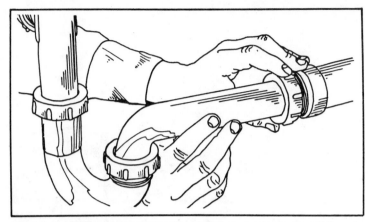

Fig. 9-1. The trap is connected to the extension pipe by means of slip nuts and compression rings.

P-trap is used for a wall stub, while an S-trap will be required for a floor stub. There are countless configurations available for this installation, together with the necessary accessories to make any kind of hookup needed or desired.

The first thing to do is determine the correct length for the stub and cut it if necessary, removing any caps that may have been installed previously for testing purposes. Attach a trap adapter to the stub. Make measurements to arrive at the proper length for the tailpipe, which is the short length of pipe that will extend down from the fixture drain flange. Slide the slip nut and flexible compression ring provided with the trap assembly up over the tailpiece. The trap is then placed in its proper position and secured loosely with a slip nut also provided. Then check the length of the trap extension pipe, trim it if necessary, and connect one end to the trap and the other to the stub trap adapter with the slip nuts and compression rings (Fig. 9-1). Adjust the trap assembly for the best fit and proper lie and snug the slip nuts down hand-tight. Run a sufficient amount of water through the fixture to fill the trap and provide a seal, checking at the same time for any leaks. If an leaking is discovered, simply tighten the slip nuts until no leakage occurs.

There may be a few variations on this procedure with regard to different types of kitchen sinks. Double-basin sinks can both be connected to one trap. All that is required is a length of pipe, called a continuous waste drainpipe, attached to the sink. An end-opening continuous waste drainpipe drops directly down from one basin, while the second basin drainpipe angles into it. Each side is connected to the sink drain flange tailpipes by means of slip nuts and compression rings. The trap is then attached in the usual manner to the bottom of the vertical portion of the assembly (Fig. 9-2). Similarly, a center-opening continuous waste assembly consists of a vertical-centered drainpipe section opening into a pair of

224

equal length horizontal drainpipes. Each one is then attached to the basin tailpipes with slip nuts and compression rings, with the trap attached in the same manner to the vertical portion of the assembly. Figure 9-3 shows this method of installation.

REPLACING AN EXISTING LAVATORY OR SINK

Replacing a sink or lavatory may be a necessity due to either damage or malfunction, or it may simply be just a matter of preference. New styles are being manufactured all the time in countless colors and designs, and you may wish to improve the overall appearance of a bathroom by replacing the fixtures. As a general rule, sinks usually come equipped with instructions for installation and all the accessory parts that will be required.

Two types of sinks or lavatories are commonly available, either wall-mounted or cabinet-mounted. By and large, wall-mounted sinks are not as popular as the cabinet type for residential use. They are usually encountered more in commercial bathrooms today. A typical wall-mounted sink is shown in Fig. 9-4.

Cabinet-mounted sinks are considered to be almost standard equipment in a residence. They can be purchased as a complete unit with the sink, cabinet, and accessories, or if desired, either the sink or cabinet may be purchased individually to come up with the desired design. Figure 9-5 shows a typical cabinet-mounted sink, which can be used both in a bathroom or a kitchen.

Before removing the existing fixture, take the time to study the method originally used to attach it to the plumbing system. In the case of a wall-mounted sink, the most common method is a hanger assembly (Fig. 9-4). The sink is supported by the horizontal portion of the assembly which is secured to the walls by means of studs on the back of the device. A cabinet-mounted sink may either be an integral part of the countertop

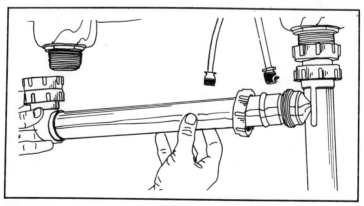

Fig. 9-2. An end-opening continuous drainpipe connection for a double-basin sink.

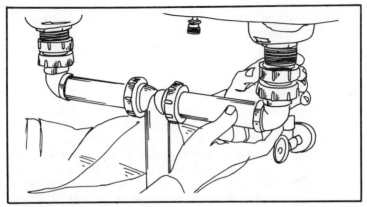

Fig. 9-3. A center-opening continuous waste assembly may also be used.

itself, or it may be a separate piece which is dropped in or out of the opening in the cabinet.

The next step is to study the water supply and drainage lines in order to determine the method of disconnection. If the faucet is not attached to the sink, as in some types of cabinet-mounted fixtures, the water lines do not have to be disconnected. In the wall-mounted types, the hot and cold water lines are disconnected by unscrewing the nuts. None of this should be attempted without first shutting off the water supply to this fixture at the appropriate point. Referring to Fig. 9-6, the drain line is disconnected by unscrewing the slip couplings (S). Also disconnect the stopper linkage (C). If there is an existing garbage disposal in the old fixture, this is disconnected at the sink drain (Fig. 9-7). The two drain lines come

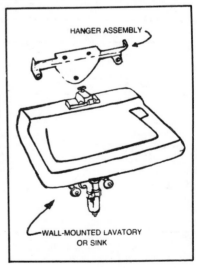

Fig. 9-4. A typical wall-mounted sink which is attached by means of a hanger assembly.

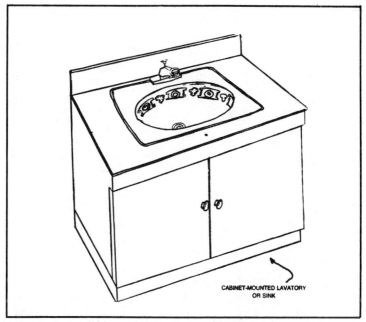

Fig. 9-5. Cabinet-mounted sinks are considered to be a more modern fixture.

together at the trap, and then a single line carries the drainage flow to the back of the cabinet. If such a fixture has sinks which lower out of the cabinet for removal, it will be necessary to dismantle the trap, the drain line, and the garbage disposal before attempting to remove the sink.

After all the appropriate disconnections have been effected, the old sink may be removed. Depending upon the age and type of sink, this may

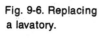

Fig. 9-6. Replacing a lavatory.

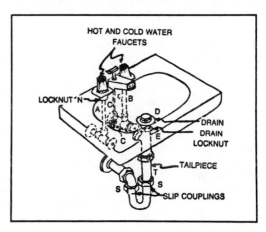

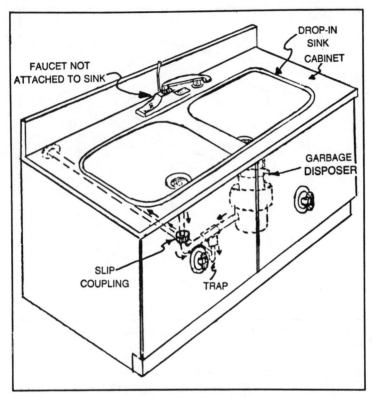

Fig. 9-7. Disconnecting a sink with a garbage disposer.

require the help of another person, since older sinks were made of heavier materials than those manufactured today. Before discarding the sink, check the faucets and drain to see if they are reusable. Even if they are not appropriate for installation with the new fixture, either because of design or configuration, it's a good idea to hold onto them. They may come in handy at a later date.

The removal of these parts is really quite simple (Fig. 9-6). The faucet is disconnected by opening the two locknuts and washers under the hot and cold water controls (N). Similarly, the drain (D) is detached by removing the locknut and washer under the stopper.

The procedure for installing the replacement sink or lavatory is basically the same as the removal, except in reverse manner. It should be removed from its container or box, cleaned, and placed in the position that was occupied by the old sink, either over the wall hanger or inside the cabinet openings. Make sure the sink is properly aligned, in that the drain opening is over the drainpipe and the faucet openings are over the hot and cold water supply lines. Apply a layer of plumber's putty on the bottom of the faucet, place the locknuts and new washers in position, and tighten them. Referring to Fig. 9-6 again, insert the drain (D) into the drain opening, place the washer and drain locknut (E) below the underside of the sink, and tighten the assembly. A little plumber's putty should be applied on the top of this washer so that it grips the surface of the sink itself. This installation will be simplified a bit if another person holds the sink in place while the necessary connections are made. Also, although most installations will follow the same basic guidelines provided here, be sure to adhere to each manufacturer's instructions if any differences are apparent.

Once the faucet and drain are attached and the sink is mounted securely in place, the water supply and drainage lines may be connected. The water supply lines are tightened up by the locknuts (N), and new washers should always be used for these connections. Connect all the drain lines which were disconnected, again using new washers.

It may be necessary to effect some minor modifications to the existing piping in order to connect up a new sink or lavatory. For example, the tailpiece tee may have to be shortened or lengthened to provide the proper fit for the trap. Similarly, the traps may have to be realigned either by tightening or loosening the slip couplings. None of this is very difficult, and instructions can be found in Chapter 8.

Once the assembly is completed, the water supply is then turned on and the faucet opened to test for leaks. If any leaks are discovered, turn off the system and make the necessary repairs by either simply tightening up the assembly or possibly using plumber's tape.

CLEARING DRAINS

Blockages can occur in almost any part of a drainage system. Some common causes may be the gradual accumulation of grease, dirt, and similar materials in the lines, or indiscriminate disposal of food, coffee grounds, rags, paper, hair, fruits, vegetables, etc., into sinks, toilets, or other fixtures. The methods used to clear the blockage will depend upon whether it is the main drain or an individual fixture that is stopped up.

With this in mind, the first step is to pinpoint the exact location of the blockage. Before beginning an in-house inspection, if the system is connected to a municipal sewer, check with neighbors to see if they are experiencing the same difficulties. The problem may be related to either an overload at some point or possibly a power failure, causing the utility's system to back up into residential systems. If this is the case, a call to the proper authorities informing them of the problem will probably be all that is needed. It will be their responsibility to make the necessary repairs to their system, and if necessary, yours as well. Depending upon the extent of the breakdown, it may take the utility some time to restore normal operation.

If the problem is not related to the municipal sewage system, an inspection of the entire system is in order. To do this, go to the location of each plumbing fixture and open the outlets one at a time. This will include all showers, bathtubs, and sinks, but *do not* flush any toilets during this test, as this will cause them to overflow. If all the fixture drains are working except one, then that particular drain is the one that is clogged. If it is found that two or more fixtures are not draining properly, this will probably mean that the point of blockage is at the main house drain.

Clearing the Main House Drain

At this point, you are probably realizing the significance of installing the various cleanouts in the system; without them, there would be no access points to allow for clearing blockages. A house main drain may have more than one cleanout, depending upon the installation. In a system with only one cleanout, it is usually located at the highest point of the main drain, and this will be the access point. If there is more than one cleanout, it will be necessary to go a little further and determine more specifically the exact location of the blockage. To do this, remove the plug of the cleanout nearest to the house sewer and be sure to place an empty bucket under it. If water starts dripping into the bucket, this means that the drain between this cleanout and the soil stack is clear and the blockage is located towards the sewer side. If no water is present, open the next cleanout towards the soil stack and repeat the procedure. In this manner, it will be possible to pinpoint the exact location of the blockage. Generally, it is quite unusual for a blockage to occur in the actual soil stack.

Once it has been determined which cleanout is to be used for the clearing process, place a bucket under it and open up the cleanout plug to allow for all the water to drain out. A number of devices can be used in this next procedure, depending upon the extent of the blockage. The one to start with is a typical water hose, with the nozzle removed. Be sure to clean anything to be used before insertion is attempted, so as not to cause any contamination or spread any disease bacteria.

The hose is inserted through the cleanout and into the drain until its entry is stopped at the point of the blockage. Seal the opening by placing wet rags all around the hose. Keeping the hose firmly pressed against the blockage, let a little water flow into the system. As the material clogging the line begins to be moved by the flow of water, increase the flow gradually until it becomes apparent that the water is moving freely. Once the material is completely broken up, use a forward and backward motion of the hose to clear the drain of any small amounts that may still be present. If this procedure has proved to be successful in clearing the blockage, turn off the water and remove the hose from the line.

If this method proves to be unsuccessful, a device called the *hand auger* should be used. As before, place a bucket under the cleanout and insert the auger into the drain until the point of blockage is reached (Fig. 9-8). Turn the handle of the auger, remembering to always turn it in the same direction, until the obstruction is broken and the auger is moving freely. The auger is then removed and the hose inserted. Turn the water on, and if the water flows freely into the cleanout and the drain, the blockage has been cleared. In the event this does not occur, the hand auger

230

may be inserted once again for a few minutes more. In most cases, you will succeed on the second try.

If, however, the blockage persists, it may be caused by roots of trees having entered the drain either through its joints or through the house sewer itself. Blockages such as this will require the use of an electric auger. These are usually available for rent at a local equipment rental store. An *electric auger* consists of the same type of flexible cable as a hand auger, but it is also equipped with a flexible blade with quite sharp teeth. These teeth, when turned around and around inside the drain, will prove quite successful in cutting through even the most difficult stoppages caused by tree roots. Electric augers come in a number of various designs and configurations, so make sure proper operating instructions are obtained from the renting dealer. These instructions should be strictly adhered to, and it is wise to read them over at least once before attempting to use the device. One thing to remember about electric augers is that even though they are usually quite long and somewhat flexible, they will not perform well if they must pass through a number of fittings and turns. Thus, it is important to work from the cleanout closest to the actual blockage.

Once the blockage has been removed following the same basic procedures given here for use with a hand auger, remove the device and insert

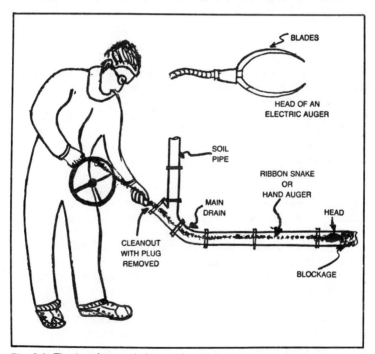

Fig. 9-8. The hand auger is inserted until it reaches the blockage.

the hose. Wash the inside of the drain thoroughly in order to dislodge any material that may remain inside the drain. Replace the cleanout plug. To test the system to insure that the repairs have been successful, test the flow by flushing a toilet several times and running water through each of the fixtures individually. If the system is still not functioning properly at this point, it is time to seek professional help. Do not be disheartened if this is necessary; rather, pride yourself on the fact that you did everything you could with the tools available for use by a do-it-yourselfer.

If blockage is determined to be somewhat extensive, do not attempt to clear the blockage with chemicals. The reason is that if the chemicals do not produce the desired results, and it becomes necessary to use any of the above methods, you will be exposed to these toxic materials as they exit the system into the bucket provided for this procedure.

Clearing Fixture Drains

If it is found that an individual fixture is blocked, the procedure given here will usually work. Each fixture is connected either to the soil stack or to the main drain system through what is commonly referred to as the fixture drain system. There is always a trap in the line and a connection to the vent pipe. Any part of this drainage system can develop a blockage, although in most cases it will not be too far from the fixture itself. Generally, it will almost always be on the horizontal portions of the drain and seldom on the risers.

Blockages at a sink are sometimes caused by a collection of dirt or other materials right at the strainer or stopper. If the blockage is the result of the strainer, it can easily be removed by unscrewing the two nuts which attach it to the basin. If it is not attached with screws, a little prying with a knife or a screwdriver will loosen it to allow for removal. Clear the drain and the strainer and replace the strainer.

If the drain is equipped with a stopper, it can usually be removed by simply turning it counterclockwise and lifting it up at the same time. As with a strainer, clear the drain and replace the stopper by turning it in the opposite direction. Some stopper designs are a bit different and cannot be removed in this manner. Figure 9-9 shows the workings of such a stopper. By raising the lifter, the rod turns about the pivot ball and lowers the stopper to close the drain opening. When the lifter is pressed down, the stopper pops up. To clean a drainpipe with this type of stopper, it will be necessary to disassemble the mechanism.

The first thing to do is place a bucket under the installation to collect any discharge that may occur. Referring to Fig. 9-9, loosen the adjustment screw (S) and remove the nut (N) from the pipe. This will make it possible to remove the rod, thus freeing the stopper and its vertical rod. The stopper can now be removed and cleaned. Make sure to clean the drainpipe as far down as is reachable. Replace the stopper and install the rod by inserting it properly at both ends. Replace the nut and tighten the screw. The drain can be tested by turning the faucet on. This procedure will usually prove successful.

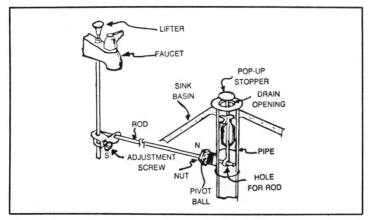

Fig. 9-9. To remove a stopper, the lifter is raised first.

If, however, the blockage is still present in the drain, the next thing to use is a plunger. As shown in Fig. 9-10, there are two types of plungers, either flat or molded. The molded one is suitable for rounded surfaces such as those in toilet bowls, while the flat type works best on flat surfaces found on sinks and shower or bathtub drains.

To use a plunger, first remove the stopper or strainer, as applicable. If the stopper cannot be taken out easily, simply raise it to its full open position by pressing the lifter at the faucet. Lubricate the base of the plunger with a thick coating of petroleum jelly. Open the faucet and let the sink or other fixture fill with water until it is 2 to 3 inches above the drain opening. Plug the overflow opening, if there is one, with a damp cloth. The reason is to prevent any leakage of pressure when the plunger is applied to the drain. In a situation involving a double sink, both the opening not being plunged and the overflow hole should be plugged tightly in some manner.

Fig. 9-10. A plunger may be either flat or molded.

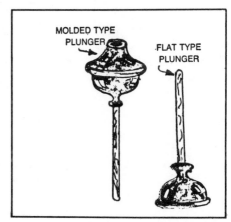

Cover the drain opening with the plunger and press its handle swiftly up and down several times. Lift the plunger after a few strokes and see if there is any indication that the drain is removing the water in the fixture. If no flow is apparent, repeat the process a number of times. It may take a bit of time to remove blockage. Once the drain has begun to work, open the faucet and let the water run for a time to clear any loose material that may remain. If the drain continues to perform satisfactorily, the stopper or strainer may be replaced and the job is complete.

If the material has been only partially dislodged from the drain, the use of a chemical drain cleaner may finish the job. However, this should not be used on a drain which is completely clogged. The chemical drain cleaners available on today's market come in a number of configurations, either as some form of liquid or in a crystal form. To use either type, remove the stopper or strainer, and place a plastic funnel over the drain opening. Drop the required amount of cleaner into the drain, as indicated by the instructions on the container label. Remove the funnel and let the chemicals ferment for a few minutes before turning the faucet on. The pressure of the water flowing through the drain combined with the action of the chemical cleaner should clear the remainder of the blockage. Rinse the funnel used in this procedure thoroughly and store it away for use in cleaning drains.

If a plunger fails to remove blockage in a sink, it will be necessary to open up the trap assembly and clean it. Dirt, grease, hair, paper, and any articles which may accidentally fall into the sink can get caught in the trap and will ultimately result in clogging. Figure 9-11 shows a typical trap assembly. In most cases, there is a nut at the bottom of the trap. Be sure to place a bucket under the trap before removing the nut, which is turned in

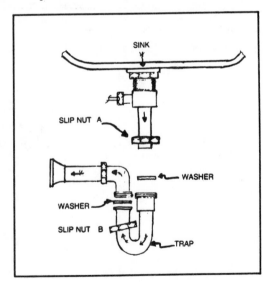

Fig. 9-11. Removal of a trap assembly.

the direction indicated by arrow A in Fig. 9-11. If a chemical cleaner was used previous to this procedure, be sure to provide some sort of protection for your hands, as the chemicals in these cleaners are quite toxic.

A wire clothes hanger which has been straightened out or any suitable length of wire of sufficient strength may be used to clear the trap. Bend one end into a hook of a size that can be inserted into the trap opening. The hook is then placed in the opening and moved back and forth. This will cause the material blocking the trap to fall down into the bucket. Repeat the procedure several times, working with it until the blockage has been cleared. Using a small piece of wet cloth or possibly a brush, clean and wipe the trap opening to remove any remaining material. The plug can now be replaced and the water turned on.

If the trap being worked on is not equipped with a nut on the bottom, the trap assembly will have to be dismantled. Again referring to Fig. 9-11, use a wrench or a pair of pliers to loosen the slip nuts (A) and (B), and take out the trap. Two layers of adhesive tape should be wrapped on any chrome pipes and fixtures before applying any tools to their surfaces. This is done to prevent damage to their finish.

Remove the deposits inside the pipes with a piece of wire and thoroughly clean the inside with a piece of rag or a brush. If the washers appear to be worn out, they should be replaced. Reposition the trap correctly, insert the washers, and tighten the slip nuts. Turn the water on, test the system for satisfactory operation, and the job is complete.

In the event that none of these methods have removed the blockage, this is probably an indication that not only the trap but the sink drainpipes—the vertical or horizontal or both—are blocked. To correct this, it will be necessary to use a hand auger. The vertical portion of the drain can be tackled by inserting the auger through the stopper opening. In order to clear the horizontal portion beyond the trap, it will be necessary to disassemble the trap as previously explained.

Figure 9-12 shows two types of augers commonly available for use in clearing drains. The first is a short, flexible hand auger which is very convenient for use with toilets and sinks. The other is a coiled-spring snake or flexible drain auger. This type has a long length of flexible wire terminating in a bulbous spring head. If inserted into a drain and turned, this auger will clear up even extensive blockages.

To clear the blockage in the vertical portion of the sink drain, remove the strainer or the stopper. Insert the hand auger and push it down until it strikes against the obstruction (Fig. 9-13). Press it quite hard, turning the handle in one direction only, and then suddenly release it by pulling back. Press and release the auger several times until the obstruction is removed and the auger is moving freely back and forth inside the drain. Now test the system by opening the faucet. If water has started flowing normally, the blockage has been cleared. Run the hot water faucet for a few minutes to clean the drain of any sticking matter or grease. The stopper or strainer can then be replaced.

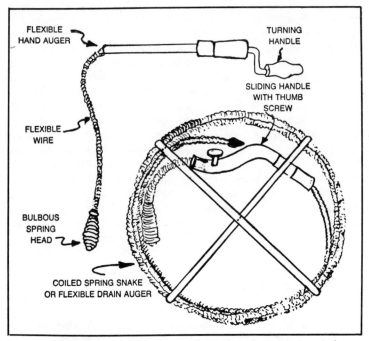

Fig. 9-12. Two types of augers are available for use in clearing drains.

If this procedure does not prove successful, this means the blockage is probably located on the horizontal length of the drain beyond the trap. To clear this, open the trap assembly in the same manner as described previously, remembering to wear gloves if a chemical drain cleaner was used earlier. Insert a hand auger inside the drain (Fig. 9-14). Turn the handle in one direction only and push it in until the obstruction is felt. Press it forward and pull back several times until it moves unobstructed. Make a note of the exact length of the auger that has gone into the drain, and withdraw it. Replace the trap, fully tighten the nuts, and open the water fully. If all is well, the job is done. If not, open the trap once again and insert the hand auger to a deeper point than before. If the blockage is encountered, repeat the process as before. If not, this means the obstruction is somewhere further down the line, and a flexible drain auger will be required. If this is either not possible due to unavailability or proves unsuccessful, it will be necessary to seek professional help.

SUMMARY

The thing to remember when either installing a new sink or repairing or replacing an old one is that a variety is found in sinks made by different manufacturers. The instructions given in this chapter will give you a basic understanding of what will be involved in a typical installation. Some

236

details may vary, depending upon the individual fixture, but the theory will remain the same. Be sure to obtain the necessary instructions for installation when purchasing a fixture from a dealer.

Most of the problems involving blockages in drains or traps closest to the fixtures themselves can be attributed to carelessness. Anyone who has had the experience of unclogging a drain can attest to the unusual items found to be the cause of the problem. Sometimes these items find their way into the system quite accidentally, and these occurrences cannot be helped. Once you become the person called upon to make the repairs as a do-it-yourselfer, and you realize a plumbing system must be treated with the respect it deserves, you will probably be more careful in the future. Likewise, you will be sure to stress this point to the other family members as well.

The parts from old fixtures, if in working condition, should be retained for possible future use. The same may hold true for the fixtures themselves. Older sinks and lavatories, if not damaged, may sometimes be sold for a nice sum, depending upon the materials used in their manufacture. Some salvage yards may be interested in purchasing these fixtures. This may also apply to lengths of pipe that may have had to be replaced, especially copper, which can be sold to help finance the cost of the new installation. In this situation, those who buy this material usually pay a set price per pound and then melt the piping down for resale to manufacturers who reuse it to produce their finished items.

None of the information provided herein is really very complicated, and as long as these guidelines are followed and manufacturer's instruc-

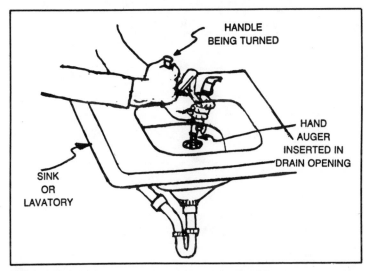

Fig. 9-13. A hand auger may be used to clear the blockage in the vertical portion of the drain.

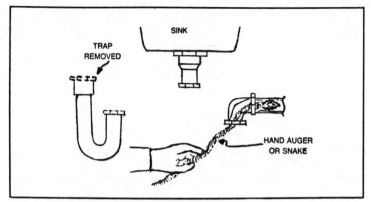

Fig. 9-14. If the blockage is on the horizontal length of the drain, it will be necessary to open the trap assembly.

tions adhered to, you should find it an easy and rewarding task to perform the necessary installations and/or repairs on your own. If any difficulties are encountered, go back over your work and examine it carefully, checking for any possible errors, before seeking professional assistance. Most problems can usually be discovered on this second check, thus avoiding a costly bill from a plumber. Remember to observe the proper safety precautions. With this in mind, you can perform a task that will result in an operating water supply at the turn of a faucet.

Installation and Repair of Toilets

As with sinks and lavatories, toilets come in a multitude of designs and prices. Those with lower prices will generally be basic fixtures and will provide long, trouble-free service. It may be desirable to invest the extra money for the more expensive toilets, which provide extra comfort and convenience and present a more eye-appealing appearance.

Although the inner workings of a toilet are quite durable and made to last a long time, the exterior is not made to handle abuse and may chip or crack if not handled carefully. If the fixture is not to be installed immediately, it is recommended that it be kept in its enclosure in an area where it will not be disturbed.

The installation of a toilet will be much easier when the home is still in an unfinished state or if a new bathroom is being installed. The water supply and drainage lines can be run through the walls and floors without too much difficulty. If replacing an old fixture, it will be necessary to possibly uncover some areas during the installation process. To make this as uncomplicated as possible, the new units should match the dimensions of the ones being replaced in regard to the inlets and outlets provided on the fixtures. In situations where this is not possible or desirable, there will be a bit more carpentry and finish work required.

TYPES OF TOILETS

There are a number of different types of toilets, all of which provide for the same function. A typical domestic toilet consists of a flush tank and a bowl with a cover. The flush tank receives its water from the water main. The actual mechanism of a toilet will be discussed later in this chapter, but the function of a toilet is to receive and remove waste. The efficiency of this unit is generally measured by its quiet operation and the extent to which the amount of contamination and nuisance it generates is kept to a minimum. As a rule, the larger the water surface provided for by the bowl,

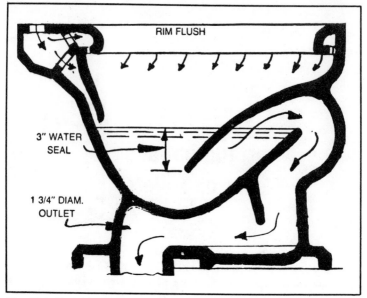

Fig. 10-1. Typical washdown toilet in which the flushing action is unassisted.

the better the sanitary conditions. Each of the difficult types of toilets available today provides a variation on the flushing action.

Figure 10-1 shows a typical washdown toilet. This is the oldest type, has been around for a long time, and will probably continue to be around for an even longer time because of its low price. The actual flushing method consists of a direct washout action unassisted by jets or vortices provided on newer models. A washdown toilet is much noise and less efficient, causing more frequent clogging as a rule. Also, a large surface area of the bowl stays out of water, causing contamination and nuisance.

A reverse-trap toilet is pictured in Fig. 10-2. This type of toilet is constructed to provide an opening at the inlet to the trapway through which a jet of water streams out with a certain amount of force (see point B). This helps the siphonic action in cleaning out the solid wastes much more efficiently than a washdown toilet. There is much less contamination and fouling, too, because a larger surface area of the bowls stays under water. Due to the jet action, clogging is less likely to occur. Though somewhat noisy, this type is the least expensive among siphon-type toilets.

Siphon-jet toilets are essentially the same as the reverse-trap toilets. Except that the jet is a bit larger and more efficient. As shown in Fig. 10-3, this type of toilet will provide for less fouling, since a much larger surface of the bowl remains under water. The outlet or the trapway passage is larger in diameter also, thus making for a much quieter operation and less chance for clogging.

Siphon-vortex toilets are the most efficient, attractive, and expensive. The flushing action is almost silent and fouling is really minimal, since most of the bowl surface is covered with water all the time. In appearance, a toilet of this type is really the same as that of a jet toilet. The flushing action is a bit different, though, as compared with the action on the other types already discussed (Fig. 10-4).

The shape of the toilet bowl itself will also depend upon individual preferences. The bowls come in two shapes: *round rim* and *elongated rim*. Many people feel the elongated rim bowl looks somewhat more elegant and is more comfortable to use than the round bowl. Also, it stays clean with little effort, since the water surface area is much larger and most of the bowl surface remains submerged.

HOW A TOILET OPERATES

The interior of a flush tank is composed of a number of components which may seem confusing at first glance. In actuality, however, the mechanism itself is really quite simple. Once the parts are properly identified, it will be easy to determine the cause of a malfunction and effect the necessary repairs.

Figure 10-5 shows the interior of a typical flush tank. The *trip handle*, also called the flush handle, is located on the outside of the tank. This is usually metallic and is the method used to begin the flushing action. When pushed down (or up), it raises the trip lever, which in turn lifts a chain (the upper lift wire) and the lower lift arm. This works through a guide arm attached to the overflow and ultimately lifts the flush ball. The flush ball is made of soft rubber and normally rests on the flush valve seat at the bottom

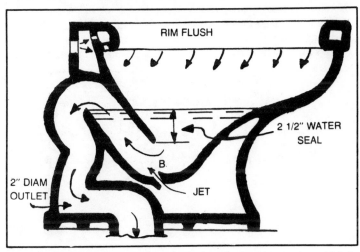

Fig. 10-2. A reverse-trap toilet provides a better flushing action which is jet-assisted.

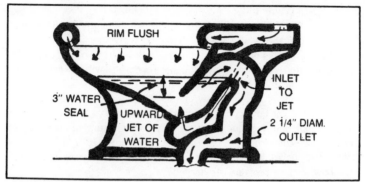

Fig. 10-3. In a siphon-jet toilet, the jet is both larger and more efficient.

of the tank. As the flush ball is lifted, this causes water to flow out through the outlet to the toilet bowl. Due to the large size of the opening, this will provide a large volume of water flowing at a quite rapid pace, causing the contents of the toilet bowl to be washed away. Consequently, the water level in the flush tank itself drops rapidly. The rubber flush ball will continue to float until the water in the tank drops to a level below its opening. When this occurs, the flush ball will obviously no longer be able to float and will drop back into place on the flush valve seat, causing the flushing action to cease.

Another device which performs a useful purpose in the workings of a toilet is the *float*. This device is normally made of either rubber or some thin metal. Under normal conditions, it will float at the surface of the water and is attached through a metallic float arm to the float valve assembly at the top of the supply pipe. When the tank is completely full of water and the float is at its topmost position, the float valve will remain closed. As the flushing action begins, the float gradually drops as the water level drops, thus opening the float valve. When the float gets to the lowest position at

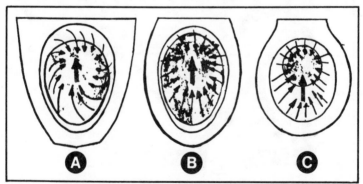

Fig. 10-4. A siphon-vortex toilet, when compared with other types, provides a circular flushing action assisted by a jet.

the bottom of the tank, the float valve is fully open. This allows for the admittance of water from the supply main into the flush tank through the tank refill tube. As the tank starts filling up, the float rises with the rising water level and causes the float valve to gradually close. When the float reaches its topmost position again, the valve will close completely, thus preventing any more water from entering the tank. This, then, accurately describes the process which takes place each time the trip handle is operated.

The flush tank has a supply line which connects to the water supply line. When the float valve opens, water rushes from the supply pipe into the tank refill tube, which in turn fills the tank. There is also a flexible bowl refill tube through which a small quantity of water flows down to the overflow tube and on to the toilet bowl. This flow, which takes place toward the end of the flush operation, restores the water seal of the toilet itself. There is another function of the overflow tube which serves a useful purpose. If, due to some defect in the mechanism, the float valve does not close fully even after the tank is full, the incoming water simply flows down the overflow tube to the toilet and on to the house drain without flooding the bathroom.

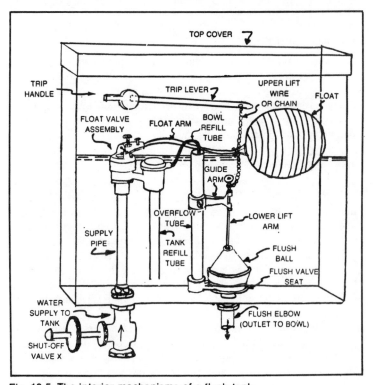

Fig. 10-5. The interior mechanisms of a flush tank.

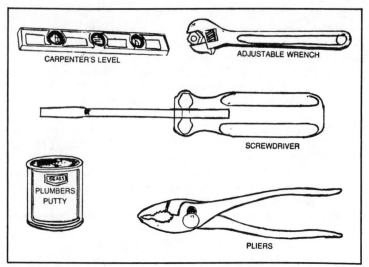

Fig. 10-6. Only a basic assortment of tools will be necessary for most toilet repairs. Shown here are an adjustable wrench, a screwdriver, a pair of pliers, a carpenter's level, and some plumber's putty.

COMMON TOILET PROBLEMS

Most of the problems involving toilets are not too difficult to repair and will require only a small assortment of tools. Figure 10-6 shows the basic tools that might be needed for such repairs. These consist of a pair of pliers, an adjustable wrench, a screwdriver, a putty knife, and a carpenter's level. Following is a discussion of the more common malfunctions encountered and instructions for their repair.

Toilet Runs Continuously

A toilet may run continuously for two reasons, either a defective float or a defective flush ball. The first step is to determine which of these mechanisms is the cause of the problem. Remove the top cover of the tank and set it down in an area where it will be in no danger of being damaged. Note the water level. If the top of the overflow tube is under water, this indicates the float needs to be either repaired or replaced. On the other hand, if the water level is ¾ inch or more below the top of the overflow tube, the defect is in the flush ball. Ideally, the water level should be about ½ inch below the top of the overflow once the tank is filled and the inflow ceases.

To repair the float, grasp the float arm and lift it up slowly to the point where it won't move any further. Be sure not to apply too much pressure, as this will cause permanent damage. This procedure is shown in Fig. 10-7. If this causes the running water to stop, you have pinpointed the cause for the malfunction. It may be that the float is touching the upper lift wire or the side of the tank, and this is a simple repair. Apply a little lateral pressure to the float arm until the float is unobstructed and is a reasonable distance away from the tank wall or the lift wire. If nothing is obstructing the float, unscrew it from the float arm by turning it in a counterclockwise direction. Shake it to check for possible water inside the float itself. If there is, the float should be discarded and a new one installed. The new float is simply screwed onto the float arm by turning it in a clockwise direction. If, once installed, the float remains immersed about ½ inch in the water, everything else is functioning satisfactorily, and no further repairs will be necessary.

If the problem still persists, bend the float arm until the mechanism functions properly. Flush the toilet by turning the trip-arm, and the water should shut off when the tank level is about ¾ inch below the top of the overflow tube. If this does not occur, continue to bend the float arm in the required direction until the desired result is obtained.

Referring back to the first step in determining the cause of the malfunction, if the water did not shut off when the float arm was raised, the problem lies with the float valve assembly. Figure 10-8 shows the normal operation of this mechanism. There are many variations in design, and each different type of toilet will not work exactly as shown. The general principles will remain the same. In some cases, the assembly comes as a sealed unit. These units usually have a rather lengthy life, but once they begin to malfunction, they should be replaced rather than attempting to make repairs. When replacing one of these units, consider the purchase of a new type of mechanism which is rather expensive. Called a plastic flush control, this unit has a self-contained, sealed mechanism which provides for quiet operation and a more precise water level control.

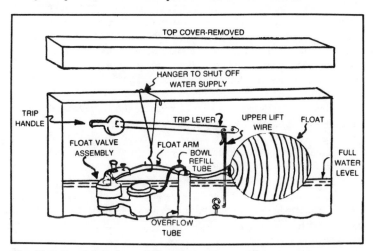

Fig. 10-7. Procedure for making repairs on a toilet which runs continuously.

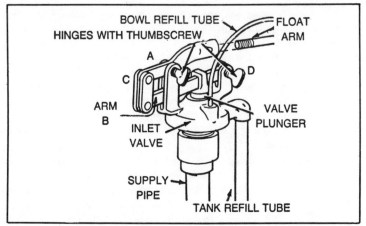

Fig. 10-8. A typical float valve assembly which is operational.

If, however, the toilet has the unsealed type of valve assembly, it can be repaired. After removing the tank cover, close the valve on the supply line under the tank. Turn the trip lever to flush and empty the tank. Remove the thumbscrews and take out the arm, indicated as (B) in Fig. 10-8. This will release the valve plunger, which should then be removed. If the washers are damaged or worn out, get exact replacements from a local supply store. Install the new washers, the valve plunger, and after replacing the arm (B) and the float arm, tighten the thumbscrews. Open the shutoff valve and flush the toilet to check the operation. The toilet should function normally at this point.

If normal water level in the flush tank is ¾ inch or more below the top of the overflow tube and the water is running continuously, the defect lies in the flush ball system. Depending on the situation, this mechanism may be either repaired or replaced completely. A number of things may be causing the problem. The flush ball may be damaged or worn out, the ball seat may be on uneven, or the lift arm may not be properly attached to the flush ball.

Before attempting any kind of repairs, the flush tank should be emptied and should remain empty during the whole operation. If the tank has a shutoff valve, the supply can be stopped by closing this valve, after which the tank should be flushed out by turning the trip lever. If there is no shutoff valve on the toilet, this can be effected by tying a string which is weighted down in some manner to the trip lever on the outside of the tank. Similarly, another method is to place the hook of a metallic clothes hanger around the trip lever, with its ends bended on both sides of the tank. This will firmly hold the trip lever in its upper position, keeping the inlet valve closed.

Once the tank is empty, inspect the flush ball assembly. This mechanism is shown in Fig. 10-9. If the lift arm seems loose, try to tighten it by turning the flush ball counterclockwise. If this does not tighten the arm, the flush arm is defective and should be replaced. Check the lower surface of the flush ball that sits on the ball seat. If it is worn or damaged, water would be leaking through the opening and causing the toilet to run continuously. In this case, the flush ball should be replaced. This is done by unscrewing it from the lower lift arm, which should present no difficulties. When purchasing a replacement, be sure to get an exact duplicate. Replacing defective mechanisms is not too difficult, since most of these devices are standardized to a great degree.

To install the new flush ball, hold it above the seat and screw it onto the lift arm by turning it counterclockwise. Make sure that the new flush ball sits evenly over the ball seat. Operate the trip level several times and observe whether the flush ball falls down and grips the ball seat evenly. If it does not, the flush ball guide arm should be adjusted. If the lift arm is simply a chain, adjust its length until the flush ball starts to sit evenly on the seat. If the lift arm is attached to the overflow tube through a guide arm, unscrew the tightening screw and turn the guide arm up or down or

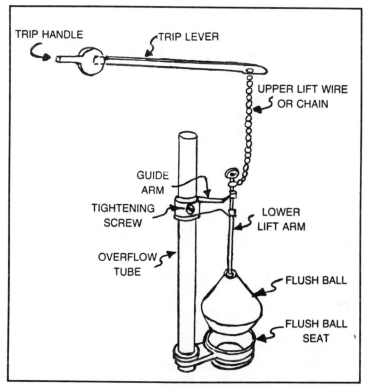

Fig. 10-9. The flush ball assembly is attached to the trip lever by means of a chain or wire.

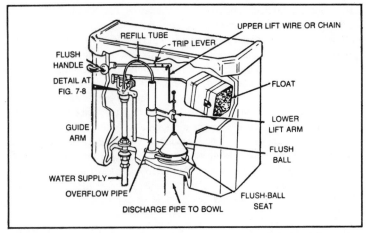

Fig. 10-10. The refill tube may sometimes become dislodged from the overflow tube.

sideways, whichever is applicable, until proper positioning is obtained. Once this is achieved, the screw is retightened, and the flushing operation is repeated to check for satisfactory operation.

During this procedure, it is suggested that the whole assembly be cleaned before replacing the tank cover. This is done by raising the flush ball up by pressing the trip lever. The ball seat can then be cleaned with either steel wool or possibly an emery cloth. Both the inner surface and the top rim should be cleaned until they feel smooth to the touch.

Remove the string or the coat hanger or turn on the shutoff valve, whichever is applicable, and check the operation of the toilet once again. If all is functioning properly, the tank cover may be replaced. Since all of the workings that could cause this type of malfunction have been either repaired or replaced, any problems still encountered in regard to continuously running water at this point are probably the result of an error somewhere. Go back over your work once again, find the error, correct it, and normal operation should result.

Toilet not Flushing or Flushing Sluggishly

As with the previously discussed repairs, remove the tank cover and set it aside in a safe place. The first thing to check is the refill tube, which may have become dislodged from the overflow tube. Referring to Fig. 10-10, simply put it back in place. It may be necessary to bend the tube somewhat to make it stay in its proper position. If the refill tube is found to be in place, check the flush handle itself, which may have loosened up, causing the trip lever to malfunction. It is a quite simple procedure to tighten the nut on the inside of the tank with an adjustable wrench, making sure to hold the trip handle firmly in place.

If both the refill tube and the flush handle are functioning properly, a further check will be necessary. The most common causes of a non-flushing toilet are either a broken upper lift wire or chain, or an unscrewed lower lift arm. If the lift wire or chain is broken, this portion of the assembly should be replaced. To remove the malfunctioning mechanism, unscrew the lower lift arm from the flush ball by turning it counterclockwise. Slide it out of the guide arm and then remove the broken chain or wire from the lower lift arm and the trip lever. Keeping the new connecting wire or chain in place, insert the lower lift arm through its hook and through the guide arm until it reaches the top of the flush ball. Attach the upper end of the new connecting wire to the trip lever. Turn the flush ball counterclockwise around the lower end of the lift arm and screw it on tightly.

Adjust the length of the upper lift wire until the hook at its lower end is positioned about ¼ inch below the hook at the top of the lower lift arm. Flush the toilet at this point to check the repair work. If all is well, the tank cover can be replaced.

If the cause of the non-flushing is the result of a loose or unscrewed lift arm, try to tighten it by turning the flush ball counterclockwise. If this method fails, both the flush ball and the lift arm should be replaced. After performing this operation, it will be necessary to adjust the length of the upper lift wire or chain, as previously explained. Again, test the toilet by flushing. If the flush ball drops after the tank is fully emptied, the defect has been corrected.

The flush ball may drop before the tank has been completely emptied. In the case of a wire upper lift assembly, simply adjust the length of this wire as explained previously. If the toilet being worked on has a chain, disconnect it from the trip lever and reduce its length to provide for a lesser degree of slack when reconnected between the lever and the flush ball. Check the flushing operation and repeat the process of reducing the chain's length until the toilet is functioning properly. Figure 10-11 shows a

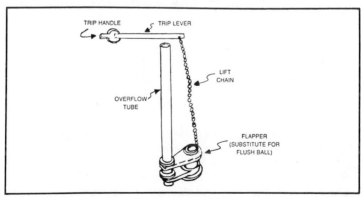

Fig. 10-11. In some instances, a flapper may be substituted for a flush ball.

249

flapper, which can be substituted for a flush ball in some instances. Instructions for the installation of a flapper are usually provided by the individual manufacturer and are not given here since they may vary somewhat depending upon design.

Toilet Tank Leaks

If water collects on the floor directly under the flush tank, this is an indication that there is a leak somewhere. This leak may be coming from around the base screws at the tank bottom which connect it to the toilet bowl. On the other hand, the washers on the inlet pipe may be either defective or worn out. In order to determine the exact location of the leak, wipe the underside of the tank with a dry cloth or sponge and watch for the appearance of water.

If the inlet pipe is leaking, it should be opened to allow for replacing the worn or defective washers. The only tool that will be needed to make this repair is an adjustable wrench. First, shut off the water supply by closing the shutoff valve near the inlet pipe if there is one. If not, close the one on the branch line or the main supply line. Remove the tank top, turn the trip lever to flush out and empty the tank, and wipe the bottom dry with a cloth. Place a bucket under the inlet pipe to catch any discharges of water. As shown in Fig. 10-12, use the adjustable wrench to loosen nuts 2 and 3. Remove the inlet pipe. Holding the ball cock assembly in place, loosen nut 1 and remove it, along with washer 2 directly above it. Remove the spud washer 1 from the top and lift up the ball cock assembly. Place the new spud washer 1 in place and insert the ball cock assembly through it. The new washers 1 and 2 can now be installed. Again holding the ball cock assembly in place, tighten this nut without applying too much pressure as the tank bottom may crack as a result. Replace the inlet pipe in its original position and tighten nuts 2 and 3. Replace the tank top and open the shutoff valve. Check under the tank to see if the leakage has stopped. If proper procedures have been followed, no water should be apparent. However, if the situation has not been corrected, it will be necessary to go through the whole process once again, making sure all connections were made properly.

If the leakage is present around the connecting screws of the tank at its bottom, either the washer or the nut and screws may have been corroded by the chemicals in the water. Again, remove the tank cover. Referring to C in Fig. 10-12, hold the bottom nut in place with an adjustable wrench and tighten the screw at the tank bottom with a screwdriver. Wipe the area around the nut with a cloth or sponge to ascertain whether or not this procedure has eliminated the problem. If the operation proves unsuccessful, it will be necessary to replace the nut, washer, and screw. In this case, disconnect the water supply at the shutoff valve or wherever appropriate and empty the tank by flushing. Wipe the bottom of the tank and remove the screw, washer, and nut. Again, these should be replaced with parts of the same size as those which are faulty. Be sure to install them in the proper order, and take care not to tighten them to a point which cracks the tank. Replace the tank cover and open the shutoff valve. Once water is present in the tank again, check the underside for leaks.

Tank "Sweats"

The temperature of the water in a toilet tank is considerably lower than that of the heated air in a building. This air is usually quite moist as well. When it comes in contact with the cool air outside of the tank, its moisture condenses, which causes the tank to sweat. A simple method of correcting this inconvenience is to line the inside walls of the tank with ½ inch thick foam rubber padding or some other insulator. Referring to Fig. 10-13, remove the top of the tank and shut off the water supply at the appropriate point. Before emptying the tank, mark the full water level on an inside wall of the tank. Flush the toilet and completely dry the inside of the tank's walls and bottom with a piece of cloth. After wiping down the walls and bottom, allow another 5 minutes drying time before proceeding any further.

Measure the foam rubber padding so it will cover the four walls of the tank up to a height of about 1 or 2 inches above the previously marked water line. Apply a layer of an appropriate resin glue *only* to the four walls, not to the bottom of the tank. Be sure to follow the directions provided with the glue being used regarding application instructions. Install the padding over the resin glue, pressing it firmly into place to insure that the

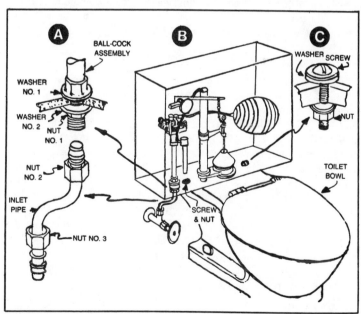

Fig. 10-12. To repair leaks, it will be necessary to remove the inlet pipe.

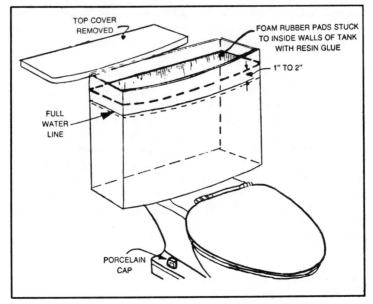

Fig. 10-13. To prevent sweating, the tank may be insulated in this manner.

glue grips it and there are no loose spots. Let the glue dry for the time indicated with its instructions. Once the glue has dried the required length of time, the tank cover may be replaced and the shutoff valve opened to let water into the tank. This should do the job nicely.

Toilet Tank or Bowl is Cracked

If it is apparent that a tank or bowl has developed cracks and water is leaking, it is probably time to replace the fixture. If time is involved in the ordering of a new fixture, a temporary repair method is given here. Purchase a tube of silicon caulk from a local supply store. Shut off the water supply to the toilet, empty the tank, and wipe the area around the crack(s) with a piece of cloth or sponge. Following the manufacturer's instructions, apply the sealant. Remember that repairs will be needed on the inside as well if the break is significant or if sizable pieces of the material have come loose. First, apply the tub caulk to the outside and let it harden for the length of time prescribed in the instructions. Apply pieces of reinforcing tape to cover the cracks and the areas around them. Finally, apply the tub caulk to the inside of the crack, again allowing sufficient time for the sealant to dry before the shutoff valve is opened.

REMOVING AND INSTALLING A TOILET

The actual procedure for removing an old toilet and installing a new one in its place will vary depending upon the location of the toilet and its proper connection to the rest of the plumbing system. A first-time installation will follow the same basic steps given here also, except that the work will involve some design and layout work in advance of the actual installation. You will use a screwdriver, an adjustable wrench, a carpenter's level, and a putty knife. In addition, you will need the following supplies and/or materials: wax gasket or putty seal, rubber spud washer, supply pipe washers, and shims of different thicknesses, which can be made of either wood or metal.

Removing and Installing a Floor-Mounted Toilet

The first step is to disconnect the water supply at the shutoff valve if applicable. If there is no individual valve for the toilet, this is a good time to install one. A valve at a point close to the toilet will allow for future repairs to be made without disconnecting the whole system or a portion of the system. Remove the tank cover and store it in a safe place. Trip the flush handle to empty the tank, and use a sponge or a piece of cloth to soak up the remaining water. Remove as much water as possible from the bowl by bailing it out with a container of some sort. Disconnect the water supply pipe at the bottom of the flush tank.

If you want to only remove the bowl without disturbing the tank, check to see if the connection between the two parts is made through a large diameter pipe which is called a *spud* (Fig. 10-14). Most older installations have spuds with slip nuts at both ends. To disconnect the bowl from the tank, simply loosen the two slip nuts and slide them onto the spud itself. This will necessitate the use of a special wrench called a *spud wrench*, which is designed for use with nuts of considerable size. The spud

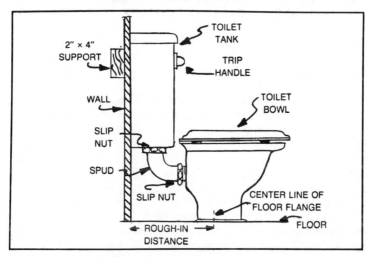

Fig. 10-14. In some situations, the tank and the bowl are connected by means of a large diameter pipe called a spud.

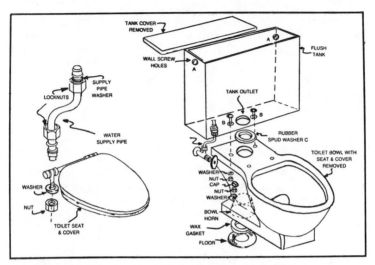

Fig. 10-15. Removing and installing a floor-mounted toilet.

can now be taken out, leaving the tank attached to the wall and allowing for the removal of the bowl as a single unit. If the tank is to be removed, too, remove the hanger bolts from inside the tank and lift the tank from the wall bracket. Generally, the newer tanks rest over the bowl with no connection to the wall.

Figure 10-15 shows a floor-mounted toilet with the tank attached to the bowl by means of two screws at the bottom. There are also two wall screw holes which indicate that the tank is attached to a wall bracket. Regardless of whether or not the tank is attached in this manner, the next step will remain the same. Remove the wall bracket screws, if applicable, from the inside of the toilet tank, and loosen and remove the two nuts and washers from the holes (B) at the bottom of the tank. Use a screwdriver and an adjustable wrench. Carefully lift up the tank and store it safety. Remove the spud washer (C).

Referring to Fig. 10-15, it can be seen that the bowl is attached to the floor by means of a pair of nuts and bolts on both sides. The bolts are held in place by a metallic floor flange which is positioned over the soil pipe. To remove the bowl, unscrew the nuts with a wrench and remove them, along with the washers. These nuts usually have ceramic caps which are set in place with a chemical compound. In order to remove the caps to gain access to the nuts, this compound will have to be pried loose with a sharp instrument, such as a putty knife. Before using this method, however, check the caps to see if they can be screwed off. Some caps manufactured are simply screwed in place. If attempts to unscrew the caps do not unloosen them, go ahead and use the putty knife to pry them off.

Once the caps and the nuts underneath them are removed, the bowl may be lifted and removed. Tap its top gently a couple of times or twist it around slightly to free it from the base. Make sure when actually lifting the bowl that the motion be straight up to prevent any of the remaining water from spilling out. After the bowl has been removed, it should be completely emptied of water, cleaned, and stored upside down in an area where it will not be subject to damage.

The whole installation process will be much easier if the new bowl has the same *rough-in dimension* as the old one. The rough-in dimension is the horizontal distance between the finished wall and the center of the floor flange. The dimension of the new tank can be smaller than the old one if the tank is not attached to the wall but is merely resting on the bowl. Before setting the bowl in place, clean up the floor flange by removing the old putty seal or wax gasket from it. If the old bowl is to be reused, clean up the inside of its horn, which is shown in Fig. 10-15. Install a new putty seal or wax gasket inside the bowl horn, pressing it down firmly until it remains in place. Lifting the bowl up and holding it just above the floor flange, align it with the two bolts provided for the actual connection to the flange. Gently lower the bowl into place. During this procedure, it may be necessary to twist the bowl backward and forward in a horizontal direction to position it correctly over the bolts and the floor flange. Apply some pressure on the top of the bowl to insure that the gasket is gripping the flange tightly, providing the necessary watertight seal.

Before continuing any further, use a carpenter's level to check the position of the bowl. Place the level over the bowl and slide it around in different positions to see that the bowl is truly level. If any discrepancies are noted, insert the metallic or wooden shims under the bowl, keeping the level in place over the bowl to check for the proper position. It will probably be necessary to do some rearranging during this process to find the exact positioning that provides a completely level fitting.

Once the bowl is in place, install a new rubber spud washer over the bowl opening just under the tank outlet. Lower the toilet gently into place, making sure the openings are lined up. Using an adjustable wrench, install the bolts and nuts which were removed earlier from the openings (point B, Fig. 10-15). Check the alignments of the tank and the bowl in regard to the wall. If all seems well, the bowl can be tightened over the floor flange bolts by replacing the nuts and, if applicable, new washers. Be careful not to overtighten the nuts, as this may cause damage to the fixture. Replace the caps over the nuts, applying an appropriate sealing compound if necessary. If the caps are threaded, simply screw them down into place. Referring to Fig. 10-15, insert the screws through holes (A) inside the tank wall and secure the tank to the wall mounting. The tank cover may now be put in place over the tank.

At this point, the supply line may be reinstalled. Insert a new washer on the supply line at the point shown in Fig. 10-15. Holding the line in its proper position between the tank bottom and the water supply main, tighten the two locknuts with a wrench. All that remains to be done now is to install the bowl cover and seat. Again referring to Fig. 10-15, this is

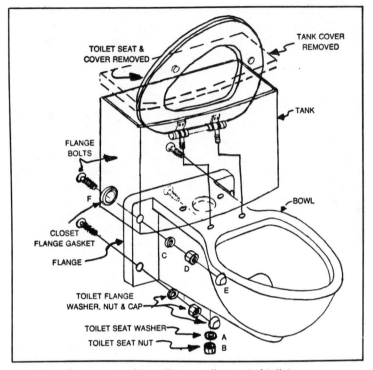

Fig. 10-16. Removing and installing a wall-mounted toilet.

done by replacing the proper washers and nuts and tightening them by hand. Do not use a wrench for this procedure, as damage to the fixture may result. Once everything is assembled, open the shutoff valve and test the toilet. If all instructions were followed correctly, the fixture should operate properly.

Removing and Installing a Wall-Mounted Toilet

The procedure for removing and installing a wall-mounted toilet is basically the same as that for a floor-mounted model, except that it is a bit simpler. Figure 10-16 shows a wall-mounted toilet with the tank resting over it. To perform this installation, the following tools and supplies will be needed: an adjustable wrench, a closet flange gasket, a putty knife, an assortment of appropriate washers, a screwdriver, and a pair of pliers.

Again, the first step is to shut off the water supply at the appropriate point in the system. Remove the cover of the tank and store it away in a safe place. Turn the trip handle to flush. Empty the tank and soak up any remaining water with a piece of cloth or sponge. Disconnect the supply pipe to the tank and remove it, as explained previously. Referring to Fig. 10-16, remove nuts and washers (A) and (B) and remove both the toilet seat and its cover. These should also be stored away in a safe place.

Note that the toilet bowl is attached to the wall by means of four flange bolts which have their heads fixed inside the wall. The washers (C), nuts (D), and caps (E) are on the bowl side of the flange. To disconnect the bowl, first remove the caps by either prying them off or unscrewing them, whichever is applicable. The nuts and washers are then removed with a wrench. During this portion of the procedure, it will be necessary to provide some means of support for the toilet bowl, perhaps in the form of another person if possible. Otherwise, the bowl will fall and may sustain damage. Once the nuts and washers are removed, the bowl can be lifted out of place and stored away for possible future use or discarded. If the bowl is to be saved, scrape its flange with a putty knife to remove the old gasket (F).

The installation of the new toilet begins by first placing a new gasket (F) in its proper place on the back of the toilet flange at the toilet opening. If necessary, press it firmly down to provide a tight seal. Lift the toilet bowl and carefully pin it onto the four bolts provided for this purpose in the wall. Use an adjustable wrench, install the four nuts and new washers. Screw on the caps or secure them by applying a sufficient quantity of the appropriate sealing compound. Restore the water supply to the tank at this point by reinstalling the supply pipe and opening the shutoff valve. Install the toilet seat and cover as explained previously, and replace the tank cover. Check the operation of the toilet. If all is as it should be, the installation is complete.

As can be seen, there is really nothing too difficult involved in the installation of a toilet. Some plumbing codes prohibit homeowners from doing this type of installation. Instead, they require that a licensed plumber perform the whole procedure. It is always wise to check with the local authorities before beginning to make sure strict adherence with codes is maintained.

Repairing a Loose Toilet Bowl

It is not unusual for a toilet bowl to come loose at its base. This is generally attributed to either misuse or a careless original installation. The repair is very simple. As was mentioned previously, the bowl of a toilet is secured to the floor by means of nuts which are usually covered by porcelain caps. These caps are either screwed or sealed in place by means of some compound, such as plaster of paris or caulk. If no leakage is present at the base of the bowl, simply remove the caps and tighten the nuts to a certain degree, being careful not to overtighten them to a point which will cause damage to the fixture. If some leakage of water is occurring, the procedure will not be quite so simple.

Shut off the water supply to the fixture, flush the toilet, and sponge out the remaining water. Disconnect the tank from the incoming supply line, unscrew the foundation nuts, and remove them. Carefully lift the toilet bowl straight up to expose the bottom of the fixture (Fig. 10-17). The seal is probably worn out, causing the leak, and will have to be replaced. A putty knife can be used to scrape the old seal from the bottom of the bowl. In purchasing a replacement, make sure the measurements are the same as the old one.

Apply a ring of putty along the rim of the toilet base (A through D in Fig. 10-18). The new wax or rubber ring is then placed inside the discharge opening. Lower the toilet carefully over the floor flange in a straight and gradual manner. After it settles in its proper position, press it down firmly and give it a slight twisting motion back and forth.

The floor flange bolts will thus meet with the holes in the base of the toilet. It is advisable to have another person help with this portion of the procedure, as most toilets are quite heavy and awkward to handle.

Once the toilet is in place, the nuts may be installed and tightened lightly. If too much force is used, this will crack the flange. Replace the caps, using whatever method is applicable. Reconnect the water supply, install the seat and cover, and the repair is complete.

If a new tile floor has been installed recently in the bathroom, the bowl may not fit properly over the floor flange. Under these circumstances, it will be necessary to use two wax seals instead of just one. These are installed one over the other to provide a completely watertight joint.

If the leak is at the base of a toilet which has a spud connection between the tank and the bowl, the procedure will be a bit different. First, remove the spud pipe by unscrewing the slip nuts. This will disconnect the flush tank. Lift the toilet straight up without tilting it and place it upside down on some thick paper padding. Remove the old seal by scraping it with a putty knife. Scrape off the setting compound remaining on the rim of the

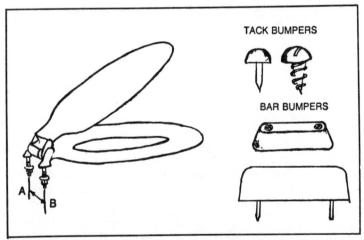

TACK BUMPERS

BAR BUMPERS

Fig. 10-17. To obtain an exact replacement for a damaged toilet seat, measure the distance between points A and B. Exact replacements should also be obtained with regard to the washers and bumpers.

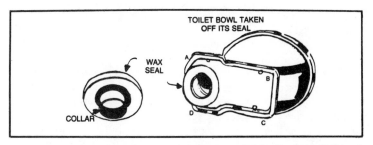

Fig. 10-18. A layer of putty is applied along the points indicated A,B,C, and D.

flange. Check the level of the floor and the floor flange with a carpenter's level. It may be that the floor has settled in places and is no longer level. Place small pieces of wood shingles near the flange bolts to serve as shims, and place the toilet back in position over the floor flange. Check the level, adjusting the shims as necessary until the bowl top is as level as possible. Without disturbing the shims, remove the toilet again and store it away. Apply a thin bead of plaster of paris along the rim of the bowl and another thicker bead along the shims. Install a new toilet seal and gently lower the toilet into position over the floor bolts and the wax seal. Check the level once again and make adjustments if necessary. Tighten the nuts lightly so as not to damage the fixture. Any excess plaster should be cleaned from the surfaces of the toilet fixture itself. Reconnect the supply line, open the shutoff valve, and check the operation of the toilet.

Replacing a Toilet Seat

Although this is not something that will need to be done all that frequently due to damage, it will be a bit more common that replacing a whole fixture. In order to obtain a replacement, it will be necessary to measure the distance between the seat bolts very carefully (Fig. 10-17). In some models, the seat bolts go through the toilet tank. If this is the case, the replacement should fit this type of installation. If the toilet being repaired is specially designed, it is wise to order a replacement from the same original manufacturer to provide a correct fit.

It is not absolutely necessary to discontinue water to the fixture during this procedure, but it is a wise move. If any supplies or even tools accidentally fall into the bowl, it will be a lot easier to restore them to their proper place if no water is present in the fixture.

It may be a bit difficult to remove the necessary nuts and bolts. Due to their location, there is a tendency for them to rust. This will be quite common when working on older installations. Apply a small amount of lubricating oil to the nuts and allow some time for the oil to loosen them up. Using the wrench, tap the nuts lightly to assist in the loosening process, being careful not to apply too much force. If a regular wrench does not provide sufficient reach, a deep-throated socket wrench might do the trick.

If difficulties are still encountered in removing the nuts, the last resort is to use a hacksaw. This is done by putting it under the hinge and sawing through the two bolts. Care should be taken when using a hacksaw this close to the fixture for obvious reasons.

Once the appropriate nuts and bolts are removed, the old seat can be lifted off and discarded. To install the new seat, insert its bolts into the holes and tighten the nuts, again being careful not to overtighten them. Since the washers used here are rubber, too much pressure may crush them, shortening their lift span considerably. This will cause a toilet seat to appear loose.

If the toilet seat is loose and its condition does not warrant a complete replacement, replace both the washers and the rubber bumpers. These bumpers can be either tack bumpers or bar bumpers, both of which are pictured at the right of Fig. 10-17. The only difference found in the tack bumpers is that some are nails and some simply screw in. Try to find exact replacements if at all possible.

CLEARING TOILET BLOCKAGE

Most clogs in toilets are the result of materials getting caught in the toilet trap. By itself, a clog such as this is not too difficult to clear. However, if this condition is combined with another defect such as a running toilet, the problem may be a bit more serious. Minor clogs are easily repaired by means of a plumber's friend or plunger. Before using the plunger, make sure there is sufficient water in the toilet to cover its rubber suction cup. If not, pour some water in the bowl up to a level just below the rim. Do not flush the toilet at this point, as this will undoubtedly cause the bowl to overflow.

Apply a thick coat of petroleum jelly all around the base of the suction cup. Place it over the bowl opening, keeping the handle as vertical as possible. Work the plunger up and down vigorously until the water begins to drain out, indicating the stoppage has been cleared (Fig. 10-19). This

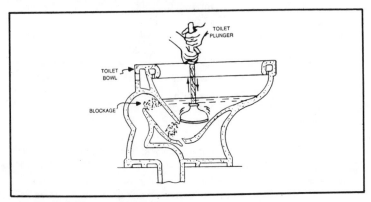

Fig. 10-19. Clearing a blocked toilet.

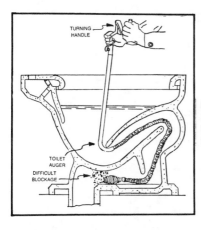

TURNING HANDLE

TOILET AUGER

DIFFICULT BLOCKAGE

Fig. 10-20. A toilet auger is preferred over a drain auger. To clear the blockage, insert the auger into the trap and turn the handle.

procedure will work under most conditions, providing the blockage is not a severe one.

If the blockage is a stubborn one and the plunger does not succeed in clearing the fixture, the next step is to try an auger, preferably a toilet auger rather than a drain auger. The drain type will usually work, but it is likely to scratch the surface of the bowl. Insert the auger into the trap and start turning the handle, pushing it further and further until the obstruction is felt. This procedure is shown in Fig. 10-20. Always remember when using an auger that the handle should be turned in only one direction. After reaching the blockage, pull the auger in and out, turning the handle continuously, until the obstruction is removed and the auger starts moving freely. At this stage, never flush the toilet to check if the blockage has been removed. Instead, bring a pan of water from a nearby sink and pour it into the bowl. If it flow freely into the trap, this is an indication that the trap has been cleared. If the water stays in the bowl, repeat the procedure once again until the water flows into the trap satisfactorily. The toilet may be flushed two or three times to wash away any materials not dislodged during the clearing procedure. Clean the auger thoroughly before storing it away. In cases where the blockage is quite extensive, a plumber's auger may work where the toilet auger failed. This is used in much the same manner as explained previously.

If none of these methods provide satisfactory clearing of the blockage, it will be necessary to remove the bowl from the floor, following the directions provided earlier in this chapter. Once the bowl has been removed, place it upside down on a padding of some sort, possibly old newspapers or rags. The obstruction can now be worked on directly through the discharge opening. It may be that the earlier methods simply pushed the blockage on further into the waste line. Clear the blockage by working with a snake through the soil pipe opening that is now exposed. If it is not possible to obtain an appropriate auger, a wire coat hanger may be used as a substitute. This is done by cutting the hanger at the two spots

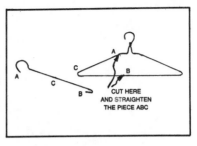

Fig. 10-21. If it becomes necessary to remove the bowl from the floor, a tool can be made from a wire coat hanger.

shown in Fig. 10-21 and straightening out the portion A-C-B. Bend one end to form a hook about 1½ inches to 2 inches wide, and bend the other end into a hook about ¾ inch wide. Insert the narrower book into the trap area and work it around to remove the blockage. Although this device will not be quite as effective as the toilet auger, it will do the job under most circumstances. If none of the methods described in this chapter clear the lines, it is time to seek professional help.

SUMMARY

Under normal circumstances, most people don't take on the task of installing a new toilet that often. These fixtures are usually quite durable and will last for many years if not abused. They are built to withstand some degree of abuse, since they are probably the most used fixture in a residence. The most common reason for replacing an old toilet is for remodeling purposes when it is desirable to change the whole appearance of the bathroom. Under these conditions, be sure to stick with the old measurements if at all possible and suitable to the new design configuration. This will make for a much simpler installation, since no carpentry work will need to be done to allow for the new fixture.

If a bathroom is being completely gutted, this is a good time to check over all the pipelines, both the water supply and drainage systems, for any wear and tear. If the house is an older one, and past experience has shown that there are some inadequacies present in the system, it might be a good idea to install newer, more modern piping. New fixtures may present a nicer appearance, but an improperly or inadequately functioning water distribution or drainage system will most certainly detract from the proper functioning of these fixtures. Remodeling and improving an existing plumbing system is covered in another chapter of this text for those who wish to correct inadequacies in their present system or to upgrade the original materials used.

The main points to remember in the installation of a toilet are to always disconnect the water supply and take the whole process slowly, checking all work as you go. This will save both time and money, since a careless job will always result in having to go back over everything to make the necessary corrections. Do it right the first time and save yourself a good deal of frustration.

Repair and Installation of Faucets

11

A *faucet* can quite simply be defined as a fitting which controls the flow of water at the end of a water distribution line. The actual device is available in dozens of styles and configurations. Generally, most are quite durable and are built to last a long time. The repair of a faucet is quite simple, and many homeowners today are able to perform this type of repair without the need of a professional plumber. The only difficulty that may be encountered is in locating specific replacement parts for a faucet mechanism, particularly if the fixture itself is quite old. In this case, replace the whole faucet assembly rather than trying to substitute mismatched parts.

Although an amazing amount of styles, shapes, and designs can be had when purchasing faucets, they can really be divided into two types in regard to actual operation: *compression* and *non-compression*. The two are easily distinguished from each other. Figure 11-1 shows some typical non-compression faucets. As can be seen, these have a single knob or lever which is connected to both the hot and cold water supply lines. The knob or lever regulates the proportion of hot and cold water, as well as its volume. A compression type faucet has separate hot and cold handles with washers (Fig. 11-2). As either handle is turned toward its off position, the washers compress a seat, which closes off the water supply. On the other hand, when turned in the opposite direction, water is allowed to flow from the supply pipe to the faucet. A compression faucet may have either one or two faucets, although those with two faucets are considered the less popular choice. These are usually found in older homes. More and more installations are being done today with the non-compression faucet.

COMPRESSION FAUCETS

The compression faucet is sometimes considered to be the more complicated device. Its appearance tends to make you think it is more

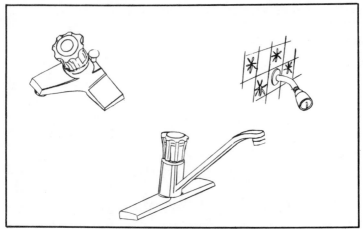

Fig. 11-1. Non-compression faucets have one knob or level which uses washers to control the flow of water.

complex than it really is. Figure 11-3 is an exploded diagram of a typical compression faucet. Most of the problems encountered with this type of faucet can be handled with little or no difficulty as long as everything is done carefully and correctly. Likewise, only some basic tools will be needed to effect the repairs. A screwdriver, a pair of long-handled pliers, and an adjustable wrench will suffice for almost all faucet repair jobs. If you already have a set of open-end wrenches, however, these will come in handy, although it is not necessary to go out and purchase a set solely for use in repairing a single broken faucet. Likewise, a *basin wrench* may be useful when working in cramped quarters, such as under a sink or lavatory basin. The next portion of this chapter will deal with the more common causes of faucet problems and how to deal with them.

Clogged or Dirty Aerator

An *aerator* is a small strainer located inside the end of the faucet assembly where the water exits into the sink or other fixture. A clogged or dirty aerator will not only reduce the quantity of water flowing out of a faucet, but may also make the flow turbulent and quite uneven. As time goes on, fine particles suspended in the water will deposit inside the small openings of the aerator and the filter, causing the assembly to become clogged. Before attempting to make repairs, be sure to shut off the faucet handle(s) completely.

Figure 11-4 shows the aerator assembly. Try to remove the aerator by turning it in a clockwise direction by hand. If this is not successful, cover the outer surface of the aerator with a few layers of adhesive tape to protect its finish, and then try to remove it using a pair of pliers. This will almost always work. Check the condition of the washer; if it is worn out, get a new one. Make sure to get an exact replacement. This may necessitate taking the whole assembly to the plumbing supply store in some cases. Remove the aerator from inside the filter and clean the fine screen by holding it about 9 inches below the faucet opening. Let the water run over it with full force for a couple of minutes. If necessary, clean it further with a toothpick. Remove the filter from the other end of the assembly and clean it with a toothpick or toothbrush. Take care to remove all material lodged in both the aerator and the filter.

The assembly is reinstalled in reverse, turning in a counterclockwise direction this time. Test the faucet. If all is well, remove the adhesive tape. The job is completed.

Spout Drips

A faucet which is either dripping or leaking sometimes goes ignored for quite some time. This is very unwise, since not only is a certain amount of water being wasted, but this continuous drip can ultimately cause stains and ruin the appearance of a sink or bathtub. A *drip* should be considered to be different than a *leak*. A drip occurs when the faucet handle is turned off and some water continues to flow from the spout. A leak exists when water is observed seeping from the stem or the base of the faucet when it is in operation.

A drip may be caused by a number of defective or worn-out parts. The most common cause is usually a defective washer, with the second most

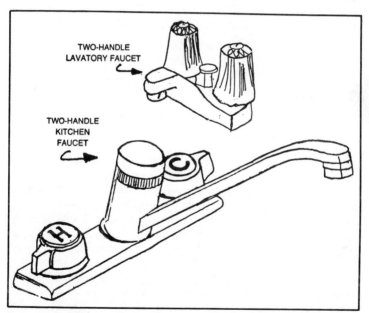

Fig. 11-2. Compression faucets have separate hot and cold water handles and may have one or two faucets.

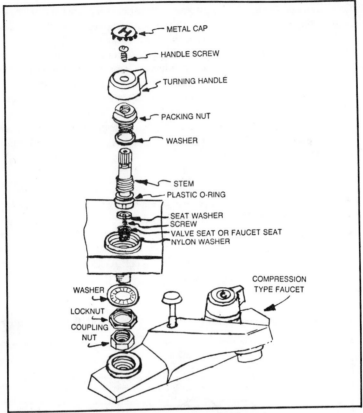

Fig. 11-3. Exploded diagram of a compression faucet.

common one being a worn-out valve seat. Since the water in the supply lines is always under pressure, all seals must be watertight. Otherwise, some of the water under pressure will simply pass right through them and drip out of the faucet. In order to determine the cause for the leakage, it will be necessary to dismantle the faucet assembly.

In a single faucet mechanism, the water supply should be shut off at the individual shutoff valve or at the branch or main supply line before you begin to disassemble the faucet. On a double faucet with separate hot and cold water controls and a common spout, determine which faucet is causing the leakage at the spout. The test is simple if there are separate shutoff valves for each faucet. Close one of the valves and see if the drip has stopped. If it has, then that particular faucet is leaking. Try the other one, since both may be leaking. In a situation where no individual shutoff valves exist, there is no way to identify and isolate the problem. Under these circumstances, there is no alternative except to repair both the cold and hot water faucets.

Shut off the water supply at whatever point is appropriate. Referring to Fig. 11-5, remove the faucet handle. The top screws are often hidden under the round plastic or metal caps marked H or C (hot or cold). If the caps are the screw-on type, unscrew them. Otherwise, simply pull them off. The handle screw can now be removed with a screwdriver, and the turning handle can be pulled off. Remove the packing nut by turning it in a counterclockwise direction. This portion of the assembly is shown in Fig. 11-3. Also, remove the stem along with the washer. Take out the seat washer by opening the bottom screw with a screwdriver.

Examine the washer and the screw. If they appear to be worn out in any way, obtain and install exact replacements. Also, examine the valve seat. If the washers were defective for any length of time, this will be damaged. When a worn-out washer is not changed immediately after the drip is noted, the spindle rubs against the faucet seat every time it is turned. Another cause for a worn-out seat is corrosion of the metallic faucet seat by the presence of chemicals in the water, or the physical accumulation of such chemicals around the seat, preventing the washer from sitting tightly over it in the closed position.

To repair an uneven and defective faucet seat, a special tool called a reseating tool will be needed (Fig. 11-6). This tool is not very expensive. To make this type of repair, the packing nut of the faucet is slipped over the tool's stem and screwed back over the top of the faucet. Tighten the nut until the cutting edge of the tool touches the valve seat. By turning the handle of the tool back and forth several times, the faucet seat is ground. Be careful not to overdo this procedure, as it is possible to completely destroy the valve seat. Any grindings should be flushed away.

If the faucet seat is irreparable, having been destroyed by too much grinding, or if a reseating tool is not available, the whole seat should be removed and replaced. A special tool called a seat removing tool is used.

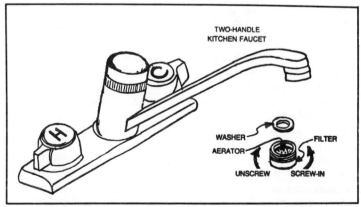

Fig. 11-4. If the aerator cannot be removed by hand, cover it with adhesive tape and try removing it with a pair of pliers.

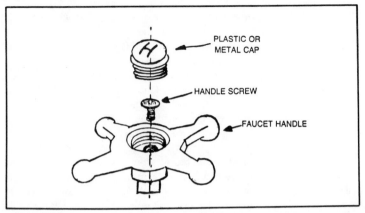

PLASTIC OR METAL CAP

HANDLE SCREW

FAUCET HANDLE

Fig. 11-5. To repair a drip at the faucet, it will be necessary to remove the faucet handle.

Be sure to obtain an exact replacement for this seat. A mismatch will not work. After the necessary replacements and/or repairs, reassemble the faucet in reverse manner, taking great care to follow the proper order. Any errors here will cause the faucet to malfunction. The water supply can now be turned on and a test run. If all work has been done carefully and correctly, the drip should no longer be present.

Stem and Base Leaks

Most of these leaks are caused by worn-out washers at some point in the faucet assembly. If a leak is observed around the faucet handle, it is called a *faucet leak*. A worn-out washer at the bottom of the stem or damaged packing in a packing nut on top of the stem will result in what is called *stem leak*. A *base leak* is caused by a worn-out washer at the base of the packing nut. Referring to Fig. 11-7, there are two types of packing nuts; one is equipped with a washer (C) at its base, while the other type is stuffed with packing material (B).

To correct any of these leaks, first shut off the water supply at the appropriate location. Use a few layers of adhesive tape to cover those surfaces of the faucets which will come in contact with the wrenches. Open up the faucet assembly as explained previously. Remove the turning handle and take out the packing nut by turning it in a counterclockwise direction. If this causes the stem to come out, separate the two with a wrench.

If the packing nut is the type which is stuffed with packing material, remove the packing. If it is the other type, take out the washer or *O-ring*, as shown in Fig. 11-7, from the bottom of the stem. Get exact replacements and install them in place of the worn-out parts. A small quantity of heat-resistant grease may be applied to the washer also. Assemble the stem and the packing nut. The stem should be inserted inside the packing

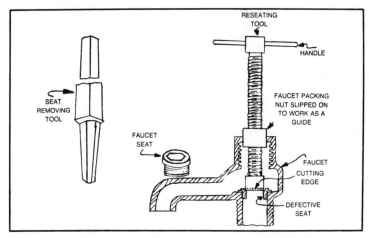

Fig. 11-6. A reseating tool is used to repair an uneven or defective faucet seat.

nut if it is not already inside the faucet. Tighten the packing nut with a wrench. Mount the turning handle and install the handle screw. Snap on the caps or screw them on, whichever is applicable.

It is always wise to check all the washers in the assembly. If any are found to be worn out, replace them. Turn the faucet to the off position and reconnect the water supply. Open the faucet and check for any leaks. If none are present, the procedure is completed.

Aligning Faucet Handles

If it becomes apparent that the turning handles of a faucet are not in proper alignment, it is a good idea to straighten them out, if only for the sake of appearance. This is a simple procedure.

First, bring the handles to their off positions to shut down the water. Remove the caps and the screws from the tops of the faucets. Take out the handles and put them on so that they are symmetrical. Install the screws. Snap on or screw on the caps, whichever is applicable. Test for proper alignment by turning the two handles so that they are in positions which permit freedom of movement for maximum flow through the spout.

NON-COMPRESSION FAUCETS

The modern non-compression faucets do not have washers. Instead , a rubber diaphragm or metal-to-metal contact serves to control the flow of hot and cold water. This type of faucet will have only one spout or faucet (Fig. 11-8). Repairs are a bit simpler. Depending on the control mechanism, the single-handled faucets can be classed into three categories: *ball faucets, valve faucets,* and *cartridge faucets.* It will not necessarily be possible to identify a faucet simply by looking at it. In order

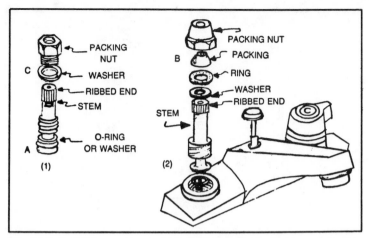

Fig. 11-7. Repairing a stem or base leak, usually caused by a defective washer.

to determine what type of non-compression faucet is on a particular fixture, it will be necessary to open the mechanism. Although each type comes in many different styles and design configurations, they all operate in basically the same manner. If at all possible, refer to the manufacturer's instructions during the actual repair procedure. If these instructions are unavailable, the guidelines given in this text should be sufficient.

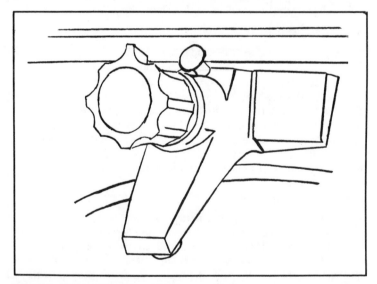

Fig. 11-8. Non-compression faucets are a bit easier to repair and tend to break down less often.

Ball Faucets

Figure 11-9 shows a dismantled ball faucet. In actual operation, as the handle is turned, this causes a ball within the faucet to rotate. The ball controls the flow of hot and cold water. If the faucet develops a drip or a leak, or is otherwise defective, this type of assembly is easily dismantled.

First, shut off the water at the individual shutoff valve or at whatever point is appropriate in either the branch or main supply line. Use some adhesive tape to provide protection for the faucet's finish. Snap off the cap from the top, or unscrew it, whichever is applicable. This will bring the handle button into view. Unscrew it and remove the handle. In some designs, there are no caps. In this case, the handle has a screw around its base which should be opened. The handle can then be removed. Remove the lower cap by rotating it counterclockwise. The cam assembly and the ball will now be exposed and should be lifted out, along with the seal assemblies (the seats and springs). In some models, there is a swing spout, which should be pulled out after moving it back and forth (sideways). Normally there will be an O-ring at its base, which is also removed. Examine all the parts for either wear or corrosion from the chemicals in the water. Anything that seems to be worn to any degree should be replaced.

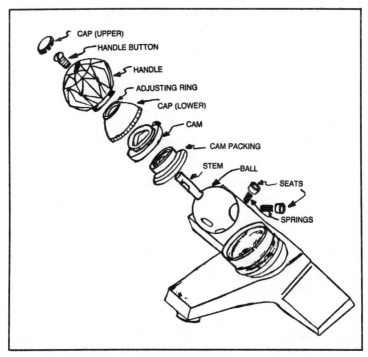

Fig. 11-9. A ball faucet is controlled by means of a ball assembly, which is easily dismantled if repairs are necessary.

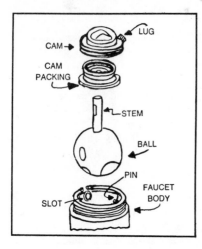

Fig. 11-10. A ball faucet is assembled in reverse manner.

Using exact replacements, begin reassembling. If the faucet has a swing spout, start by installing the O-ring. Apply a small amount of grease to this part and then push down the swing spout, rotating it until it sits in its correct position over the slip ring. Install the two seat assemblies inside the faucet body in their proper location. The slot on the ball should go over the pin in the faucet body, and the lug on the side of the cam should go into the slot body (Fig. 11-10).

Install the ball, the cam packing, and the cam. Place the lower cap on top and tighten it by turning in a clockwise direction, first by hand and then with a pair of pliers. The handle is now placed over the stem, and the water is turned on. Inspect the area around the stem for any leaks. Do not place the upper cap in position yet. If no leaks are present, remove the handle and position the cap on it. Tighten it and remove the protective adhesive tape applied earlier. Finally, place the handle or knob over the lower caps again. Tighten the handle screw or button. If everything has been done properly, the leak should no longer be present.

Valve Faucets

As was mentioned previously, there are no washers in a noncompression faucet. Figure 11-11 shows a typical valve faucet. In this type of arrangement, the hot and cold water flow is controlled both by water pressures and the rigidity of a spring contained in the valve assembly. Referring to Fig. 11-12, the mesh screens, or strainers, protect the valve assembly from being damaged or clogged by the fine suspended matter which is often present in water. These screens should be checked for any possible deposits periodically. To clean the screens, they can be flushed under running water with the help of a toothpick or a fine toothbrush.

If a leak or a drip develops in a valve faucet, turn off the water supply at whatever shutoff valve is appropriate. Open the faucet to allow all the water present to drain out. Referring to Fig. 11-11 again, use adhesive tape to cover the connecting ring as a protective covering. This ring is loosened with an adjustable wrench by turning in a counterclockwise direction. The swing spout can now be lifted. Take out the O-ring and remove the faucet body cover. You should now see plugs on either side. Using the adjustable wrench again, unscrew the plugs, one at a time, and take out the gasket, the strainer, and the valve assembly, Take out the valve seat with the help of a valve seat removal tool or an allen wrench. Examine the strainer and clean it if necessary. If the strainer has sustained any damage whatsoever, it should be replaced with an identical one. Also examine the gasket, valve assembly, valve seat, and the O-ring for either damage or corrosion. Get replacements if necessary.

To reassemble the faucet, install the valve seat using a valve seat removal tool. The valve assembly and the strainer are installed next. Place the gasket in its proper position and screw in the plug. The faucet body cover is replaced after all else has been properly installed. The final step is to reinstall the swing spout. First, apply a small amount of heat-resistant grease to the O-ring and slip it on the spout. Drop the spout carefully into the opening at the faucet body. Using an adjustable wrench, tighten the connecting ring by turning it in a clockwise direction. Remove the adhe-

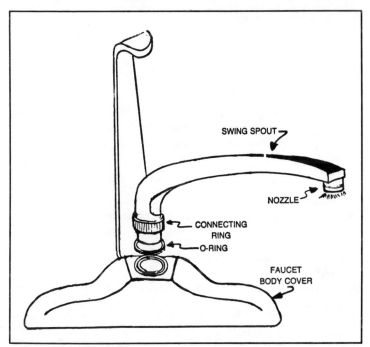

Fig. 11-11. In a valve faucet, flow is controlled by water pressure and a spring in the valve assembly.

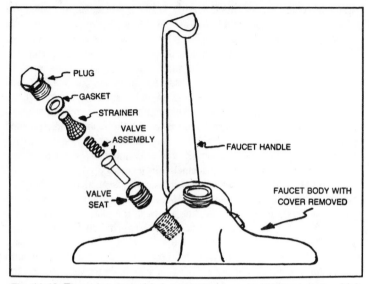

Fig. 11-12. The valve assembly is protected by a screen or strainer which may become clogged. To clean the screen, it should be removed and flushed under running water.

sive tape applied earlier as a protective covering, turn the water supply on, and check for any leaks or drips.

Cartridge Faucets

This type of faucet consists of a special cartridge assembly equipped with either ports or holes which control the flow of hot and cold water. Cartridge faucets can be had in a number of designs and configurations, but all have one common feature. The cartridge is held in position by means of a retaining clip, which can be either external or internal. Due to the fact that the actual cartridge assembly is a single unit, this type of faucet is a bit easier to repair. The assembly can be removed and replaced quite easily without having to dismantle the unit. Some cartridges can be repaired by a local plumbing dealer, but these repairs are not really worthwhile, since the replacement unit is usually quite inexpensive. Also, the life of a new unit is generally longer than that of an old one which has been repaired. The retaining clip, whether it is external or internal, provides access to the cartridge assembly. Once the clip is removed, the cartridge assembly is simply lifted out. Again, make sure an exact replacement is purchased.

To repair a defective, leaking, or dripping cartridge faucet with an external clip, the only tool that will be required is a screwdriver. Referring to Fig. 11-13, shut off the water supply at the appropriate shutoff valve and open the faucet to drain out any remaining water. Remove the external retaining clip and lift out the cartridge assembly. Purchase an exact replacement, possibly bringing the old assembly to the plumbing dealer as a reference. Install the new cartridge in its proper position, snap on the retaining clip, open the shutoff valve, and turn the faucet on. If the correct cartridge has been installed in the proper manner, the faucet should no longer drip or leak.

To effect repairs to a cartridge faucet with an internal retaining clip, again disconnect the water supply by shutting off the appropriate shutoff valve and draining the faucet of any remaining water (Fig. 11-14). If the cap is present on the top of the faucet, this is removed by either unscrewing it or snapping it off, depending upon the faucet design. This will bring into view a gasket, which should be removed. The handle screw is also removed using a screwdriver, which will allow the handle to be lifted up. The retaining clip is under the clip retainer ring. Remove the ring to get to the clip. As soon as the clip is removed, the cartridge assembly may be removed and an exact replacement obtained. The new assembly is then installed properly and the retaining clip replaced. The clip retainer ring is

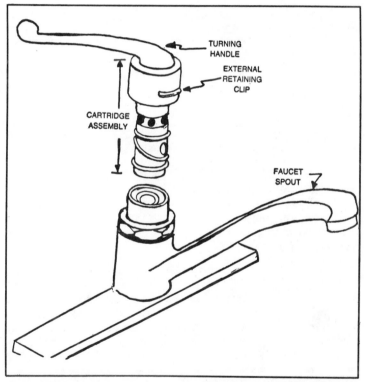

Fig. 11-13. A cartridge faucet has a cartridge assembly which is a single replacement unit. The assembly is held in place by means of an external retaining clip.

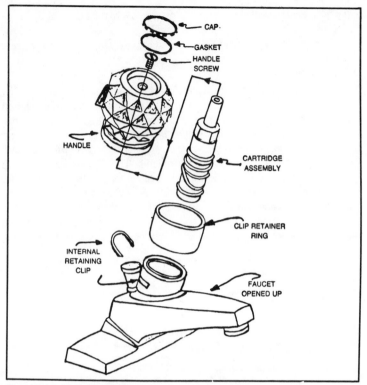

Fig. 11-14. If the retaining clip is located inside the faucet assembly, repairs will necessitate opening the assembly.

now slipped back into position and the handle installed by tightening the screw. Replace the gasket and cap, turn on the water supply and faucet, and check for drips or leaks. All should be functioning normally if proper procedures were followed.

KITCHEN FAUCETS

Plumbing fixtures in general are built to last a long time. The same holds true for faucets and their internal mechanisms. However, when a faucet assembly does begin to develop defects under normal operation, especially an older home, it may be a perfect opportunity to do some modernizing rather than attempt to repair the old fixture. By simply replacing the old faucets with new ones, the whole appearance of the fixture is changed. Newer faucet assemblies being manufactured today are not only durable but also quite sleek and elegant looking as compared to the older, more standardized models.

Before making any purchases, be sure to obtain exact measurements and dimensions of the fixture in order to aid the plumbing supply dealer in providing the proper assemblies that will fit your old sink. This may involve some special ordering from the original manufacturer of the fixture, depending upon the age of the unit. It may even be wise to bring the old faucet assembly along when planning to make the necessary purchases, if possible. If not, at least measure out all pertinent dimensions, such as the sizes of the water lines underneath the faucets and the center-to-center distance between the shanks of a two-handle faucet. If the fixture has a single-sink faucet, be sure to note whether it has male or female threads.

Replacing Single-Handle Kitchen Faucets

Figure 11-15 shows a typical single-handle kitchen faucet with one handle controlling both the hot and cold water and a separate spray assembly. The spray attachment is generally connected to the spout by means of a flexible hose. This type of faucet doesn't have washers. Instead, it will have either a ball, valve assembly, or a cartridge to provide control of the hot and cold water. These generally have a longer life than the older washer faucets.

Faucet assemblies usually come with instructions for installation. The whole procedure is not very complicated. If you are planning to replace an old double-handle faucet with a new single-handle faucet, be sure that the replacement unit will adequately cover the old holes in the sink. With a flashlight, examine the underside of the sink and measure the center-to-center horizontal distance between the studs. Also, be sure to

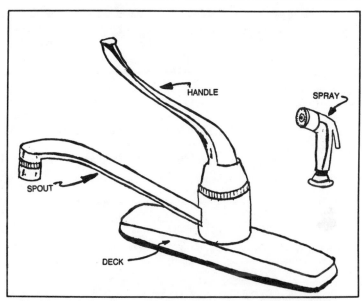

Fig. 11-15. Some kitchen faucets have one handle with a separate spray assembly, which is attached by means of a flexible hose.

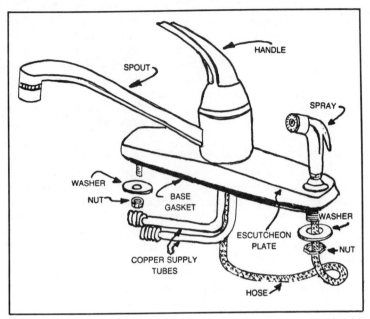

Fig. 11-16. If the sink being replaced had three holes, the new unit should have the spray assembly attached to the deck.

measure how far removed the water supply line is from the studs. Both of these measurements will be critical when choosing a replacement assembly. If planning to install a spray assembly as a new addition to the sink, these may come included with the faucet assembly or possibly as a separate unit. If the old sink had three holes, it will be necessary to purchase a new faucet with the spray attached to the deck itself (Fig. 11-16). Alternately, if there are four holes, a separate spray unit will be needed to cover up the fourth hole (Fig. 11-17). The addition of a spray assembly will increase the overall cost of the replacement, so this should be taken into consideration if expense is a factor.

The tools and supplies that will be needed for the removal of the old assembly and installation of the new one are: an adjustable wrench or channel-lock pliers, a basin wrench for use in cramped quarters under the sink, and some appropriate form of plumber's putty. Before beginning, remember to shut off the water supply at the appropriate shutoff valve and drain the faucet. Unscrew the nuts from under the faucet deck and remove them, along with the washers. If there is spray hose connection, detach it from the faucet. Remove the nut and washer from under the spray hose to release it. Once this assembly is removed, disconnect the supply tubes from the supply line and, if necessary, bend them to the position shown in Fig. 11-16. The faucet assembly can now be lifted out. It will probably have to be loosened a bit by either tapping it gently or rocking it back and forth.

278

Remove the gasket under the deck. The new assembly can now be installed.

First, place the gasket over the new deck. Remove the nut and washer from the spray hose assembly, and lower the hose through the sink hole on the extreme right. Take the hose up through the middle hole and connect it to the faucet. If the spray hose is separate from the faucet, it can be installed at this point over sink hole 4 (Fig. 11-17). The hose can now be lowered down through this hole, raised through hole 2, and connected to the faucet. The faucet assembly can then be lowered over the sink until it rests squarely on the gasket. If there is no provision for a gasket, apply a layer of plumber's putty over the sink before lowering the faucet deck. Be sure to take care not to kink the copper supply tubes when lowering the assembly and passing these tubes down through the sink hole. The faucet assembly is connected to the sink itself by positioning and tightening the appropriate mounting nuts and washers. Adjust the copper supply tubes as necessary to align them properly to meet the supply lines. Generally, the water supply lines in a residence run in such a way that when facing the mounted faucet, the hot water line is on the left. The bending of the copper tubes will be necessary, since single-handle faucets are manufactured in such a way that the right-hand side copper tube is for hot water and the left-hand side is for cold water. The tubes have to be connected properly within the faucet so hot water is received by turning the faucet toward the left, and cold water is received by turning it toward the right. The copper

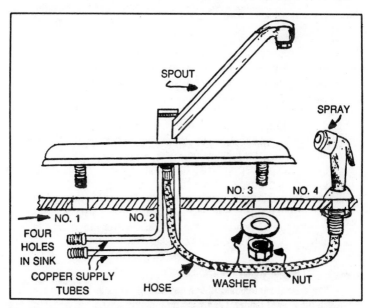

Fig. 11-17. If the sink being replaced had four holes, this will require a separate spray unit to cover up the fourth hole.

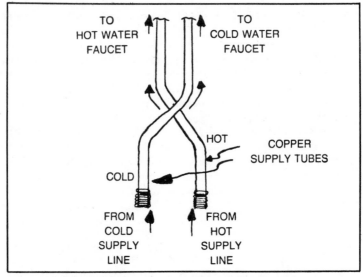

Fig. 11-18. Copper supply tubes are bent in a crisscross pattern to insure that hot water is received when the faucet is turned toward the left.

supply tubes will have to be bent in a crisscross pattern to provide water (Fig. 11-18).

A simple technique which has proven successful in bending tube without causing any damage is given here. First, grip the tube firmly at the bend with one hand, while slowly pulling toward the right side with the other hand, applying pressure at the same time with the thumb and the palm (Fig. 11-19).

Once the tubing is properly positioned, it can be connected to the hot and cold water supply lines. Apply pipe joint compound and use two wrenches during this procedure, one to hold the shank of the fitting on the copper supply tube, and the other to turn the nut on the water supply line in a counterclockwise direction. Be careful not to damage the tubes in any way while making this connection. If the supply lines are inadequate in regard to length or the connecting fittings are not exactly matched, it may be necessary to purchase some adapter tubes and fittings from the same store that supplied the replacement faucet assembly. Once the supply lines have been connected in whatever manner is appropriate, open the shutoff valve to supply water to the fixture. Go over the faucet and spray assembly to check for any drips or leaks.

Replacing Double-Handle Kitchen Faucets

The procedure for installing a new double-handle faucet will be done in much the same manner as with a single-handle assembly, and there may or may not be a spray attachment. As was explained previously, measure the distance between the studs, and the distance from the studs to the supply line, before making the new purchase. If there are three holes in the sink top, the only way to attach a spray hose will be to drill another hole. The method of installing the spray hose will remain the same. The free end of the hose is connected to the underside of the spout after running it through the middle hole.

As can be seen in Fig. 11-20, no copper tubes are present in a double-handle faucet installation. Instead, the connections between the supply lines and the hot and cold faucets are made by means of straight pieces of copper tubing. Be sure to place the black base gasket in position before installing the faucet deck. Follow the procedures described as with single-handle faucets, and you should have no trouble with this installation.

Another type of double-handle kitchen faucet is designed with the deck concealed. In this case, the deck is installed under the sink top, with only the two handles and the spray hose visible at the top of the fixture. To perform this type of installation, the handles are first removed from the deck. The next step will probably require the assistance of another person. The deck is held in place under the sink, while the handles are positioned on top of the sink and screwed down into the deck. The rest of the procedure will remain the same as explained previously under single-handle faucets.

It is not unusual in older homes to find that the faucets were attached originally in the backsplash against the wall. In this situation, the hot and cold water supply pipes will probably be coming directly through the wall. To replace this type of installation, the faucet is removed by means of slip nuts behind the backsplash. The new faucet is replaced in the same manner.

SINK STRAINERS

A sink strainer is almost always installed on a kitchen sink and sometimes on a bathroom sink. The obvious purpose is to prevent waste

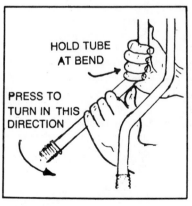

HOLD TUBE
AT BEND

PRESS TO
TURN IN THIS
DIRECTION

Fig. 11-19. A method which can be used to bend the tube without causing any damage.

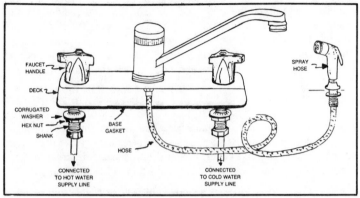

Fig. 11-20. In a double-handle kitchen faucet assembly, connections to the supply lines are made by using straight pieces of copper tubing.

material from entering the drainage line and causing blockages. Strainers are not installed in fixtures having garbage disposers. Since these devices perform a quite useful function, if they appear to be damaged in any way or are malfunctioning, they should be replaced immediately.

Figure 11-21 shows a typical strainer as it appears both assembled and disassembled (A and B). In order to replace one of the units, it will be necessary to work from underneath the fixture. This will simply involve disconnecting the strainer from the drainage line. First, open the slip nut with an appropriate size wrench and remove the small washers. Remove the larger slip and the set of washers. This will free the strainer. Lift it out and get a suitable replacement, along with new washers for the drain line connection. Some Teflon non-stick coating tape and putty will be needed during the installation of the new strainer.

To install the new device, first apply plumber's putty all around the strainer opening in the bottom of the sink. Unscrew the slip nut from the new strainer and take out the washers. Place the strainer in position over the putty layer and, working from under the sink, install the larger washers and the large slip nut. It may be beneficial to recruit another person to help hold the strainer inside the sink while tightening the slip nut. A screwdriver and hammer may be used, as long as the taps with the hammer are quite light so as not to cause any damage.

To connect the strainer into the drainage line, use the smaller slip nut and the new set of washers. Wrap the threads with Teflon non-stick coating tape and make the connection. Turn the faucet on and check for leaks. Close the strainer opening with a plug, fill up the sink almost to the top with water, and again check for any possible leaks. If the procedure has been done carefully, all should be functioning properly.

LAVATORY FAUCETS

A *lavatory faucet* operates in much the same manner as a kitchen faucet and can be either a double-handle or a single-handle type. Replacement and installation will follow the same basic procedures described earlier in this text. The only real difference between the two is that a lavatory faucet sometimes is equipped with some sort of pop-up drain plug which is operated by a lifter located on the faucet deck.

Figure 11-22 shows a typical double-handle lavatory faucet assembly. In this illustration, the water supply lines for the hot and cold water are equipped with a shutoff valve (7) in order to discontinue water when repairs are necessary. If one is not present on the fixture being replaced or repaired, it's a good idea to install such a valve.

To replace a lavatory faucet assembly, shut off the water supply and drain the faucet completely. The only tools that will probably be needed to perform this work are an adjustable wrench of channel-lock pliers, a basin wrench for working in cramped quarters under the fixture itself, and some plumber's putty. Referring to Fig. 11-23, the next step will be to remove the stopper assembly. Loosen the adjustment screw (1) at the bottom, disconnect the stopper lifter from the linkage, and pull it out. Going back to Fig. 11-22, loosen the nut (2) on the supply lines below the sink, using the

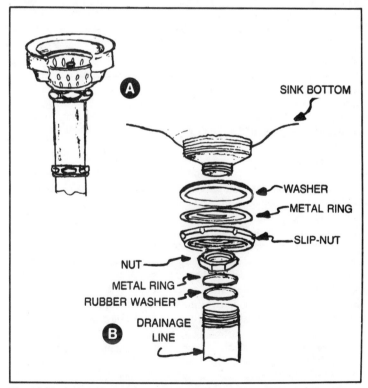

Fig. 11-21. A typical strainer unit assembled (A) and disassembled (B).

283

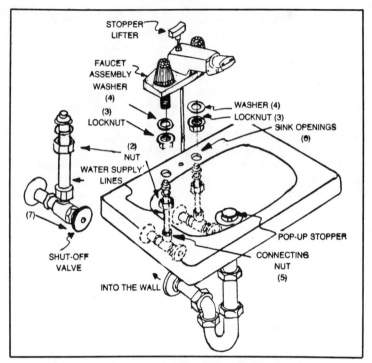

Fig. 11-22. A double-handle lavatory faucet assembly.

basin wrench. Disconnect the supply lines after slipping the nuts and washers on them. Open the locknuts and washers (3 and 4) under the faucet assembly. This will free the assembly, which should come loose with a few gentle taps.

In purchasing a replacement, be sure the distance between the inlet pipes is the same as that in the old assembly. Otherwise, the sink openings will have to be modified, which will involve a bit of work. Remove the washers (4) and locknuts (3) from the new assembly. Clean the area of the sink around the opening (6), and apply a layer of putty around the bottom edge of the faucet assembly. If the faucet assembly comes with a gasket, this should be installed first, with the assembly installed over the sink so that the shanks go down through the two holes (6). Using a wrench, install and tighten washers (4) and nuts (3). The previously disconnected supply lines should now be reconnected by means of the nut (2) and fully tightened to prevent any leakage. Install the stopper lifter by threading the lifter rod (2) through the linkage and tightening the screw (1) (Fig. 11-23). This should complete the installation. Turn the water supply on and check the work for any errors in the form of drips or leaks.

To install a new pop-drain unit such as the one shown in Fig. 11-23, the following steps should be followed:

284

■ Remove the tailpiece (9) by unscrewing it.

■ Unscrew nut (4). Mount the gasket and seal on the ball rod (5). Tighten the nut after making sure that the ball slides into the correct position to make the proper connection.

■ Remove the flange (8) by unscrewing it. Apply plumber's putty around the sink drain hole, and place the flange over it. Holding the pop-up drain body (7) under the sink, align it below the sink hole and tighten the flange over it. The ball rod should again be checked to make sure its position is correct. Install the locknut and washer, which perform the function of holding the pop-up drain body firmly to the underside of the sink.

■ Align the tailpiece (9) on the trap, and screw the tail into the body. Tighten the locknut on the trap, using a new washer.

■ Lower the lifter rod (2) through the hole in the spout. Attach it to the strip rod (10).

■ Attach the operating lever (3) to the strip rod (10) through a strip clip, which should be adjusted so that it opens and closes the pop-up drain correctly.

■ Go over everything carefully. Insert the stopper in the sink hole and test the function of the assembly to see that it works properly. Some

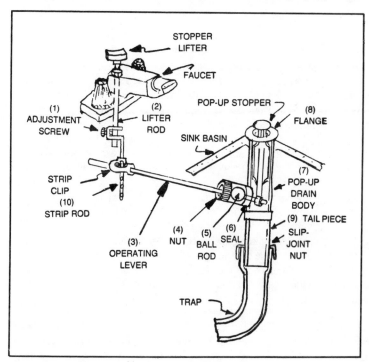

Fig. 11-23. Removal of the stopper assembly.

285

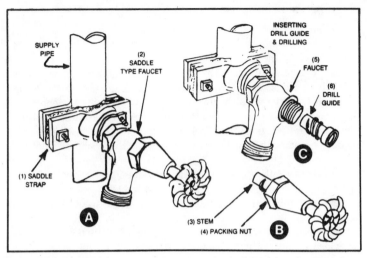

Fig. 11-24. Single-handle faucets are usually installed in a basement or other area where it is desirable to provide a water supply where none previously existed.

adjustments may be necessary at the junction of the strip rod and the operating lever.

Figure 11-24 shows a typical single faucet assembly. These come in a number of different designs and variations, but all are installed in much the same manner. These single-handle faucets are the type usually installed at some location in the house, such as the basement, where no water supply previously existed. These are very simple to install, but they will require the use of an electric drill with a ¼-inch metal bit. The faucet normally used for this installation is a saddle faucet. The new faucet should be of a size that can be suitably installed on the pipe. First, turn off the water and clean the surface of the pipe where the faucet is to be installed, possibly with some light sandpaper. Remove the nuts and bolts from the saddle strap of the new faucet. Install the faucet at the appropriate spot on the pipe by tightening the strap around the line with the nuts and bolts. Unscrew the packing nut and remove it, along with the faucet handle. Insert the drill guide into the faucet opening and, with the ½-inch metal drill, make a hole in the pipe. Remove the drill and guide, and clean the faucet to remove all metal drillings. Replace the stem and packing nut into the faucet hole and tighten them up. The installation is now complete. Turn the water back on and test the faucet.

Figure 11-25 shows another type of faucet connection which uses a saddle-type tee connector. Actually, this method is suitable for installing types of threaded pipe, not just faucets. To perform this procedure, the saddle tee is first installed at the pipe by placing it around the pipe and tightening with nuts and bolts. A drill guide is then inserted into the tee, and a hole is drilled in the pipe. The whole assembly and pipe are cleaned to remove any drillings, and the faucet assembly is screwed in place on the tee. Turn the water on and test the faucet. If all has been done properly, the faucet should perform satisfactorily.

Replacing a Hose Connection Faucet

To perform this procedure, first shut off the water supply and open the old faucet to drain any remaining water in its body. Referring to Fig. 11-26, clamp the pipe firmly with a pipe wrench while turning the faucet assembly counterclockwise with an adjustable wrench. This should loosen the faucet from the line and allow for removal of the old assembly. When purchasing a new faucet assembly, make sure that its dimensions match those of the pipeline. Install the replacement on the line and tighten it first by hand. Holding the pipeline firmly with a pipe wrench, turn the faucet assembly clockwise with an adjustable wrench until it is tightly in place. Teflon non-stick coating tape may be used in this installation by wrapping it over the threads of the pipe before the new faucet is connected, if desirable. Reopen the water supply and test the faucet for any leaks or drips.

INSTALLING A NEW OUTDOOR FAUCET

The supply line for an outdoor faucet or one to be located in an outbuilding is best installed below the frost line if winters are severe. At its point of origin in the basement, a stop-and-waste valve should be installed. This valve is equipped with a drain hole through which water from the line can be drained out. When the line is in use, this valve is normally capped. To drain the line, open the turning handle at the top of the valve and remove the cap from the drain hole. This is a very useful device, as it will keep the outdoor faucet from freezing in the winter. Be sure to

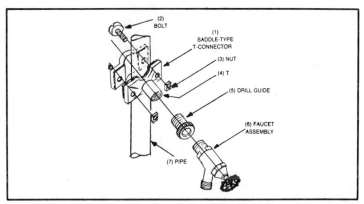

Fig. 11-25. Another method of installing a faucet is with a saddle tee connector.

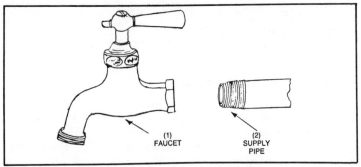

Fig. 11-26. Replacing a hose connection faucet is very simple using a pipe wrench and an adjustable wrench.

place a bucket under the hole during this procedure. Once the line is drained, the cap should be replaced.

The actual installation of the line and the faucet assembly is basically the same as discussed earlier. Since the line will only be supplying cold water as a rule, plastic, copper, or galvanized pipe can be used, at the option of the builder. Figure 11-27 shows an arrangement which will make for a freeze-proof installation on the outside wall of an outbuilding which is not heated. In this case, year-round service can be maintained by wrapping the vertical pipe with electric tape up to a point just below the frost line. In addition, the exposed length is covered with insulating tape. A thermostat installed above the ground line will control the heat and keep it at the necessary level to prevent freezing. This arrangement also includes a stop-and-waste valve in the basement of the main building where the line originates.

INSTALLING A NEW SHUTOFF VALVE

It is a real advantage to have a shutoff valve located at convenient points near each fixture. In this way, when a repair is necessary at an isolated point in the home, it will not require shutting down the whole system from the main shutoff valve in the basement. This is neither complicated nor expensive.

The actual installation will vary somewhat depending upon the arrangement of the fixture in relation to the walls and/or floors and the location of the supply lines. The basic procedure is simply to remove the section of pipe leading from the fixture into the wall or floor, install the valve on the stub-out at the pipeline, and joint the fixture to the valve with some flexible copper tubing. If the supply pipe is coming up from the floor, use a straight shutoff valve; if it projects from the wall, a right angled shutoff will be required. Some typical shutoff valves are shown in Fig. 11-28. If the pipe supplying water is threaded, a threaded shutoff valve will be used. Likewise, if the stub-out is copper, it will be necessary to purchase an adapter which will be sweat-soldered at one end and screwed on to the shutoff valve at the other end. Once the valve is properly connected to the supply line, the next step is to complete the connection to the fixture.

There are kits on the market which make this installation a bit easier. Called *Speedee kits*, they come in different designs, each serving a different purpose. Some even come equipped with the shutoff valve already jointed to the Speedee, or the flexible copper tubing used for connection to the fixture. The kit will come complete with all instructions and necessary small parts required for the connection.

The valves normally used for household puposes are generally of the globe type, which can provide either partial or complete shutoff of the water supply. Remember to shut off the water supply before installing this new valve.

REPLACING SUPPLY TUBING

When replacing an old fixture, you may find that the supply tubing is damaged. These tubes are generally made of flexible copper. To replace these tubes, shut off the water at the appropriate place. Unscrew the nuts at the upper and lower ends of one of the tubes, place an empty bucket underneath to collect any water remaining inside, and remove the tube. Repeat the process for the second tube, if necessary. Cut replacement sections of the required size and length. The connections will be of the compression type using a nosepiece compression ring and a flange nut at each end. Slip a compression ring over one end of the tube. Hold it vertically in position so that the upper end with the ring fits into the fixture connection. Place a flange nut around the tube at the lower end and slide it up to the top. Tighten the nut by hand. Slip another nut and sleeve over the lower end of the tube. Align the tube, bending it carefully if necessary to

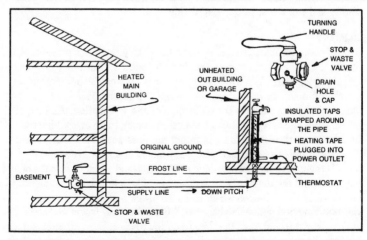

Fig. 11-27. An arrangement whereby an outdoor faucet will be able to provide year-round service with no danger of freezing.

289

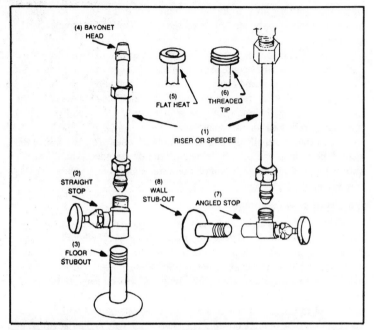

Fig. 11-28. Shutoff valves serve a very useful and important purpose in any home. Ideally, one should be provided at every fixture.

align it with the lower connections. Tighten the nut again by hand. Check over all the connections carefully. If all seem proper, gently tighten them with a wrench. Repeat this process with the other tube, open the water supply, and run a test of the installation.

SUMMARY

As costs continue to rise, so does the hourly rate for a professional plumber. This, in itself, may be the deciding factor in learning to do your own plumbing work. You will save somewhere between 30 percent and 50 percent by doing your own repairs and modifications alone, even if you don't feel able to tackle a complete design and installation. As long as proper guidelines are maintained and care is taken to do everything right the first time, the whole thing is rather painless and quite satisfying.

Practically everyone has had a leaky faucet at one time or another. It is really the most common plumbing malfunction and probably also the one that is ignored the longest. A drip may seem like a minor inconvenience, but consider that a faucet which leaks 15 drops a minute wastes somewhere around 500 gallons a year. Don't ignore a leaky faucet and hope it will go away. Instead, take the time to do what is necessary to correct the problem. Save yourself some money, both in water and in damage to the fixture itself. The whole operation will probably take less than an hour.

Showers and Bathtubs  **12**

Bathtubs are available in practically any size, shape, or color imaginable. Showers now come as completely prefabricated units which can be installed in a matter of hours. Units can now be purchased which provide a stimulating whirlpool right in the comfort of your own home.

Many different materials have been used in the manufacture of bathtubs since their inception, ranging from cast iron to fiber glass and formed steel with porcelain enamel surfaces. The average size for a bathtub is anywhere from 30 to 48 inches wide and 4 to 6 feet long. The steel tubs are lighter than the cast iron, although the latter is a bit more durable. The most popular material used today in the manufacture of showers is fiber glass coated with a special gel which protects the surface from the continual flow of water. This increases the life of the unit significantly.

Personal preference has played a larger part in the manufacture of the newer models of tubs and showers. Large size bathtubs are becoming increasingly popular. If a square tub is appealing, one can be had equipped with a seat at either one or both ends. If replacing an existing bathtub or shower installation, the decision may be hampered somewhat by the need to match the dimensions of the old fixtures. In a new home the sky is the limit, as far as both design and cost are concerned. The fancier and more unusual your choice, the higher the price tag.

INSTALLING A BATHTUB

If the new fixture fits exactly in the space occupied by the old one, the whole installation will be much simpler. It will involve only the hookup of the plumbing connections and fixtures. This will be the exception rather than the rule, since the newer fixtures are usually larger and will probably need more floor space. It may be necessary to actually move some walls, either partially or completely, to allow for the new installation. Refer to the manufacturer's instructions provided with the new tub in order to get the exact roughing-in dimensions. The framing for the tub will have to be built before installing and connecting the tub to the existing water supply lines, since this will protect the finish of the fixture.

The whole process will consist of the following steps: disconnecting and removing the old tub, building the framing for the wall(s), cutting a hole in the floor, running new pipe, installing the tub, making the necessary drain and water connections, and finishing the walls and the facing. Make sure all the necessary tools are on hand before beginning.

Figure 12-1 illustrates the type of enclosure that will be required if you are replacing an old tub with high legs and no enclosure with a new shower/tub combination. This will probably involve building a new end wall and stripping the other end wall and back wall completely of any tiles or other type of covering. The new tub will be installed directly on top of the subfloor, so any floor covering here will have to be removed too. When this is done, clean the area thoroughly. The plumbing portion of the project can now be done.

Connections will be needed for a drain line, two water lines, and the shower pipe, if applicable. It may be necessary to change the location of the existing drain hole in the floor in order to correctly match the requirements of the new tub. Installation of a 2 × 4-inch floor plate at this point is a wise move (Fig. 12-2). Examine the existing hot and cold water lines and check for any signs of corrosion or other damage. Fit these lines with any new extensions, offsets, or adapters necessary in order to bring them up to the exact point where connections will be made to the new tub and/or shower unit inside the wall. Install a vertical pipe for the shower unit, if applicable. This pipe will need at least one cross brace in the framing and some U-clips in order to position it rigidly in its proper vertical position.

Examine the drainpipe to see if any modifications are necessary for connection to both the drain and the overflow opening of the new fixture. The necessary dimensions should be included with the tub and/or shower, but it is wise to double-check by measuring them yourself at the location of the installation. In order to install the faucet unit and its stub-out securely in position, it will be necessary to notch a crosspiece. With the assistance of one or two other persons, remove the tub from its packing enclosure and temporarily position it on its supports inside the enclosure. Be careful not to strike it against anything, as damage may occur. Using the manufacturer's instructions, secure the tub to its foundation in the proper manner.

The next step is to check the position of the tub. Using a plumber's level, check the level at the top of the rim in both directions. A very slight pitch toward the direction of the drain hole may be allowed. If the tub slopes in the opposite direction, this will have to be corrected, since the tub will obviously not drain properly if left in this position. This is corrected by means of ⅜-inch padding installed between the tub and the subfloor. Shims are prohibited for use with tubs and/or showers because they provide only point supports. A tub full of water is quite heavy. The bottom of the tub would crack under the load if only shims were used as the sole means of support. Install the padding and check the level again, repeating the process as necessary, until the proper level is obtained. Once the tub is positioned properly, the unit may be secured to the studs by means of flathead brass screws (Fig. 12-2). Holes for the screws are drilled through the lip of the tub, and the screws should be countersunk into these holes.

Provide an access panel in the new wall at the head of the tub, which can be removed to gain access to the supply pipes at a later date when repairs may be necessary. The walls can now be finished, making sure they are of a waterproof material. The usual method of providing this waterproofing is to install furring strips on the studs right up to the ceiling and cover them with nailing wallboard. A coat of water-resistant sealer is then applied to the face of the nailing flange. The wallboards should completely cover it. The wallboard can then be painted with whatever color is desired. If you are installing a prefab tub and enclosure unit, the prefab walls should be installed after the tub has been secured in the proper position. Manufacturers include special clips for attaching these

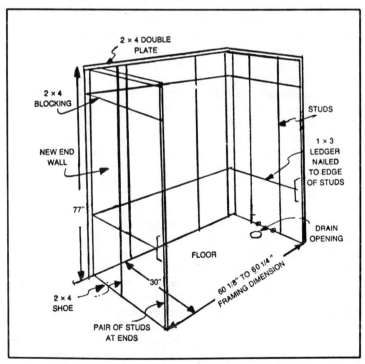

Fig. 12-1. This type of enclosure will have to be constructed if replacing an old claw-foot bathtub with a bathtub/shower combination.

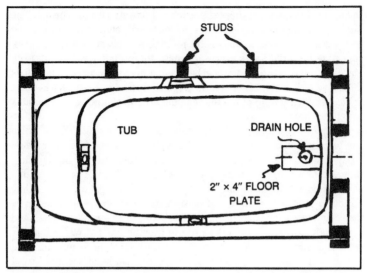

STUDS

TUB

DRAIN HOLE

2" × 4" FLOOR
PLATE

Fig. 12-2. Installation of a floor plate at the drain location will greatly simplify future repairs.

walls to the studs. A sealant designed for these purposes should be used to completely seal the joint between the plastic walls and the tub.

The next step is to install the supply and drainage fixtures themselves. Referring to Fig. 12-3 and working from the rear and underside of the tub, make the drain and overflow connections first. The supply system consists of two valves with operating handles, a divertor valve at the spout, one connecting pipe leading to the spout stub-out, and another running upward to the elbow to which the shower spout is connected. Figure 12-4 shows the arrangement of faucet handles, the valve assembly, and its connection to the cross brace attached to studs. Cover the walls above the plastic side walls with the appropriate materials, and complete the paint or tile job.

Figure 12-5 shows a feature found on most tubs which serves a very useful purpose. This is a combination stopper and overflow device. The stopper is usually operated by means of a lever or handle located between the tub's faucet handles. Depending upon which direction the lever is turned, this will open or close the drain.

The overflow portion of this device serves the purpose of preventing the tub from overflowing. Should the water reach the level of the overflow, it will simply drain through its opening into the drainage piping. The installation of this device is somewhat critical, since the waste pipe must be designed in such a manner that no water rises up to the overflow and into the tub when the stopper is in the closed position. Likewise, the piping should also prevent any water from being retained in the overflow pipe when the tub is emptying. This overflow passageway also provides an additional safeguard in keeping the trap filled with water, since it allows an amount of air in the lines.

If this device malfunctions and needs to be replaced, removal is quite simple. It is usually attached to the wall by means of two screws. To take out the system, simply remove the two screws with a screwdriver of appropriate type and size. The device should lift right out. Be sure to purchase a suitable replacement that will fit with the tub.

INSTALLING A SHOWER STALL

A prefabricated shower unit is the ideal solution in those instances where you want to install a new bathing facility within a limited space. This will work equally well in a bathroom which already has a bathtub with no

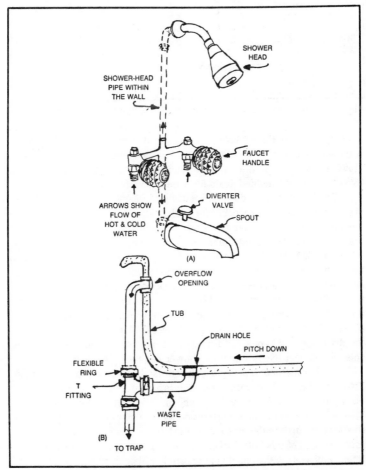

Fig. 12-3. Installation of the supply and drainage fixtures.

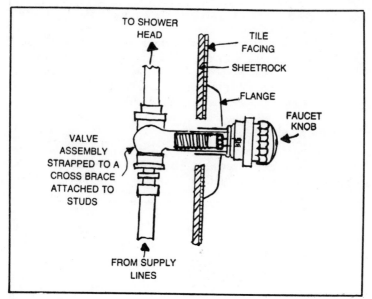

Fig. 12-4. The only visible part of the faucet assembly is the handle.

shower facilities. Figure 12-6 shows a typical prefabricated shower stall. This unit measures 30 × 36 × 75 and is considered to be adequate. However, if space permits, a larger unit may be installed, and some can even be had with molded seats for added convenience.

The installation is done in basically the same manner as with the previously described procedure for a tub. Depending upon the location of the stall in the room, one or two stud walls will have to be constructed for the enclosure. The overall dimensions will depend upon the size of the unit to be installed. The dimensions can either be measured out or taken from the manufacturer's roughing-in instructions which normally come with new units. To install the drainpipe, cut a clearance hole of suitable size in the subfloor. Referring to Fig. 12-7, stub-outs should be installed within the wall, along with the valve assembly, the shower head assembly, and the vertical supply pipe. Position the shower stall in the enclosure. Install the drainpipe and connected fittings from below through the clearance hole. The faucet handles and the shower head are then installed.

Another method of providing shower facilities is to construct one in place. A waterproof pan with a drain hole is placed at the desired location in the room, and walls are built around it. The shower head and valve assembly are mounted in one of the side walls. A fairly large clearance hole should be cut in the subfloor to accommodate the drainpipe. Determine the exact location of this hole, either from the manufacturer's drawings or by actually measuring the offsets of the pan. It will be necessary to cut some structural members if a drum trap is to be used. Since such a trap has to be

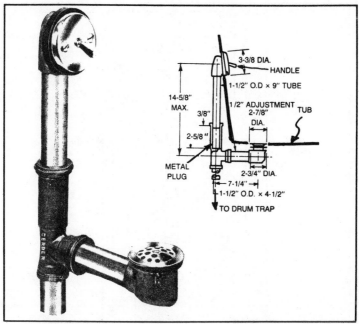

Fig. 12-5. A combination stopper and overflow assembly (courtesy Gerber Pbg. Fixture Corp).

cleaned at regular intervals, provision should be made for an easy access to it during repairs. The drainpipe is passed through the clearance hole and locked in place both above and below the pan bottom by applying plumber's putty or by using fitting washers. The lower portion of the drainpipe is

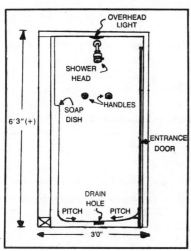

Fig. 12-6. A prefabricated shower stall is done in much the same manner as a bathtub, since it will need a water supply and a drainage connection.

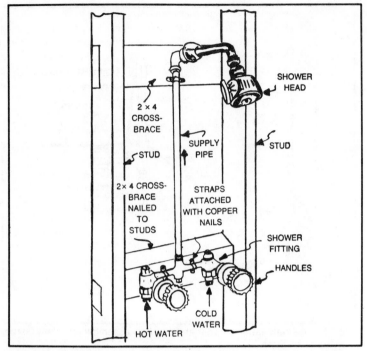

Fig. 12-7. The faucet and shower assembly are braced against studs.

connected to the drum trap which, in turn, is connected to the appropriate branch drain or soil stack. The procedure used to attach the shower head and the valve assembly to cross braces running between two adjacent braces is done in the same manner as with a prefab shower stall. Brass or copper straps are nailed to the cross braces to make this connection. The hot and cold water connections to the supply lines are also made in the same manner. Once the plumbing is completed, the walls are finished with whatever material is desired, either tile or Sheetrock, gypsum wallboard, as long as waterproofing is provided. It is wise to leave some clearance in the wall around the valve stems for easy access in the event repairs are necessary at some future time. After the walls have been finished, the valve flanges and the handles and shower head are installed. The installation is now complete. Turn the water supply on and test the arrangement.

REPAIRING BLOCKED BATHTUB DRAINS

The most common cause of blockage in a bathtub drain is an accumulation of hair and other debris, although it is sometimes necessary to replace some of the working parts of the drain assembly due to wear over a long period of time. There are a few variations in regard to the actual drain assembly, and the method used to repair each type is a bit different.

Figure 12-8 shows the mechanisms of a drain which is commonly called the *tiptoe drain plug* because the stopper can be conveniently operated with the touch of a toe. A downward movement is used to both open and close this type of drain. As can be seen, this type of drain contains no moving parts internally, thus making it the least likely to cause mechanical problems. Since there is nothing inside the drain for debris to get caught on, the chances for blockage are reduced.

If it becomes necessary to either clear the drain or possibly replace any of its parts, simply remove the screws as indicated in Fig. 12-8. Obtain the necessary exact replacement parts, and reinstall the device with the new parts. The whole process is very simple.

Older bathtub installations will probably have the type of drain assembly shown in Fig. 12-9, which is called a *trip waste drain.* This type of drain has a few moving parts located internally; thus, it is more likely to clog or break down than the drain already discussed. To operate the drain, the lever is pressed down, causing the internal adjustable rod to lift the cylinder, thus opening the drain. To retain water in the fixture, the lever is pulled up, which fixes the cylinder in the drain as a blockage to prevent the water from escaping. Aside from the usual difficulties associated with moving parts and accumulations of debris in the strainer itself, the trip waste's cylinder has a tendency to "hold" a certain amount of hair and other materials. This causes it to rest in an improper position on its seat so that

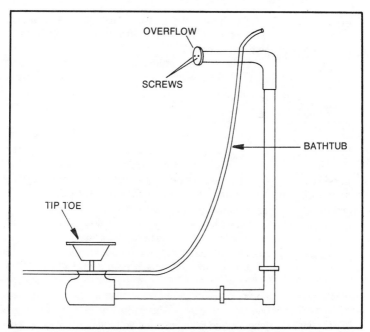

Fig. 12-8. A tiptoe drain mechanism contains no moving parts internally.

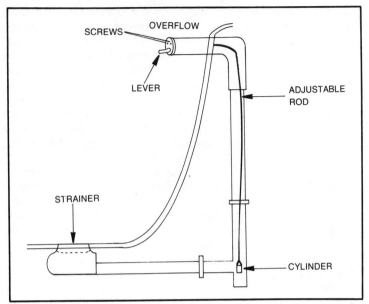

Fig. 12-9. A trip waste drain is more likely to clog or break down due to its internal moving parts.

even when the lever is pulled up, water will still drain from the tub. If this type of malfunction is occurring, the cause can usually be attributed to some debris clinging to the cylinder.

To clean the cylinder, simply remove the necessary screws in the overflow plate (Fig. 12-9). The cylinder can then be freed and lifted out for cleaning, which is done by holding it under a somewhat heavy stream of water. Before reinstalling the device, apply a coating of petroleum jelly to its surfaces and then restore it into the drain.

Another problem associated with this type of assembly is also attributed to its moving parts. The rod portion of the mechanism is equipped with threads and a quite small locknut which is used to adjust the motion of the cylinder. This adjustment must be very accurate to provide the proper downward or upward motion necessary for the cylinder to rest in its seat and retain the water completely. This rod is quite slim, usually about ⅛ inch in diameter, and quite fragile. If deterioration does occur, it usually results in the cylinder snapping loose and becoming jammed somewhere in the drain opening. If you are lucky enough to have an access panel, the repair will not be all that difficult. If the installation is an older one with no way to get to the blockage, it will be necessary to go in through the overflow plate, using a clothes hanger or some other length of strong wire. Make a hook on the end of the wire and insert it into the opening. Just feel around until it catches on the cylinder. This will require a bit of patience and luck.

The third type of drain is called the pop-up assembly (Fig. 12-10). In operation, it is quite similar to the trip waste drain in that it is equipped with a lever, an adjustable rod, and an overflow plate. Instead of a cylinder, however, the pop-up drain uses a spring which is attached to the stopper to control the flow of water. Although it will cause the same problems as the trip waste drain due to its moving parts, it is a bit more durable in some respects. The adjustment of the rod does not have to be quite as precise to provide adequate performance, so less wear and tear will be experienced in this portion of the mechanism. The spring assembly does have a tendency to collect hair and other debris, but it is cleaned quite easily. Simply remove the screws on the overflow plate, and the whole assembly can be lifted up and out for cleaning.

If a thorough cleaning of any of the types of drains discussed does not clear the drain, the problem is probably in the trap or waste line. In order to clear a blockage in either of these areas, it will be necessary to use a plunger. Fill the tub with enough water to cover the rubber suction cup portion of the plunger. Remember to tightly plug the overflow in order to create a vacuum. Place the suction cup over the drain opening and push up and down a few times. This should be sufficient to clear the obstruction, which should be evident by some of the water draining from the fixture. As the water drains out, add more water and repeat the process until the flow is continuous and rapid. If this method does not succeed, it will be necessary to use a hand auger.

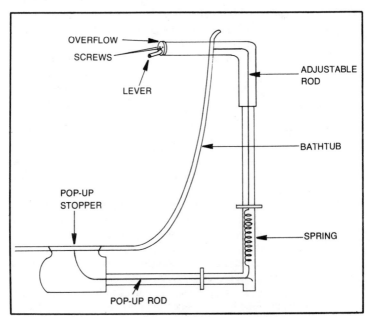

Fig. 12-10. A pop-up drain assembly is controlled by means of a spring.

REPAIRING CLOGGED SHOWERS

As with bathtubs, the most frequent material that causes blockages is hair which has accumulated either in the trap or drainpipe. In order to clear the clog, it will be necessary to remove the shower strainer. Depending upon its design, this device may be held in place with screws or simply snapped in place. In any case, once the strainer is removed, use a plumber's friend. This will usually work; if not, use a small auger.

REPLACING AN EXISTING TUB OR SHOWER FAUCET

If a tub or shower faucet assembly begins to cause problems that appear to be more than minor drips or leaks, it's time to replace the whole fixture. On the other hand, if the fixtures are old-fashioned and a new look is desired, installing new faucets with a more modern appearance may be desirable. Most faucets on tubs or showers are usually wall-mounted. The actual body of the faucet is installed behind the wall, where it is connected to the supply piping. This is where most of the problems associated with this type of project come in. Obviously, since the body of the faucet is out of view, there's really no need to replace it. If the assembly is quite old, especially in older homes, it may be virtually impossible to find faucets that will match and fit the body as they should. An exact fit is a necessity here. If the faucets for your particular type of assembly have been discontinued, the only choice left will be to replace the body. If no access panel is present in the wall, it will be necessary to get behind the wall to remove the old faucet body and install a new one. An assortment of wrenches will be required to loosen this assembly, along with the necessary carpenter's tools to remove a portion of the wall.

The first step is to shut off the water supply at the appropriate shutoff valve. If one is not provided on the line for this fixture, now's the time to install one. This is explained in Chapter 11. To get to the screws on the handles, it will be necessary to first remove some sort of decorative caps, usually each label H or C. These can either unscrew or simply snap out. The handles are right underneath and are then removed, along with any washers. Take to the plumbing supply dealer and see if exact replacements are available. If so, these are usually installed very easily, along with new washers, and the procedure is complete. Turn the water back on and check the fixture for any leaks or drips.

If it is not possible to obtain the correct replacement faucets, the whole faucet body will have to be replaced. Do whatever work is necessary to gain access to this portion of the assembly, which is connected to the water distribution system. Make sure the water supply is shut off. Then proceed to disconnect the faucet body from the supply lines. These may be either threaded or soldered. If they're threaded, unscrew them by using one wrench to hold the supply pipe and another to turn the locknut counterclockwise. If the lines are soldered, it will be necessary to use an appropriate torch to disassemble them.

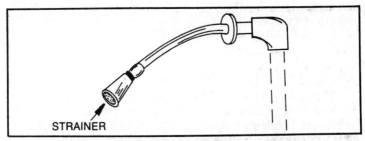

Fig. 12-11. A shower head assembly should be disassembled and cleaned periodically.

Once the lines have been disconnected in the appropriate manner, remove the old faucet assembly. When purchasing a replacement unit, try to buy one with either the same or similar dimensions, since no extra work to the supply lines will be necessary. The new faucet assembly is installed by connecting it to the supply pipes by either soldering or threading, making sure the connections are completely watertight in either case. The wall is then restored to its original condition. Be sure to provide an access panel for any possible future repairs. It is a good idea before closing up the wall to provide some sort of support for the lines themselves to insure that they can't come in contact with each other. This is the cause of banging noises often heard in the walls of older homes, and is easily preventing using some type of straps to keep them a sufficient distance from each other. Check the condition of the distribution lines and replace them if necessary.

Newer tubs and showers today generally have a single-handle control with a cartridge assembly much the same as those found on lavatory and sink faucets discussed in Chapter 11. Although this assembly may appear to look different than a sink assembly, it functions and is repaired in much the same manner. If a single-handle faucet is dripping, the cartridge assembly probably needs to be replaced. In order to reach this portion of the faucet, remove the handles, which should bring the retaining clip into view. Using some long-nosed pliers, remove the clip and simply lift out the cartridge assembly. Get an exact replacement and install it in reverse manner. This procedure should correct any drips.

REPAIRING SHOWER HEADS

If a shower head is spraying somewhat sporadically and unevenly, it probably needs a good cleaning. Figure 12-11 shows a typical shower head.

Some of the newer types have a handle attached to the head at some point to allow for adjustment of the spray. To get at the strainer, remove one or two screws from the cover plate. remove the strainer and soak it in a cleaning solution; then use a toothpick to remove any remaining particles still lodged in the holes of the strainer. Disassemble the whole shower

303

assembly and soak its parts, too, checking at the same time for any worn or defective pieces. Once all the parts have been cleaned and any defective ones replaced, reassemble the shower head in reverse manner. Turn the water on to test the spray.

ADDING A SHOWER FACILITY TO AN EXISTING BATHTUB

If you want to provide shower facilities in a bathroom which is too small to allow the addition of a separate shower stall, the project will require some carpentry and replacement of the existing bathtub fixtures. The new faucet assembly will be equipped with a special divertor lever which, when operated, directs the flow of water to either the tub or the shower attachment. Generally, the whole assembly is purchased in a single unit with all the necessary pipes and fittings included.

The installation itself is not too difficult, and the assembly will come with complete instructions. The first step will be to make the necessary openings in the wall to allow for the new supply connections to the shower head. After referring to the dimensions of the new fixtures, measure and cut the opening. Be sure to provide adequate support inside the wall for the shower pipe running vertically. The connections to the supply lines and the attachment of the faucet assembly and shower head are done as discussed earlier in this chapter.

SUMMARY

As more and more people take on the task of installing, repairing, and replacing plumbing fixtures, the industry itself has been quite helpful. The instructions normally provided with these fixtures are much more down-to-earth and simple to understand and follow. The language used makes it easy for a layman to know what to do and how to do it right. Very little difficulty should be encountered if all directions are followed carefully and correctly.

Care should always be taken in handling bathtub and shower fixtures. Their surfaces can be damaged quite easily, especially if any type of tool is applied to them without providing some sort of protective covering.

When it is necessary to remove walls or partitions to allow for the installation of a new fixture, it is wise to go over that portion of the plumbing system which is exposed by this process. If the home is an older one, the supply lines or possibly the drainage system may have served their purpose and need replacing. This is the time to make such improvements. Most older homes will not have individual shutoff valves at each fixture, and these should be added. They provide a real convenience for the do-it-yourselfer. It is always wise to look beyond the scope of the intended project to see if all is well with other portions of the system whenever possible. If something goes undetected or ignored at this point, chances are it will cause problems at some future time and will have to be repaired anyway. Save yourself some time and probably some money by taking care of it immediately.

Repair of Common Pipeline Problems

Most of the problems associated with either water supply or drainage lines are quite common and fairly easy to repair with a normal complement of tools and supplies. Even if a system is properly designed, installed, and maintained, there will more than likely be times when malfunctions do occur. This chapter will deal exclusively with problems associated with leakage, condensation, freezing, noises, water hammer, and pressure problems which can be dealt with successfully by the do-it-yourselfer.

If the system is an older one, it may be necessary to replace whole lines due to normal wear and tear. In other cases, it may simply be a case of repairing joints or adding new valves in order to make for a satisfactorily operating system. Regardless of the extent of the problem, be sure to take the time to get an overview before beginning to make repairs. Sometimes a defective portion of the system has caused problems elsewhere that may be overlooked or unnoticed if everything is not taken into consideration before beginning. This is not a time-consuming process and is well worth it in any case.

LEAKING WATER SUPPLY LINES

Piping is quite durable and is built to last for the home's lifetime; however, this will be the case under near-perfect conditions which are not always present. Some of the things which may cause these lines to become either worn or weakened are improper installation, corrosion due to chemical content of the water, or continued abuse. Corrosive chemicals will tend to weaken the lines over a long period of time. This weakened state will make the piping much more susceptible to cracking and splitting. Likewise, internal water pressure, or *water hammer*, may also trigger a leak. The definition of water hammer is a banging noise heard in a water supply line following an abrupt change of flow and a resultant surge in pressure. Very often the leakage is occurring at the threaded joints of a pipe, which can be caused by either the joint itself loosening or possibly a fracture in the pipe. If the joint has simply become loose, the leak can be stopped with a little tightening. Be careful, however, that an adjoining fitting is not loosened in the process. If tightening the joint does not stop the leak, the fitting should probably be replaced with a new one. This will usually work, unless the pipeline itself is corroded and breaks under the pressure of the tools when installing the new fitting. If this occurs, there is no alternative except to replace the entire length or a complete run of the pipeline. Small leaks at a threaded fitting can be repaired by opening the joints and remaking them with an appropriate jointing material.

If the piping is plastic, there will be less danger from freezing because the pipe is an insulator. Copper tubing, however, enlarges every time it freezes, and this process will tend to break down the tubing over a long period of time.

It is obviously better in the long run to make the necessary replacements if either piping or fittings are leaking rather than make repairs. This will cost more, but it will also preclude the necessity of repeating the same repair process at a later date, particularly if the line and/or fitting are quite old. However, if leaks occur at a time when it is not possible to get replacement parts right away, it is possible to make temporary repairs. Some examples of emergency procedures are shown in Fig. 13-1.

To repair a small leak temporarily, an appropriate type of water proof tape may be wrapped triple-strength around the pipe. Before applying the tape, wipe the pipe as dry as possible under the circumstances with a rag or piece of cloth. Start the tape about 3 inches from the leak on one side. Extend it the same distance on the other side. This type of repair should suffice to block the leak temporarily. This may not be possible in some situations where plastic piping is leaking, especially if it is on a hot water distribution line. The tape may be in danger of melting and deforming the pipe. Some special types of tape must be used when working on plastic pipe, and this tape may not be available. In this case, do not attempt to patch the leak with a substitution.

Another method is to apply a compound stick, such as *Krak-Stik*, which is available at most plumbing and hardware stores. This sealant comes in a stick, as its name implies, and is very simple to use. Rub the stick over the hole and the leak should stop almost immediately. It is not necessary to shut off the water before applying this sealant.

If the pipeline is actually cracked or has larger sized holes in it, a common approach is the clamp and patch method (Fig. 13-1). To do this, you will need a clamp and a patch made from sheet plastic or possibly an inner tube. The patch should be about 3 inches longer than the damaged section of the pipe and slightly less wide than the outer circumference of the pipe. Clamps come in standard sizes. A clamp should be purchased which will match the diameter of the pipe that is to be repaired. For covering a large area, use a type I clamp with two screws. For smaller lengths, a type II clamp will suffice. If the pipe is a small diameter one, a patch can be quickly made by splitting a short length of common garden hose and slipping it over the pipe leak after applying plumber's cement on the pipe surface. Clamps are then fitted at both ends and the repair is complete.

Yet another method of making temporary repairs to a leaking or cracked pipe is to cut out the defective length with a hacksaw and slip a length of rubber hose or flexible tubing over the ends. Then tighten the tubing with clamps (Fig. 13-1). Be sure to shut off the water supply before making this type of repair. The inside diameter of the hose should be equal to the outer diameter of the pipe. A garden hose may also be used, as long as it fits the required measurements. For example, a ¾-inch hose will fit over a ½-inch pipe, and a 1-inch hose will fit over a ¾-inch pipe.

To make more or less permanent repairs on a threaded pipe, it will be necessary to replace the entire section along with the fittings. This is

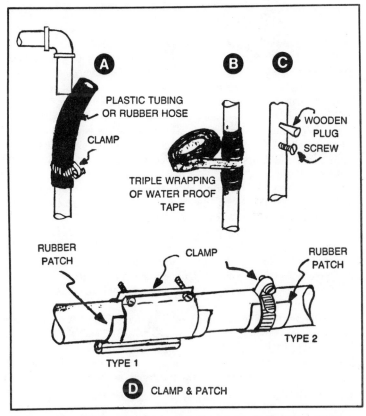

Fig. 13-1. Examples of some emergency repairs in the event piping or fittings are leaking. These should be considered as temporary.

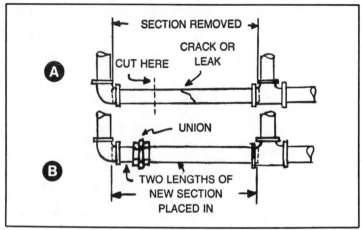

Fig. 13-2. Replacing a worn section of pipe is done by cutting the pipe at least 4 inches from fittings on either side.

really the best way to solve the problem. The method of replacing the worn section is shown in Fig. 13-2. First, shut off the water supply and then cut the pipe with a hacksaw somewhere in between the two closest fittings. Unscrew the two sections from the line. Make sure the cut is at least 4 inches away from fittings on either side. If there is a union on this line, it will be possible to simply unscrew the lengths on either side without having to cut the line. Make a new section of pipeline with two lengths of pipe and a union, so that it is of the exact size as the old pipe. Replace this section in the line, first tightening at one end and then at the other. The repairs are now complete.

If leaks or defects occur in a section of pipeline which is located either inside walls or under floors, it will be necessary to gain access to them. This may require removing part of the floor or walls before repairs can be done. Sometimes the leaky pipe can be uncovered at two accessible points near the fittings. It can then be cut at these points. The leaky portion can be removed and discarded. The best replacement piping in this case will be flexible copper tubing, since it can be placed in the walls or floors quite easily. To join the copper tubing to the existing plumbing, use whatever method is appropriate.

To repair joints in copper pipes, first shut off the water supply and heat the joint to a melting temperature with a torch. Touch the wire solder at the leaking part of the joint and fill it with molten solder. To avoid damage to any other fittings in the immediate area, wrap them in a damp cloth. This method should work the first time, but it may be repeated if necessary. Make sure the joint is given enough time in between steps to cool down.

If a section of copper pipe is cracked or crushed by a heavy object, the damaged section will have to be replaced. The new piece is attached to the line by means of slip couplings, which are small lengths of tubing with an internal diameter slightly larger than the outer diameter of the copper pipe on which they are to be used. The couplings are slipped over the ends of the old pipe. After the new section of pipe is put in place, the couplings are centrally spaced over the joints. Solder is melted using a propane torch. The joint is made in the usual manner.

LEAKING DRAINAGE PIPES

The only major difference between making repairs on water distribution lines and drainage lines is that the process will be a bit simpler. The reason is that drainage lines do not operate under any pressure. To repair a leak, clean the area first with some soapy water and allow it to dry thoroughly before wrapping with some electrician's tape. This will be effective on small and medium size drains and will stop leakages even at the joints. If the drainage pipe happens to be made of cast iron with leaded joints, which are common in older homes, the leakage will probably be at the joint itself. This can be repaired by caulking the joint with a caulking tool. Additional lead yarn may be added if necessary. If a caulking tool is not readily available, use a hammer and chisel. Don't cut too deeply into the lead during this procedure. Place the chisel inside the lead, and apply a few light blows with the hammer. As the lead spreads out, this will cause the leak to stop.

LEAKING DUE TO BURST PIPES

The first thing to do in the event a pipeline bursts or a joint opens up completely is to shut off the water supply at the main shutoff valve in the basement in order to protect the structure from flooding damage. In this type of emergency situation, there is really not time to call a plumber, so it is important that everyone in the home know the exact location of at least this one valve in the system. A shutoff valve is shown in Fig. 13-3.

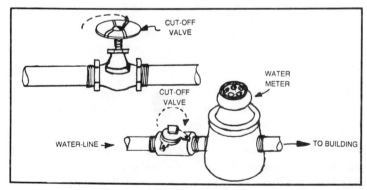

Fig. 13-3. Every member of the family should know the location of the main shutoff valve in the event of an emergency.

Depending upon the extend of the flooding, it may be advisable to shut off the electricity, since some electric appliances can short-circuit if they get wet. All family members should be familiar with the location of this master switch.

In some instances, the shutoff valve is located on the main water line just outside the house. In this case, it will be either a *gate valve*, which is operated by hand, or an L-shaped rod, which is turned with a wrench. A few clockwise turns will suffice to shut off the supply. Since these valves are seldom used, they may become corroded due to moisture. It is a good idea to check them once or twice a year. Make sure they are in good working condition by operating them once or twice. A little lubrication will also help to keep them working properly.

Once the water and electricity (if applicable) have been shut off, go to the location of the pipe which is damaged. Examine everything to determine the extent of the damage. Many times you will find that either the pipe has sprung a few good-sized leaks, or a joint has burst open. Repairs of this sort are done in the same manner as described earlier. In some instances, however, a section of the line maybe so corroded and deteriorated that there is no alternative but to replace it with new pipe. The same procedures described earlier will be followed if this is the case.

ALLEVIATING CONDENSATION

This condition is quite common in older homes with metallic pipelines. Another name for it is *sweating*, and it is caused by condensation of the moisture in the air as it comes in contact with the cold surface of the water line. The dripping can damage not only floor and wall coverings, but it may also damage the structural supports of the building such as studs and joists.

The reason condensation occurs is that the cold water in the pipe never remains stationary and thus does not have time to get heated to the room temperature. One way to effectively deal with this problem, which is frequently experienced in basements during the summer months, is to install a *dehumidifier*. These devices also help stop condensation on the cool surfaces of walls and floors. A cruder method of dealing with sweating pipes is to keep a few collecting pans under the areas of drips, but this does not really solve the problem.

Another very effective method of alleviating this problem is to use anti-drip insulation tape, which comes in self-sticking rolls. Both the pipe and the fittings should be fully covered with this tape, which is available at most plumbing supply stores. Alternately, the pipe may be coated with a special paint containing cork granules. Some of the self-sticking tapes come with these granules used in their manufacture. There is another type of tape which may be used. It is also available at most plumbing stores and is an asbestos, non-self-sticking tape which must be attached by means of special asbestos cement. These compounds are applied in the same manner as paint, with the help of a brush. As these materials are not reputed to

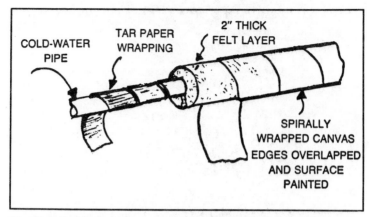

Fig. 13-4. Exposed pipelines may be insulated effectively in this manner.

be all that durable, it will probably be necessary to apply several coats to achieve the desired results.

FROZEN PIPES

In most cases, pipes can be prevented from freezing by providing an appropriate method of protection at the time of installation. If temperatures go below freezing in the winter, all exposed pipes should be insulated in some manner. Likewise, all pipelines running underground should be laid well below the frost line. This depth will vary somewhat from area to area, so be sure to check this at the time of the original installation. Figure 13-4 shows one method of insulating exposed pipelines. The pipes are first wrapped with tar paper and then covered with an insulation which generally consists of a 1½-inch to 2-inch thick felt layer running continuously so as to not leave any air gaps between adjacent lengths. Finally, the felt layer is covered with spirally wrapped canvas until its edges are overlapped, and its outer surface is painted.

If an exposed pipeline was not installed with adequate protection against freezing, enclose it in a rectangular wooden box which is then tightly packed with sawdust, mineral wool, ground cork, or some other similar material. The thickness of the insulation should be more than 2 inches to provide adequate protection.

Figure 13-5 illustrates a method of insulating a vertical riser pipe by enclosing it inside an earth tile. The pipe is wrapped with a layer of felt and placed inside the tile, leaving an annular space or air gap which also helps in the insulating process. The top portion of the pipe just below the floor is enclosed in a wooden box with an insulating material such as sawdust packed in between.

During the original installation, water supply lines should be placed inside inner walls, whenever possible. Lines running to outdoor faucets will be an exception to this, and these should be insulated in an appropriate manner. The faucet assemblies themselves should be of the frost-free variety and should also be drained and shut off sometime in the fall.

Even if all the proper precautionary measures have been taken, there is still a chance of freezing and subsequent bursting of pipelines. For example, if a heating system shut down due to a power failure, all the lines inside the home would be in danger of freezing if the temperature got cold enough. The reason that pipes burst when the water inside them freezes is the water expands when it turns into ice. If there is no place for the ice to go when it expands, it will simply continue to push out and eventually cause the pipeline to burst. This will not necessarily occur at the point where the freezing occurs. For instance, ice may start forming at an exposed section of the pipe and progress toward the inside until it meets an obstruction such as a bend, a rough spot, or a valve. As the movement is stopped, the pipe at that point is subjected to high stresses, causing it to burst at that particular spot. This is why a pipe may mysteriously burst inside a kitchen which is not cold enough to cause it.

There are several methods which can be used to thaw frozen pipelines, and several schools of thought on which approach is best. Regardless of the method, all rely upon an external heat source for thawing. A quick and inexpensive way is to use electric heating by attach-

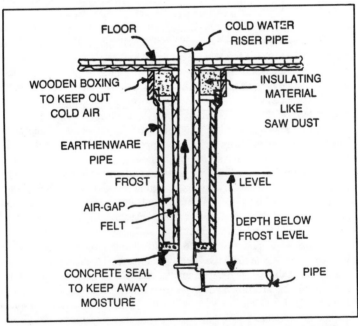

Fig. 13-5. A vertical riser pipe may be insulated by enclosing it inside an earth tile.

ing the terminals of a welding machine or a pipe-thawing transformer to the frozen ends (Fig. 13-6). The heat is continued until the ice melts and the flow of water resumes. If using this method, be sure to follow instructions, as this may be hazardous if done improperly and without taking the proper precautions.

Another method that can be used if the line is accessible is to wrap the pipe with cloth, heavy towels, or burlap bags. Pour very hot water over it. Be sure to place something under the pipeline to catch the overflow, and be careful not to expose any parts of your body to the hot water. Thawing should always be started at one end of the pipe, never from the middle. Otherwise, the formation of steam inside the pipe, or even the expansion of heated water trapped by ice, may lead to a burst. Start from the faucet side of the frozen section and work toward the supply end, keeping the faucet open to allow for the escape of steam or melted water. This will also serve as an indication of when the ice has melted. Be prepared to shut off the water supply if the pipe has ruptured.

Yet another approach would be to use a propane torch, as long as proper precautionary measures are taken. A fire can easily start if you happen to ignite some combustible material such as wood, cloth, or paper, or if a wooden wall is running somewhere near the line. Asbestos cloth may be used as a protective shield. The flame should not be played on one spot too long, but it should be kept in continuous motion. Also, do not apply

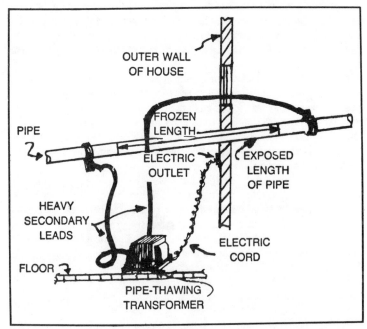

Fig. 13-6. Frozen pipes may be thawed with a pipe-thawing transformer.

too much heat at the joints or they may open up, causing additional problems. Some people prefer to use a slower but effective approach such as holding and electric iron against the pipe or possible using an electric heater. These methods are also considered to be safer.

Another very good method which is used to thaw drain pipes as well as water supply pipes is to run boiling water into the pipe if a section of it can be removed. This operation can also be started from the outlet. A small pipe or a garden hose may be used for a horizontal pipe. For a vertical pipe or for the end of a drainage pipe, either a garden hose or a length of rubber tubing will have to be used (Fig. 13-7). To perform this procedure, insert the tubing into the pipe until it strikes against the ice obstruction. On the drain pipe, it will be necessary to remove the trap. Attach a funnel to the other end of the hose and, after raising it, start pouring very hot water through the funnel. A bucket should be placed under the end of the pipe to collect any backflow. Take care not to get scalded by the hot water. This process will gradually melt the ice. As the ice is melted, the hose will have to be pushed further into the pipeline until the whole obstruction has been cleared satisfactorily.

PIPELINE NOISES

Each common noise associated with pipelines has a different cause and will require a different method of alleviating the noise. Some people simply choose to live with this disturbance until it becomes quite loud and annoying. If noises are detected in a plumbing system, it is wise to find the cause and make the necessary repairs at once, since it may be attributed to a defective part which can cause further damage if ignored for any length of time.

One of the most common noises heard is the whistling and chattering at faucets, which can be caused either by a loose or worn-out washer or possibly by poor hydraulic design in the case of the less expensive faucet assemblies. The first step is to shut off the water supply and open the faucet. Take out the assembly. Check the washer and its seat. If necessary, change the washer. Instructions pertaining to this type of repair will vary depending upon the type of faucet.

In the more poorly designed faucets, the water passage area is insufficient. This condition tends to cause a high velocity of flow. When this high velocity water changes direction near the faucet outlet, it generates air bubbles which subsequently collapse, creating vibrations and noise. Depending on the frequency of vibration, the noise will be chattering, whistling, or perhaps even pounding. In old plumbing systems where Fuller-type faucets are still found, sound noises are the result of either a loose ball inside or a worn-out ball shaft and spindle eccentric, which will cause the entire assembly to rattle. If this is the case, the whole faucet should be replaced.

If a banging or squeaking noise is heard, the pipes are probably striking against something solid. The first thing to do is determine the

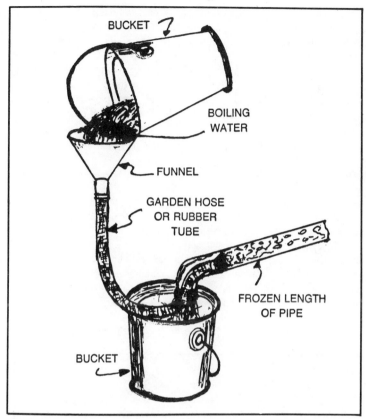

Fig. 13-7. If a section of the line can be removed, the pipes may be thawed by pouring boiling water into them.

location of the problem by turning the water off and on, watching for any movements on the pipeline. If the noise comes from a length of pipe located behind the wall or below the floor, it is not necessary to tear anything out to make the repairs. This is what most people think, which is why they choose to live with the disturbance. Actually, it is quite simple to alleviate the noise. Stops placed at each end of the line where it comes out of the walls or floors will effectively eliminate the banging.

Figure 13-8 illustrates some methods of stopping pipeline noise. If the noise is coming from the vicinity of a tube strap or U-clamp, it is probably caused by the pipe being loose and banging against the wall. A patch of rubber or a small piece of garden hose slit lengthwise and inserted around the pipe will silence the noise. Alternately, if the pipe is banging against a masonry wall, place a wooden block between the pipe and the wall. Secure it to the wall with masonry nails. A pipe strap is used to attach the pipe to the block.

To eliminate noise on a long length of horizontal pipeline, use some sort of firm support such as pipe straps, hangers, or possibly even shelf brackets. Be careful not to fix these pipes too rigidly, as temperature changes cause expansion and contraction. The pipe should be able to move freely to a certain extent. When securing a pipe with an anchor or a bracket, always place a rubber piece between the pipe and the clip. Sometimes a water pipe and a drainpipe running next to each other vibrate and rattle. It may be a good idea to solder them to each other.

A line that is either corroded internally or possibly inadequately sized for its intended purposes may tend to cause noise problems. This is because the water is being restricted somewhat. The only two choices are to replace the line or insulate the outside to cut down the noise. Obviously, the former is preferred for the obvious reason that the latter does nothing to eliminate the problem at its source. A pipeline which is not functioning as it should, especially if it is not of the proper size, should be replaced before further damage occurs from overtaxing.

Most of the problems discussed here will not occur in drainpipes because the speed of the flow is not quite as high; nor is the system under pressure. One noise which is somewhat common in drainage lines is a sucking sound when waste water goes down a sink or washbasin into the drain. This is usually caused by a clog in the vent pipe if one is present. If one is not part of the system at this particular point, this could also be the problem. The condition may lead to sewer gases entering the home.

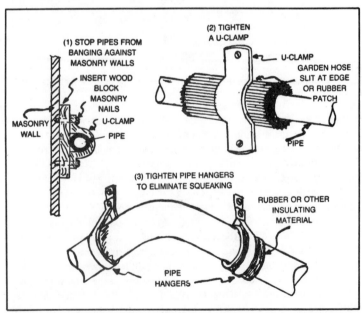

Fig. 13-8. Some of the methods by which pipeline noise can be eliminated.

Purchase an anti-siphonage trap from a plumbing store, and install it on the drain. This should effectively deal with both the noise and the sewer gases. If the drain appears to be clogged, refer to the instructions provided in this text on clearing blocked drainage lines. The easiest method to use will be to clean the drain with a plumber's snake.

WATER HAMMER

Most people attribute any noise in their plumbing system to water hammer, which is really not correct. This type of disturbance is quite common and has much to do with pressure.

When a faucet or valve is closed abruptly, the flowing water is brought to a sudden rest. The momentum of the water is converted into shock waves which travel back and forth between the valve or faucet and the point where the pipe connects to a larger pipe or tank, or until the line changes its direction. In addition to causing a hammer noise which can usually be clearly heard anywhere in the house, the energy of this wave also vibrates and shakes the line violently. This vibration may rupture the pipe at its fittings or even crack the fittings themselves. This is why the use of malleable iron fittings is recommended on water pipes. In hard rigid pipes made of steel, brass, or iron, the water hammer noises and stresses are much more pronounced than in flexible pipes like plastic, copper, or lead which tend to dampen and absorb the waves or pressures. The situation should not be ignored, as further damage may occur.

The problem of water hammer may be alleviated somewhat by installing pressure reducing valves or *regualtors,* or by bracing the pipelines. This will work to some extent, but the real solution is to install air chambers where necessary. An air chamber is simply a small length of vertical pipe capped at the top and attached to the line near the valve or faucet. Two such devices are shown in Fig. 13-9. These air chambers are usually installed inside the walls. They contain air which is compressed when the water hammer wave hits it, thus providing a sort of cushion which considerably reduces the impact and noise and prevents pipes from shaking and banging.

An air chamber may be a continuation of the riser pipe (Fig. 13-9A), or it may be offset (Fig. 13-9B). Certain requirements should be maintained regarding the proper size for each individual situation. The minimum chamber length is 18 inches. The minimum capacity is equal to 1 percent of the capacity of the pipeline on which the water hammer problem exists. The size of the air chamber pipe should be one or two sizes larger than that of the pipeline itself.

One problem which may be encountered even with the use of air chambers is if the water pressure is quite high and the flow velocity in the pipes is equally high, air may be sucked out of the chamber, causing a vacuum. In this situation the air chamber will be either partially or completely filled with water, thus losing its effectiveness. When water hammer noises occur on a system which has been quiet in the past, the

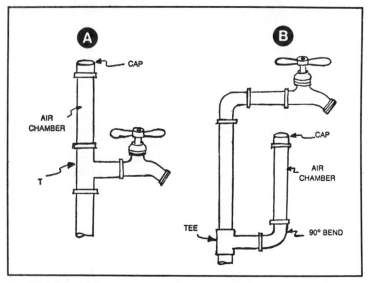

Fig. 13-9. Water hammer is alleviated by the installation of air chambers.

problem can usually be traced to an air chamber which has filled with water. To alleviate this, turn off the water supply and let the lines drain out completely through the faucets. The air chambers will fill with air once again and should perform satisfactorily. The supply can then be opened again. The only other problem that may occur with these devices is that they may become clogged with either impurities or chemicals present in the water supply. To clean the chamber, remove the cap, remembering to shut off the water first. Since the air chamber is a bit larger in diameter than the pipe line, it can withstand a certain amount of clogging, more so than the piping, before it needs to be cleaned out. However, this does not mean that the malfunction should be ignored, but rather than it will hold up a little longer than the pipeline in the event blockage does occur.

If no air chambers are present in the system, and it is not practical to install them, there is another device which may help to alleviate the problem. Figure 13-10 shows a coiled air chamber which is really a coiled piece of metal which can be fitted directly on top of the faucet or valve, thus avoiding the need to open walls or floors to install an air chamber. Although this device is not quite as effective as an air chamber, it will perform adequately in controlling water hammer and is recommended for use where it is not desirable to install air chambers. These can be purchased at most plumbing supply stores and are not very expensive.

It is possible to make your own chamber if you wish. Again referring to Fig. 13-10, replace the bend (A) with a tee and attach an 18-inch length of pipe of the same size as the water line, or somewhat larger, to the top of the tee. Close this pipe with a cap, and it will now perform in almost the same manner as an air chamber. Remember to shut off the water supply before opening the line. Once the cap is in place, water can be restored.

LOW OR VARYING WATER PRESSURE

Problems with water pressure are quite common and may be attributed to any number of things. It may be that the pressure varies enough to be an inconvenience during different times of the day or year. Alternately, it may be caused by the water supply not being delivered at an adequate pressure. Other reasons for low pressure are a clogged service pipe, in adequate sizes of water lines in the home, or a clog or obstruction on an outside line. To determine the cause of the low pressure, if it is already known that the supply being delivered is not at fault, it will be necessary to use a device called a water pressure gauge. If the problem is due to low pressure from the utility supplying the water, the only alternative is to install a booster pump inside your home, probably in the basement.

To use a water gauge, place it on the service pipe a little upstream from the meter (toward the curb). Shut off the supply to the house by closing the shutoff valve. If the gauge shows adequate pressure (50 psi or more), this means the public water supply is not at fault. Either the service pipe or the plumbing system is causing the problem. Open all the faucets in the house, as well as the shutoff valve, and let them run. If the pressure at the gauge drops down quite a bit, the service pipe is either clogged or inadequately sized. If, however, the pressure does not drop considerably and the pressure at the fixtures is still not adequate, then the water pipes inside the house are either clogged or too small.

Take the gauge inside the house and attach it to the main line. Observe the pressure first without any water running through the fixtures,

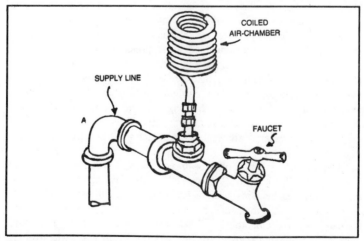

Fig. 13-10. If it is impractical to install an air chamber inside the walls, this coiled air chamber may be installed directly on top of the faucet.

and then do the same with all of them open. Note the drop in pressure. If the drop is insignificant, move the gauge further along the line in the direction of the faucets and other fixtures until you find the point at which the pressure drop is significant. The line between this point and the fixtures is either inadequate in size or clogged. This procedure may point out more than one area where the problem is occurring, so be sure to continue through the whole system even when one spot is found. Once the problem areas are located, it will be necessary to either clean the lines or replace them with pipes of larger sizes, whichever is applicable.

If it becomes noticeable that the hot and cold water mixture at a fixture suddenly changes either in temperature or volume when another fixture is in use nearby, the pipes are either inadequately sized or are worn due to corrosion or scaling. It may be that the lines were not increased in size at the time new fixtures were added to the system. An increased number of fixtures adds new lines, fittings and bends, thus resulting in more friction loss and subsequent low pressure. Regardless of the reason, the best solution is to replace the pipelines with larger sized ones. A temporary solution is to install an automatic mixing valve at the location of the fixtures where low pressure is a problem. This will in no way increase the water flow in inadequately sized piping and only serves to alleviate the problem in one given location. Although this will work, it is best to make the necessary replacements even though this is more expensive.

SUMMARY

It is important to take the necessary precautions during an original installation of a home plumbing system to insure that the pipelines will be protected from freezing temperatures. Plan the system so that no lines are installed in exterior walls, especially those which are not insulated. Always remember to take into consideration any possible future additions of fixtures, appliances, or whole new bathrooms. These will tax the present system if the piping is not adequately sized to handle the extra load. It will also have an effect on water pressure, since more lines, fittings, and directional changes will undoubtedly be involved in these additions. Careful planning will take care of all these things satisfactorily. Even if these precautions make the original installation a bit more expensive, they are well worth it in the long run, since future additions and alterations may necessitate ripping out interior walls and floors. This would not be necessary if allowances were made while the structure was in an unfinished state.

Older homes will most often be prone to pressure problems, since the piping is probably inadequate or corroded. Modifications to a plumbing system which has been in operation for many years will probably be a bit more expensive than the same procedures on a newer home. In trying to alleviate one problem, others will be encountered. A minor repair may grow into a whole house overhaul very quickly. In any case, whatever is needed to make the system function satisfactorily should be done, since the purpose is to provide convenience and continued good service.

Miscellaneous Indoor and Outdoor Installations 14

Some of the more common outdoor installations include pools, sprinkler systems, and gazing pools. The plumbing for each of these is a bit different, but their installation is certainly not beyond the scope of the do-it-yourselfer. Obviously, the main thing involved in an outdoor installation of any type is the supplying of water from the basement to the subsystem itself.

Appliances such as garbage disposers, dishwashers, and automatic clothes washers are becoming more popular today and can be considered almost standard equipment in most homes. They are real time-savers. The installation, repair, and general operation of these units is not really difficult at all, but if handled improperly or carelessly, serious accidents can result. Care should be taken when working with units powered by electricity. Make sure the appliance is unplugged before beginning to work on the interior of the unit. The interior mechanisms of these appliances may not all be the same. They all do operate, however, in much the same manner. Be sure that the unit purchased comes with an instruction booklet, which should be used to supplement the information provided in this text.

AUTOMATIC WASHERS

Washing machines are available in many different sizes, colors, and operating mechanisms. The different types of mechanisms are really beyond the scope of this book and will not be discussed in detail, but the four kinds are *front-loading, top-loading, vacuum-cup,* and *agitator.* The basic principle of operation of these kinds of clothes washers is really the same. The machine generates a motion between the wash water and the soiled clothes. This motion makes the water run through the clothes so that foreign matter, stains, and dirt attached to the clothes are softened, detached, and washed away.

To install an automatic washer, you will need a water supply, a drain, and an electrical connection. Obviously, in planning for an installation, try to find a location that will make access to both water and electricity as convenient as possible. Also, be sure to check with local authorities as to any regulations applying to the installation before beginning. Codes will most often specify the types of piping that are allowed and the proper sizes which have proved adequate for satisfactory operation.

Figure 14-1 shows two different types of hot and cold water connections for the installation of a portable washer. In one system the rubber hoses are connected directly to the sink faucets; in the other, separate shutoff valves of the threaded type are used to make the connections. Obviously, the latter alternative provides for a better and more independent control. New washers come equipped with special hoses already attached. These hoses are simply attached to cutoffs exiting from the walls.

To drain off the waste water produced by the unit, a standpipe must be connected to a drain located somewhere nearby. The top of the standpipe should be a few inches higher than the maximum water level in the washing machine. It is not necessary to make a permanent connection between the standpipe and the drain hose of the machine. It can be installed simply by inserting the hose into the standpipe and securing it in place so that it does not fall out and cause flooding. This type of open connection is a safety device. It prevents a cross connection between the waste water line and the water supply pipe by creating a vacuum breaker.

If the washing machine is being installed in a separate laundry room or space set apart in the house, the plumbing work will be minimal. All the plumbing lines in this case were probably installed at the same time as the rest of the system. The water supply lines are provided with air chambers which prevent water hammer noises. Otherwise, as the washer shuts on and off several times during one complete cycle, there would be a considerable amount of water hammer noise. If, however, the plumbing system did not include allowances for the installation of an automatic washer, and it is necessary to run the lines, be sure to provide these air chambers.

To arrange a washing machine drainpipe, first note the location of all openings, spaces, or indentations in the back panel of the machine. This space can be used to house the drainpipe, so that the machine will cover the pipe and still set back against the wall. After making the necessary calculations, bring a 1½-inch drainpipe line up through the floor in the appropriate position. Terminate the pipe somewhat above (or below) floor line with a trap assembly attached to the end of the pipe. Attach a tubular S-trap to the adapter, and adjust the trap unit to the most convenient position. Insert a short stub of drainpipe or a short tailpiece into the top of the trap. Insert the crook of the washing machine drain hose into the top of this pipe stub (Fig. 14-2). There is no need to seal around the hose connection with putty or any other material; nor is this desirable. The trap itself provides a water seal. If the top opening is completely sealed up as well, no top venting will take place. The machine may not discharge properly.

The installation of an automatic washer in a more or less permanent position will require a bit more effort beyond just the plumbing and electrical hookup. Some washers are bolted down to the floor. If the floor is concrete, it will be necessary to drill holes in it with a star drill and install bolt anchors. The foundation bolts are set into the concrete during the pouring process (Fig. 14-3). In wooden floors, the bolts should go right through the floor and be anchored to the wooden beams below level. If this is not practical or possible, attach a large sheet of plywood securely to the wood floor by means of a few screws, and bolt the washer to the sheet.

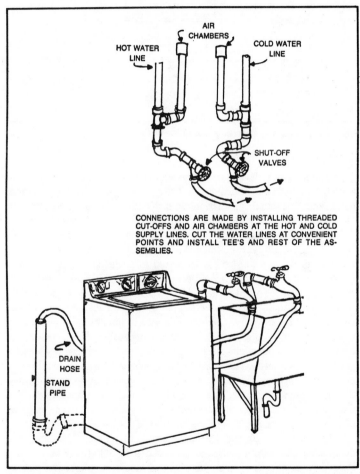

Fig. 14-1. A portable washer may utilize supply lines at an existing fixture or may be connected to its own supply lines.

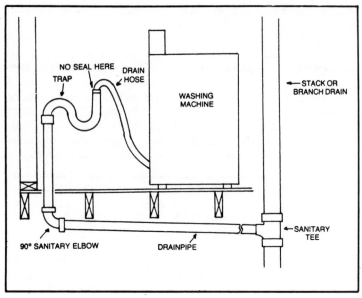

Fig. 14-2. A washing machine drainpipe installation.

DISHWASHERS

Although more homes today are equipped with *dishwashers,* they are usually installed as an afterthought. The required plumbing will go hand in hand with the actual installation. A dishwasher is truly a time-saver. Operation is basically the same in all units. Hot water is mixed with a dishwashing detergent which flows over the items to be washed. After the wash portion of the cycle is complete, the soiled water either drains out or is pumped out, and the load is subjected to a rinsing action with clean water. Finally, a heating element dries out the load. This last step of the cycle may also be used, if desired, to warm up serving dishes. Dishwashers are usually only connected to the hot water distribution system, and up until recently they needed a supply of hot water at a minimum temperature of around 140°F to clean adequately. Manufacturers are coming out with new units which will clean satisfactorily with water not quite so hot, sometimes around 120°F. As people turn down the thermostat on their hot water heaters to decrease electrical or gas consumption, they have found the water supply not hot enough for this sole device in their home. Similarly, until recently, most dishwashers carried some sort of statement which disallowed the washing of pots and pans in their unit. This has also changed, and it is now possible to put almost anything in a dishwasher and have it come out clean. Care should always be taken to make sure an item is dishwasher safe, and most new dishes and plasticware carry a label stating whether or not they can be washed in a dishwasher.

324

The main operating parts of a dishwasher are a motor, an impeller placed at the bottom, a watertight tub cover, draining and filling apparatus with a pump, dishracks, and a basket for silverware. In addition, there are a number of controls providing for only a single operation, if desired. As more and more people look for ways to conserve, the dishwasher is cut off after the rinsing cycle to avoid the time-consuming and electrical drying portion of the operation.

The installation of a portable dishwasher is really very simple. This type of unit usually is equipped with wheels, making it convenient to place it near a sink during operation. The dishwasher will also come with a water hose which is connected to the hot water faucet of the sink by means of a special adapter. The drain hose of the dishwasher is positioned so that it will discharge into the sink. Hence, the unit is really using facilities already provided for in another installation. The soiled water is pumped from the dishwasher into the sink by a motor-driven pump which is housed inside the unit itself.

To install a permanent dishwasher, it will be necessary to run a hot water supply to the proposed location for the unit. Permanently installed dishwashers are equipped with flexible water hoses with coarse-threaded, female screw couplings at the end. These couplings are connected in an appropriate manner to the hot water line in the easiest fashion by means of

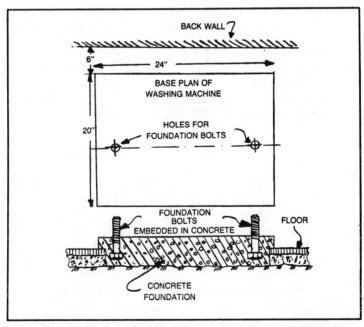

Fig. 14-3. To permanently install a washer, it should be bolted down in an appropriate manner.

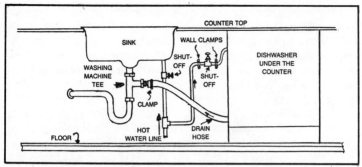

Fig. 14-4. Installation of an under-the-counter dishwasher.

a faucet which has garden hose threads on the spout. Such a faucet is inexpensive and easily installed. Just cut away a piece of pipe from the hot water line and install a tee. The faucet is attached to this tee by means of a nipple.

Figure 14-4 shows a typical under-the-counter dishwasher installation. The hot water supply comes from the line leading to the sink nearby. If at all possible, this is an ideal location for the obvious reason that less plumbing work will be required. A shutoff valve is provided on the connect-

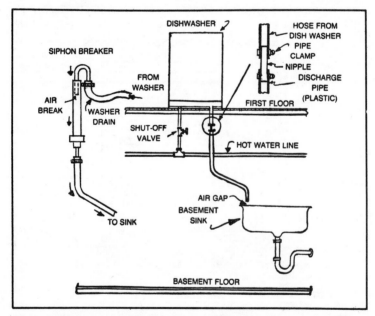

Fig. 14-5. If it is not possible to install the dishwasher near an existing sink, it may be connected to another sink or tub in the basement. It is advisable to install a siphon breaker.

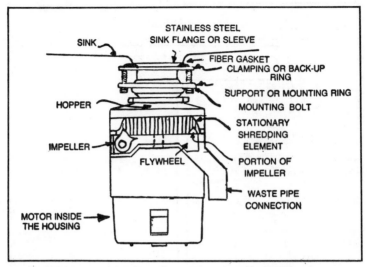

STAINLESS STEEL
SINK FLANGE OR SLEEVE

SINK

FIBER GASKET

CLAMPING OR BACK-UP RING

SUPPORT OR MOUNTING RING

HOPPER

MOUNTING BOLT

STATIONARY SHREDDING ELEMENT

IMPELLER

FLYWHEEL

PORTION OF IMPELLER

WASTE PIPE CONNECTION

MOTOR INSIDE THE HOUSING

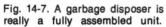

Fig. 14-6. Interior mechanism of an automatic garbage disposer.

ing line, which is secured to the wall by two clamps. The drain hose from the machine is connected to the sink drain and discharges the waste on the upstream side of its trap. A special washing machine tee will be used in this installation. First, open up the connection between the trap and the tailpiece of the sink. Swing the trap to one side. After removing the tailpiece, fit the washing machine tee in place. Assemble the drainpipes. The result will be a connection with male water hose threads to which the drain hose of the machine is attached and secured with a clamp. It may be necessary in some situations to reduce the length of the tee by cutting, but this is a simple task.

Fig. 14-7. A garbage disposer is really a fully assembled unit.

327

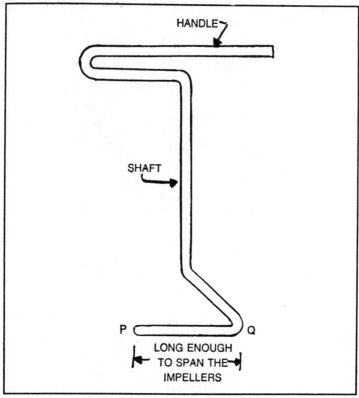

Fig. 14-8. A tool can be made using a 3-foot length of ⅜-inch drill rod, which is inserted to turn the flywheel.

If the drain hose does not have a screw-thread female coupling, attach one by using the appropriate water hose fittings and necessary hardware. If it becomes apparent that there is a more than ample length of hose, do not cut it; just coil it up and tie it with a length of string. If it ever becomes desirable to move the machine to another location, a longer piece of hose may be necessary.

If it is not possible to install the dishwasher near a sink, the drain connection can be made to a sink or tub in the basement. To effect this type of installation, a garden hose or similar length of tubing may be used to provide additional length to the drain hose equipped with the unit. To connect the two, use a rigid copper nipple and secure it by means of two clamps. The hose is connected to the basement sink in the same manner as discussed previously (Fig. 14-5). It may be necessary to install a siphon breaker such as the one in Fig. 14-5 to prevent water from being siphoned out of the dishwasher. This problem is attributed to the long drain line and is solved through the use of the breaker.

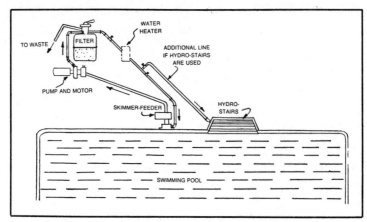

Fig. 14-9. Basic operation of an in-ground swimming pool.

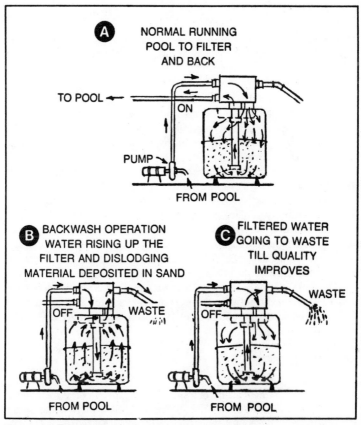

Fig. 14-10. The filtering action is composed of these three operations.

329

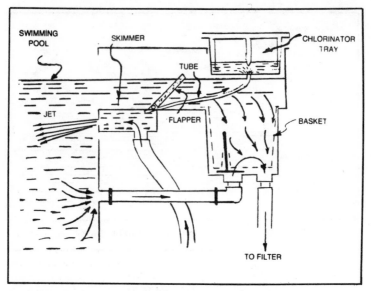

Fig. 14-11. The skimming action provides excellent circulation and cleaning.

If the kitchen sink is already equipped with a garbage disposer, it may be possible to connect the dishwasher directly into the disposer's drain line. Some disposer units are equipped with a special fitting to allow for this type of connection. To install a dishwasher in this manner, first make sure no electric current is present in the disposer unit. Using a screwdriver or an appropriate tool, insert the pointed end into the disposal unit from the sink opening. Pry loose the knockout plug which seals the fitting. The dishwasher drain hose can then be inserted over the end of the fitting. It is held securely in place with a clamp.

In a situation where it is not possible to discharge the drainpipe from the dishwasher into an existing sink's drain, it will be necessary to connect it directly to a drainage line in the house. This will require the installation of a trap and possibly a vent pipe. The trap may be of a large size, and the washer drain may hang inside it. It is preferable that a permanent solid connection be made between the washer drain and the trap. In case the sewer and house drain back up, a solid connection will prevent any waste matter from flowing out.

GARBAGE DISPOSERS

The advantage of having a *garbage disposer* is that it provides a quite sanitary and convenient means of disposing of wastes. In rural areas and in localities with no pickup service, disposers rid their owners of unpleasant garbage in a most pleasing manner. Before installing a disposer, make sure

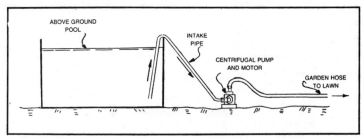

Fig. 14-12. To drain an above-ground swimming pool, it will be necessary to use a centrifugal pump with an electric motor.

that your present septic system is capable of handling the extra waste. This again stresses the point that future additions should always be taken into consideration when installing a new plumbing system.

Garbage disposers come in many different designs and models, along with a wide range of price tags. Those in the medium price range will be adequate enough to deal with most food wastes including vegetable and fruit peelings, egg shells, steaks, roasts, melon rinds, celery, carrot tops, fats, greases, small bones, tea leaves, pieces of bread, coffee grounds, etc. More expensive models with high-powered motors and heavy duty shredding mechanisms are capable of dealing with medium-sized bones, corncobs, sea food shells, and similar hard materials. Food waste disposers are not designed to handle the following: rags, tea bags, wires, strings, large bones, aluminum foil, cardboard, rubber, broken pieces of dishware, glass, bottle caps, cigarette filters, etc. If any of these types of materials are put into a garbage disposer, it is likely to just stop functioning. Also, such items as corn husks or artichoke leaves may get wound up in the mechanism and cause the unit to stop.

Figure 14-6 shows the interior mechanism of a typical garbage disposer. A disposer consists of a shredding device which is run by a motor. The shredder chops the food wastes into very small pieces. The upper portion, sometimes called a *hopper,* is mounted inside the sink drain and replaces the strainer and the flange that normally come with a sink. The lower portion of the disposer houses the motor, which is mounted inside a metal housing. The water seal between the upper and lower portions serves to protect the motor from water damage. The blades or impellers, which are attached to a flywheel, batter the garbage and throw it against the stationary shredding element and cutter surfaces. The speed of the rotation of the flywheel is 1700 revolutions per minute (rpm) or more, depending upon the power of the motor. The waste pipe connection (the tailpiece) is located a little above the water seal and is connected to the discharge line through a P or S-trap, depending on whether the line runs through the wall or the floor.

Most garbage disposers come equipped with complete installation and operating instructions, but a few of the rules common to all disposers

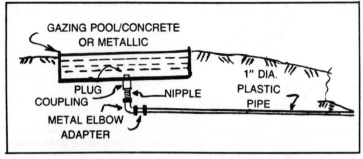

Fig. 14-13. A gazing pool is best installed on high ground with a downward slope for ease of draining each fall or for cleaning purposes.

are pointed out here. First, a disposer should not be switched on until water is flowing into the unit. Second, this water should always be cold. The reason is that cold water solidifies the fat and grease content of the waste which can then be easily shredded and washed down into the drain line. If hot or warm water is run into the unit, the liquid or semi-solid fats will stick to its inside walls and to the shredding element itself. This will cause either clogging or jamming.

Figure 14-7 shows a complete garbage disposer unit as it appears prior to installation. These units come as a fully assembled unit ready for installation. They are designed to fit the sink and drain openings without making any alterations in them, and they seldom need much plumbing work. Since both electrical and plumbing alterations will be a part of the actual installation, a check with local authorities is recommended to avoid incorrect or defective work which may cause future problems.

Installation

The first step in installing a garbage disposer in an existing sink is to disconnect and remove the sink drainpipe and the old sink flange, including any gaskets or sealing material. Apply a heavy bead of ordinary plumber's putty around the edge of the sink opening and the underside of the sink flange unit which came with the new disposer. After inserting the flange from the top into the sink opening, press it down firmly until it grips the putty layer. Do not turn it around, as this may cause the seal to break and result in a leaking installation. remove any excess putty. Working from the underside of the sink, install the fiber gasket and the clamping or backup ring with its flat side up on the lower portion of the flange. The next step is to install the support ring or the mounting ring, which has three holes with mounting bolts. All these parts should be held above the groove in the flange. The snap ring is slipped over the flange and moved up until it snaps and fits into the groove. Tighten the three mounting bolts in turn to apply a uniform pressure until they hold the clamping ring and the gasket firmly against the flange.

The main disposer unit is now raised to the sink mounting assembly. Turn the body flange counterclockwise until it engages inside the mounting assembly. Insert a screwdriver into a lug and turn the body flange toward the right until it locks in place. The unit can now be attached to the plumbing system. This is simplified by the fact that most of the necessary components will come with the unit. This is the case with the discharge tube. If this is not supplied, a straight tailpiece can be used. The trap is connected either to the disposer discharge tube or an appropriate length of tailpiece by means of a slipnut. Follow the plumbing code requirements for any and all connections. Be sure to adhere to the manufacturer's instructions, since design and installation procedures will vary somewhat from unit to unit. Once all the connections have been properly made, apply power to the disposer and check all the joints for any signs of leaking.

In regard to suitable electrical requirements, the disposer should be connected to a 120 volt 60 Hz ac power supply in most cases, although some units are specially manufactured for 200 volt service. Before making the connection, turn the power off. Open the disposer base by removing the electrical cover. Pull the electrical wires out. The *BX* or *Romex cable* for the toggle switch should be connected to these wires. Remember to refer to the manufacturer's instructions for making these electrical connections, and be absolutely sure the unit is grounded in the proper manner. The electric cover of the disposer unit can now be replaced, and the unit tested for satisfactory operation. If all instructions have been followed carefully and correctly, the unit should function normally.

Repair

A garbage disposer is a sturdy and durable unit designed to provide trouble-free service for many years if operated properly. Even if the disposer is subject to a minor amount of abuse occasionally, this should not do any permanent damage to the unit. Obviously, this does not mean that the disposer can withstand consistent abuses and carelessness. Always be

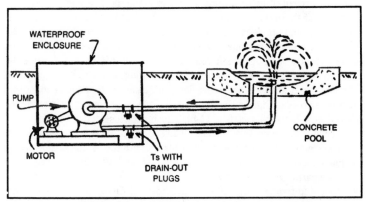

Fig. 14-14. Installation of a fountain with a pump underground.

sure to use cold water when running the unit, since this will prevent clogging of a serious nature. It is not unusual for a disposer to have a tendency to retain small amounts of grease or other materials, which will cause the unit to give off an unpleasant odor. This can be alleviated in a simple manner.

A drain cleaner may be used to clear a disposer of anything which has become lodged on its walls. However, be careful when using this method, as some cleaners are not specifically meant to be used in garbage disposer units. The cleaners approved for use with disposers contain a petroleum compound which will effectively deal with clogs and grease and will also act as a deodorant. Be sure to read the label of the cleaning agent before using it.

Another method which has proved successful in periodic cleaning of disposer units is to fill the disposer halfway with ice and run it until the cubes are shredded and the noise subsides. Then throw in a few pieces of lemon rind and run the disposer again for a minute or two. This type of treatment will not only remove any materials lodged on the walls of the unit, but will also produce a deodorizing effect.

Since a garbage disposer is really a sealed unit, it is not recommended that you attempt to disassemble it in the event of a major breakdown. Under these circumstances, it is suggested that an authorized repairman be contacted. More common problems such as a jammed flywheel, motor failure, slow grinding or shredding, water leaks, drain stoppages, and unusual noises during operation may be tackled, being careful to shut the unit off before making any repairs. If it becomes necessary to call a professional and the problem seems to be a major one, it is wise to compare the cost of repairs with the cost of a completely new unit. Obviously, if the two prices are about the same, the best choice is the new disposer.

One of the more common problems with a garbage disposer is that it makes a humming noise when turned on, but does not actually start. This is usually attributed to something jamming inside the unit and is caused by turning off the unit before the materials inserted at last use were completely ground up. A hard, solid object may remain stuck in between the shredding element and one of the impellers, or a prohibited item may have inadvertently been thrown in. Making sure the unit is off, reach inside the unit and try to locate what is causing the jamming. Remove it. Take out the disposer circuit fuse. To turn the flywheel, a special tool designed for this use is available (Fig. 14-6). Alternately, one can be made using a 3-foot length of ⅜-inch drill rod (Fig. 14-8). This is then inserted and the flywheel is turned in the reverse of its normal direction. If this is not possible, turn off the disposer and pry the impeller blades with the handle of a broom or some similar piece of wood. As the particles are loosened from under the moving parts, the impeller will move freely. This procedure may have to be repeated two or three times to provide satisfactory movement. Once the impeller appears to be freed, remove the broom handle and press the reset button, which will always trip if the unit becomes jammed. Run a test of the unit. The result should be a properly functioning disposer.

If the unit does not start when turned on, the problem can probably be traced to one of three things: a blown fuse, a defective switch, or a stuck switch plunger. Both the fuse and switch are replaced easily, and a stuck plunger is simply freed and cleaned. If the shredding and grinding action seems to be obstructed, this is usually caused by a stuck impeller, worn-out cutting edges, or an excess feed of hard solid waste such as bones. First, shut off the power and check the impellers. If they appear to be jammed, try to free them by removing the materials. Then apply a small amount of light oil in their pivots. If the impellers do not seem to be damaged, and they function properly even though the actual shredding action is impaired, the stationary shredding element needs to be changed.

If leaks are the problem, this is probably due to either worn-out gaskets or loose plumbing joints. If a motor seal is leaking, water may enter the motor and burn it out. Gaskets are easily replaced, as well as the plumbing connections. However, the service manual for the particular unit should be consulted before attempting to make any repairs to the motor seal.

SWIMMING POOLS

Although not every home is equipped with an in-ground swimming pool, it is worth discussing since the number continues to increase each year. A pool installed in the ground is a permanent installation, and strict codes governing proper operation and maintenance are in existence. These codes specify that the water must be properly circulated, filtered, purified, skimmed, chlorinated, and provision for periodic drainage must be installed. The actual installation of a pool is not really within the scope of the average do-it-yourselfer, and the work in most cases is done by professionals skilled in this type of installation. However, it is worth noting the different components that make up an in-ground swimming pool and the manner in which they operate.

In operation, the pump lifts water from the pool through the skimmer and delivers it to the inlet side of the filter, where it is purified by passing through a sand and gravel bed. The filtered water rises through the central pipe and enters the pool in a purified state. Normal filtering action consists of a number of specific steps, each designed to perform a set function. See Fig. 14-10. The three operations are the normal filtering action (A), the backwash process (B), and the waste process (C). During a normal run, the water goes down through sand and gravel and rises through the central tube. In the backwash cycle, it enters at the top of this tube and rises through the filtering media. At the start of operation each season, it takes a while for the filter to produce water which has been adequately purified. Again referring to Fig. 14-10C, it can be seen that until the quality of the water has improved, it does not return to the pool. Instead, it is drained off by means of a drainage system.

The skimming action is shown in Fig. 14-11. All water from the filter passes through the skimmer and shoots into the pool in a jet-type action. This provides both excellent skimming action and circulation. There may be two return lines to the pool, one leading to an inlet in a side wall of the pool or possibly to the stairs, and the other connected to the underside of the skimmer where the filtered water enters the pool through a nozzle.

In areas of the country where winters are severe, it may be necessary to empty the pool at the end of the season. Some towns will not permit the draining of pool water into roadside drains and/or sewers, and a septic system is really not capable of receiving this amount of water all at once. Under these circumstances, the only solution is to drain the pool gradually over the lawn so that the water can be absorbed slowly. The pool is equipped with a waste line in its plumbing system near the filter. To drain the pool, attach a garden hose or possibly a sprinkler system to this line. Let the pump run for a few hours every day. As the water seeps into the lawn, it will have the effect of improving the quality of the grass and will eventually be absorbed into the ground water table.

In the case of an above-ground swimming pool which is not equipped with a pump, it will be necessary to purchase or rent a small centrifugal pump with an electric motor in order to empty the pool. Attach an intake hose to the inlet end of the pump, and attach a garden hose to the outlet end. In order to operate the pump, remember that it will have to be primed. Once the intake hose is attached to the pump, fill the hose completely by pouring water into the other end. Holding the far end closed in some manner, lower it into the pool, open the end by removing the obstruction, and start the pump. This procedure is shown in Fig. 14-12.

Regardless of whether the pool is an in-ground or above-ground type, it is recommended that any chemicals be discontinued at least a few days before draining the pool onto the lawn. Some of these chemicals may possibly cause damage to plant life. Before beginning the actual procedure, run a simple check to make sure the chemicals are dissipated by filling a sprinkling can or some other container with pool water. Empty the contents of the container on a small, out-of-the-way area of the lawn, preferably in the evening, and check it the following morning. If the grass seems to have suffered no ill effects, the water is safe.

A check with the local authorities should provide all the necessary information and rules governing swimming pools. One such rule is that no cross-connection is permitted between this system and the municipal or community water supply, even if check valves or other protective devices are a part of the installation. Also, the pool should not be filled directly from the public supply. An open tank should be filled up with water first, and this water is then transferred to the swimming pool. Be sure to follow all guidelines carefully to make for a safe, clean, and functional installation.

GAZING POOLS

The ideal location for an installation of this type is on high ground with a downward slope angling away from the pool. This will provide the easiest type of pool to drain (Fig 14-13). The gazing pool is connected to a pipe buried underground at a suitable pitch so that the pipe's lower end comes out of the ground near the lawn or at some other suitable spot. A 4-inch plastic pipe will work well for this line. The plug in the center of the pool sits on a coupling attached to its bottom. The pipe is connected to the coupling through a metal-to-plastic adapter, a 90-degree elbow, and a small nipple. If the gazing pool is installed on level ground, this will necessitate the use of a pump in order to provide drainage when necessary. Generally, this will only apply in areas where winters are severe. To drain the pool, the plug is taken out and the line is left open. This prevents the accumulation of water or melting snow, which would cause damage to the sides of the pool if it freezes. In areas where winters are quite cold, it is recommended that a fiber pipe be used in place of plastic pipe.

FOUNTAINS

A fountain serves no useful purpose. It is simply a visual improvement to the exterior of the property. This installation is really quite simple. All that is needed is a pump, a motor, two plastic pipelines with tee connections, a small waterproof enclosure for the pump and motor, and the actual fountain assembly. Figure 14-14 shows a typical fountain installation.

Purchase a fountain pump unit from a plumbing supply store or possibly from a specialty house that deals in these types of installations. If this is not possible or desirable, an old washing machine pump or an ordinary centrifugal pump with a matching motor can be used with some minor modifications. Locate the pump so that its axis is a couple of feet below the bottom of the pool such that it will not have to be primed each time it is operated. Electrical power to the motor and to any night lights at the pool should be carried by waterproof cable installed in a trench at least 1½ feet deep. A check with local authorities is recommended in order to comply with both electrical and plumbing rules and regulations. Alternately, a flexible waterproof cord may be used as long as it can be attached safely to an outdoor electrical outlet.

As with swimming pools, fountains use the same water over and over again. They may need to be drained and refilled either during the season or at least once before winter. The two pipelines should be pitched toward the pump in order to make the draining process a bit easier and faster. To drain the pool, the tees are simply opened.

SUMMARY

Each individual unit manufactured varies somewhat from a similar unit from another company. Refer to the instruction manual provided with the appliance. Since these units do require electricity in order to operate, caution should be exercised when either installing or making repairs on them. *Always* be sure no power is present in an appliance before any work is attempted. If the plumbing portion of the system is to be worked on, the water supply should be shut off at some appropriate point. With a new installation of a dishwasher or washing machine, it is wise to provide a shutoff valve at some point near the appliance for ease in making future repairs.

Swimming pools and the like may be considered by some to be beyond their reach economically at present. As the demand increases, the prices of these types of installations will decrease as they have done in the past, making a pool well within the scope of more and more people. Fountains really serve to enhance the beauty of a lawn or yard and are also becoming more popular. An installation of this type is well within the reach of most homeowners, since most of the work can be done quite inexpensively.

Improvements and/or Modifications to an Existing Plumbing System

15

As the prices on new homes increase due to higher construction and material costs, more people are opting for older homes or simply staying in their home rather than purchasing a new one. There always have been those who prefer homes which were built 70 or 80 years ago, simply because of their design and quaintness. Unfortunately, a lot of people don't take into consideration the cost of making the necessary modifications to an older home when making the original purchase. This may make the improvements more expensive than a brand new house on the market today. It's always a good idea to have a professional come in and look over an old house before making any final decisions.

Much money involved in improving an existing plumbing system can be saved by doing the work yourself. It may simply be a case of adding another bathroom facility or additional drainage capabilities, or it may involve installing all new supply lines. Depending upon the job, it may be necessary to do some remodeling in the form of tearing down interior walls to get at the old plumbing. Whatever the case may be, keep in mind that not only will this work improve the functioning of the plumbing system itself, but it will also improve the appearance of the home and increase the value. Knowing this, you can weigh the cost of the improvement or modifications against these factors and make a decision at the onset as to how extensive the work should be.

PLANNING PLUMBING EXTENSIONS

Before doing anything else, sit down and sketch out the proposed improvement. This sketch should show the proposed locations of the fixtures, the routes along which the pipelines will be laid to connect to these fixtures, and the lengths and sizes of pipe to be used. On small and simple jobs, it will suffice to make a plan showing the dimensions of the floors and the positions of the walls. More often it will be helpful and desirable to work out an elevation, too. This sketching will also be quite helpful in pointing out any important structural features which must be taken into consideration in the laying of pipelines and the location of fixtures. The same basic principles of planning and laying out a new plumbing system apply to an old one as well. The system should be convenient and yet not overly expensive. There should be a minimum amount of breaking through the walls and floors whenever possible.

One of the main difficulties encountered is actually finding the needed space for the proposed additions. Obviously, the more reconstruction, the higher the cost. It may be a good idea to compare the cost of making the necessary modifications to the existing space with the cost of adding on the whole room needed to house the facilities. Generally, it is easier to make outside extensions if space is available. If this is done, the additions can be planned and constructed in the same manner as in a new home, with the added advantage of more freedom in laying out the plumbing system. Even though this may be much more convenient and easier on the person doing the work, chances are that the addition of a whole room will be the more costly of the two. If it is necessary to economize as much as possible, you will obviously have to work with what is available.

Refer to Fig. 15-1. At the top is a drawing of the house as it is when considering the options available. As can be seen, the floor of the home has two bedrooms with no bathroom nearby. To provide the bathroom facility, two alternatives are presented here. In plan A, the closets attached to the bedrooms have been converted into an access hallway leading to the new bathroom, which has been extended outside of the existing building. This bathroom has been provided with a tub, toilet, lavatory, and linen closet. The soil stack is placed in the new wall close to the new toilet, and the house drain runs straight to the house sewer outside. In plan B, the space occupied by the closets has been extended by installing a new partition wall, and the bathroom has been provided for in this way. This has the effect of reducing the size of one of the bedrooms. The soil stack can either be placed at position X inside the new partition wall, or outside at position Y and boxed in. A new house drain will connect it to the existing main stack or to the existing house drain.

To add a new soil stack in a single story house, it is a simple task to install the stack inside a partition and carry it up and out through the roof for venting. In a multi-story house, the stack will generally run through partitions on the upper floors. If this is not possible or practical, it will be necessary to box in the stack (Fig. 15-2). If space provides, one can even be boxed in a closet, which will reduce the changes in the appearance of a room.

When making improvements or additions, try to plan new fixtures in such a way as to make them back-to-back with existing fixtures. It will be possible to make the necessary connections to existing lines without having to run new lines. When adding a new toilet which will have to connect to a soil stack, try to position it somewhere in the home so that a new soil stack will not be needed. Generally, it is much easier to install horizontal runs of pipes than vertical ones, particularly in the basement or attic. The pipe can go under or over the joists. Figure 15-3 shows the method for running new lines under floors which will necessitate raising the level of the floor. To run new vertical lines, it will be necessary to remove the facing material from the walls to expose the framing. The same rules will apply as far as any notching to studs and/or joists that held true in a new plumbing system. The water pipes, being smaller, run in the same space with the drainage pipes. The hot water line, unless insulated, is separated from the cold water line by at least 6 inches. When making improvements, it is also better to use either plastic or copper materials, since the connecting of piping made of these materials is much easier and less space is required to do the work.

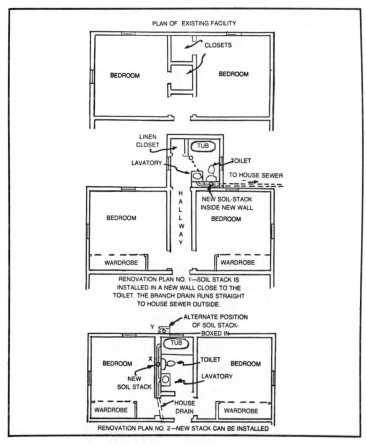

Fig. 15-1. Two options are available in this example in order to provide an additional bathroom.

341

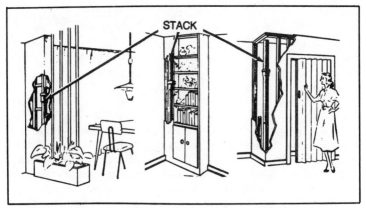

Fig. 15-2. A new soil stack may be installed in a partition, closet, or bookshelf in order to keep the appearance of a room the same.

LOCATING OLD PIPES

The first step will be to locate the old pipes inside the walls and/or floors. New pipes can be connected to an old line where it is exposed, such as in an attic or basement. However, most of the time it will only be possible and desirable to make connections to pipes concealed within a wall. In old houses where plumbing has been altered and remodeled more than once, the process of locating the piping may be a bit tricky. A metal detector will be a great help if the piping is known to be either iron or steel construction. The device will not respond to copper or plastic at all. One approach that may prove helpful in locating the lines in question is to follow the pipeline from its exposed location in the basement or attic and work up or down from there. With the help of another person, use a hammer or some other dull instrument to strike the riser while the helper positions himself or herself on the upper floor and follows the sound.

Another method is to shut and open the flow of water in the line and listen to the noise. A more exact method is to measure the distances of the exposed portion of pipe (for instance, in the basement) from two side walls, go upstairs, and measure back the same distances from the same two walls. This should pinpoint the location of the pipe inside the wall. An easy approach in trying to locate the DWV pipes is to go up on the roof and observe the pipes' positions in relation to the walls. These can then be located on the lower floors. Remember that all of these methods will probably only provide approximate locations. If the lines are cast iron or galvanized, it is possible to determine the exact location by driving long test nails into the wall until they hit the pipe. Obviously, if this method were used with either plastic or copper pipe, the lines may accidentally be punctured and flooding would occur.

Before proceeding with the actual job at hand, be sure to check with the local authorities to find out what is permitted, the kinds of pipe and joints allowed for various purposes, the type of supports required, etc. Also, using the sketches made previously, make a trip to the local plumbing store and order all necessary materials and supplies before starting. It is a real nuisance to have to stop at midpoint to run to the store for something that should have been on hand at the onset. If cost is a factor, take the time to shop around and get prices from more than one dealer before making the necessary purchases. The same holds true as far as purchasing new fixtures and appliances. With all the necessary information and materials on hand, you are now ready to begin.

ADDING NEW WATER SUPPLY LINES

Pitch is not critical when laying water lines. They can be laid horizontal, with a negative pitch, or even vertical, with the direction of flow upward. On the other hand, it is already known that since drainage pipes do not run full or under pressure, they are installed at a slight downward pitch to create the necessary flow velocity. Normal pitch for drainage pipe is ¼ inch per foot of pipe length.

To get back to the water supply lines, regardless of the type of material used for the original pipelines, the addition of a new line is quite simple. The first step is to determine a convenient point from which to start the run. The most convenient place is at a junction of two lines which are connected with an elbow. If one line has a long stretch or consists of flexible tubing, it will be easy to torch-heat the elbow and bend the line to slip the elbow off (Fig. 15-4). The ends of the stubs while still hot, should be wiped clean to remove all traces of solder and then coated with flux. Clean a new tee fitting and slip it onto the ends of the pipes. Make the joint by heat-soldering. The new line can now be run. Rigid copper pipe is the better choice in areas where the line will be exposed, since it is less likely to sustain damage. It is also wise to install a stop valve on the new line to permit work on subsequent lengths without interruption of the supply on other lines. If this is not done, remember to install a cap on the end of the new during its installation at the end of work each day.

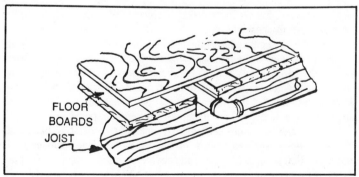

FLOOR
BOARDS
JOIST

Fig. 15-3. A method by which new lines may be run under floors.

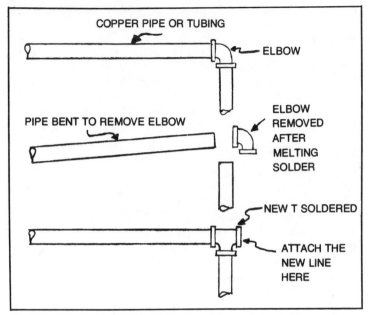

COPPER PIPE OR TUBING

ELBOW

PIPE BENT TO REMOVE ELBOW

ELBOW REMOVED AFTER MELTING SOLDER

NEW T SOLDERED

ATTACH THE NEW LINE HERE

Fig. 15-4. If the line is flexible tubing, it may be torch-heated for removal.

If it is not practical or possible to remove an elbow and replace it with a tee, or when a line is to begin at a point on an existing straight run of pipe, a small length of this straight run should be cut out and replaces with a similar length having a tee fitting attached to it (Fig. 15-5). Joints at both ends are made with the help of slip couplings which are slipped over two small lengths of pipe on either side of the tee. Once the pipes are set in place and aligned properly, the couplings are slipped over the joints and soldered.

If the project involves the addition of rooms or the expansion of an attic, the new pipes should obviously be installed before the inside walls are constructed. Joints and connections which will remain inside the walls, or otherwise inaccessible, should be carefully soldered so as to remain watertight for the entire life of the house. The advantage of using flexible tubing is that it needs a very small number of fittings and can be bent in any direction desired. In fact, it may only be necessary to provide fittings at the beginning and end of each run, which is both convenient and inexpensive. However, in most of the jobs a combination of the rigid copper pipe and flexible tubing is used. The tubing is not used in exposed locations.

If the improvements necessitate connecting to existing lines with threaded pipes, such as iron or steel, the method used will be a bit different. Referring to Fig. 15-6, fit together a union, a nipple, and a tee fitting of the same inside diameter as the line XY. Measure out the distance PQ between the center of the union and the center of the outer hub of the tee fitting. From the line XY, cut off a portion BC exactly equal to the length PQ. Unscrew the lengths XB and CY from their joints, and thread their ends AB and CD. Screw on the tee fitting at CD and put back the line CY into the system. Screw the nipple into the tee, and the union onto the nipple. Tighten up the assembly. Take the other section of pipe XB, screw it into the union, and put it back into the system. By turning the union around, tighten up and complete the job. The new line is connected starting from the end N of the tee fitting.

ADDING NEW DRAINAGE LINES

If the existing drainage plumbing in the home consists of plastic or copper pipes, additions and alterations will be much easier. With iron or steel pipes, the work is more difficult mainly due to the type of joints and fittings that will be required. A new drainage line from a fixture will have to be connected by means of a suitable fitting to the already existing soil or waste stack. If its length is substantial, a *revent pipe* will be run from the drainpipe of the fixture to the vent stack above.

The first step is to determine what path the new line will follow. A number of options will probably be available. For a new fixture, such as a bathtub, the drainage pipe will probably go down vertically through the floor, running inside a wall until it reaches the underside of the basement

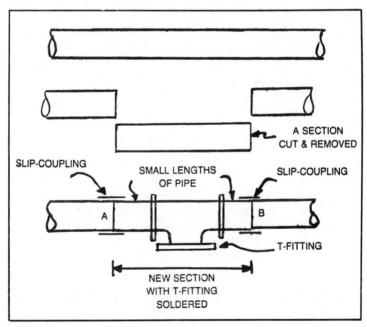

Fig. 15-5. To add a new fitting for the connection of another line, a small length is cut and removed.

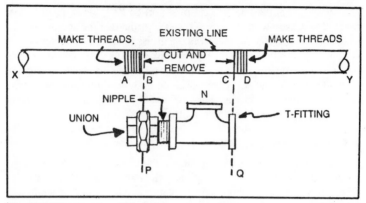

Fig. 15-6. The process will be a bit more lengthy if working on iron or steel pipes.

ceiling. From that point it will run across to the soil stack, or it may run vertically for some distance and then horizontally across a ceiling until it reaches the soil stack. Sometimes it may discharge into a waste stack instead of the main soil stack. In most cases, a vent pipe connection will be necessary.

Be sure to remember when cutting the pipes to allow for overlap length at fittings and for the dimensions of the fittings themselves. Also, be sure to adhere to the code insofar as the size and type of piping used. If no code exists, use the following measurements as guidelines: sinks, basins and bathtubs, 1½-inch to 2-inch drainpipe; toilets, 4-inch drainpipe.

The drainage connections for fixtures in a bathroom are shown in Fig. 15-7. The toilet should always be located next to the soil stack. From the drum trap of the bathtub, the drain line runs horizontally into the wall, turns right, and is installed in grooves cut in joists. Next, it joins the vertical waste pipe through a Y connection. The upper portions of the soil stack and the waste pipe act as vent pipes in this installation. A revent pipe connects the tub drain line to the vent pipe. This provides for proper venting and prevents the drum trap seal from being broken as a result of sewer gas pressures. A sink connection is made to the waste pipe. A cleanout is provided on the small horizontal pipe connecting the waste pipe and the soil stack, which will be invaluable for future repairs and cleaning.

Connecting a Drainage Line to a Soil Stack at a Cleanout

If a cleanout is the connecting point for a new fixture drainpipe, the job will be a bit easier. An additional stack may also be connected in the same manner. Remove the plug and attach a nipple or short piece of pipe by caulking or screwing it in, whichever is applicable (Fig. 15-8). A new cleanout is now attached by means of an appropriate elbows. The new drainpipe or stack pipe is connected by either caulking or screwing it in.

Connecting a Drainage Line Directly to a Soil Stack

A good deal of remodeling is done to provide for the addition of new fixtures, which will involve running new drainpipes from these fixtures to an existing soil stack or house drain. It may also be necessary to provide additional vent pipes. Performing alterations to both existing drainage lines and the soil stack is a bit more difficult and time-consuming than working on the water distribution system. This is attributed to the fact that not only are the pipelines themselves usually larger, but they are almost always entirely inside the walls except in the basement or attic. The actual procedure will differ depending upon the type of pipe used in the original system to which additions are being made.

Existing Stack Is Cast Iron or Steel

The connection of a new drainage line to a cast iron or steel stack is accomplished with a *sisson joint*, which is a slip fitting. Referring to Fig. 15-9, the upper section of pipe is first supported firmly by installing iron straps on at least two joints. Using a hacksaw, cut and remove a full length of pipe. After positioning a Y fitting, a sisson fitting, and a piece of pipe cut

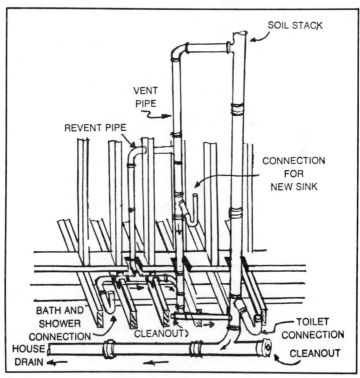

Fig. 15-7. Proper drainage connections for fixtures in a bathroom.

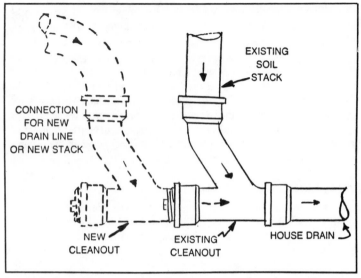

Fig. 15-8. Connecting a new drainage line to a soil stack at a cleanout.

to the required length, expand the sisson fitting and make the four joints secure by caulking. The new drainpipe is now connected to the Y.

If the addition is being done near the upper portion of the stack close to the roof, the work will be a bit easier. First, open out the topmost joint and lift the upper section a few inches. Tilt it and strap it securely in place. Referring to Fig. 15-10, use a hacksaw to make a cut at point B, discard the piece BC, and slide AB down into the lower socket. Add the Y fitting, straighten the pipes, and make the necessary joints. In situations where it is possible to lift the topmost section substantially, cutting may not be necessary. This can be done by holding the upper portion in its raised position and simply slipping the new fitting under it. The two joints are then made. The stack will now be protruding from the roof a bit more than it was before, but no ill effects will be experienced.

Existing Stack Is Copper or Plastic

The procedure to connect a new drainpipe to a copper or plastic stack is shown in Fig. 15-11. First, use a hacksaw to cut and remove the required minimum length of pipe, not a full length as was done in the case of cast iron or steel. Be sure to provide some form of support for the upper portion of the stack to prevent it from either sagging or falling out. Install a Y or tee for the new branch drain by either sweat soldering of the pipe is copper or by cementing in the case of plastic. Make a collar by sawing a coupling in half, and slip it over the lower pipe. Cut a piece of tube a little shorter than will go between the new fitting and the lower portion of pipe. Connect its upper end to the new fitting by sweat soldering or cementing, whichever is applicable. Slip the collar up and make the lower joint in an appropriate manner to provide for the connection of the short piece of pipe to the lower portion of the stack. The addition is now complete.

Adding the Fixture Pipes to the Branch Drains

This procedure is done in much the same manner as discussed in regard to soil stack connections. The only difference is that a branch drain is smaller in size and runs in a horizontal direction. Figure 15-12 shows a typical connection of a lavatory through a Y fitting. The addition of fixtures may have the effect of overloading the existing branch drains. When the system of a home is originally designed and installed, the capacities of branch drains are based on the manner of fixtures fitted at the time, plus any plans for future expansion. Each time a fixture is added, the capacity is taxed, particularly if the fixture is in the kitchen (such as a sink or garbage disposer). Its addition may increase the amounts of grease, oil, solid matter, or rinse water which needs to be drained away. If the branch drain begins to clog frequently, it should be replaced with a larger size pipe. If the overloading is presenting only minor problems, this may not be necessary. However, if it becomes obvious that drainage is hampered

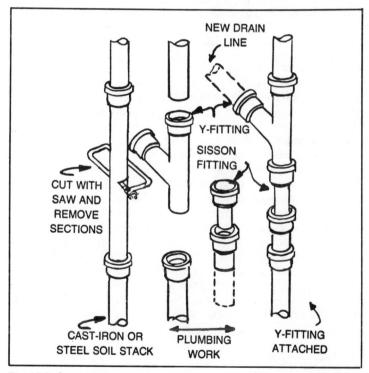

Fig. 15-9. Connecting a new drainage line to a cast-iron or steel stack.

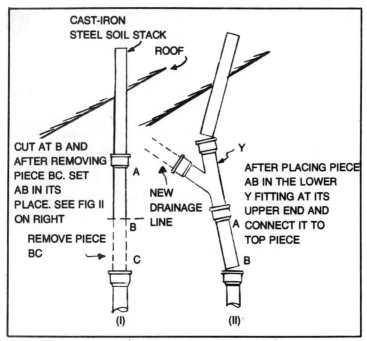

CAST-IRON STEEL SOIL STACK

ROOF

CUT AT B AND AFTER REMOVING PIECE BC. SET AB IN ITS PLACE. SEE FIG II ON RIGHT

A

NEW DRAINAGE LINE

B

REMOVE PIECE BC

C

Y

AFTER PLACING PIECE AB IN THE LOWER Y FITTING AT ITS UPPER END AND CONNECT IT TO TOP PIECE

A

B

(I)   (II)

Fig. 15-10. Connecting a new drainage line at a point near the upper portion of the stack is a bit easier.

considerably, the line should be replaced. All lines that will be expected to receive an increased discharge due to additions should have their capacities checked. The pitch should also be checked. If inadequate, it should be increased. This will provide for a greater flow velocity and less chance for materials to accumulate in the line.

When making additions to an existing system, stress should be placed on careful planning in the early stages to keep the carpentry work required at an absolute minimum. Holes will need to be cut in both walls and floors to run lines from one floor to another. In any event, some carpentry work will be necessary. The use of power tools, if economically feasible, will save you a good deal of time. The power tools that will be useful are a large saber saw with an assortment of blades, a heavy-duty angle drill with an assortment of drill bits, and a cat's paw for pulling nails.

Proper pitch must be maintained throughout the system, including the new lines to be installed. Check both the elevation and the angle of each fitting and its connection so that this pitch is properly provided for. This will keep any future clogging in the lines to a minimum.

Flexible Copper Tubing

This type of material is very popular today in home plumbing systems, mainly because it can be easily bent in any direction. Long runs can be effected with very few fittings. When it is necessary to run a length of tubing inside a wall from one floor to another, tie a long rope with a weight attached to one end. Slowly lower the rope until it reaches the hole in the wall at the lower floor. With the help of another person on the lower level, have him or her pull the weight and the rope out to a point which positions the copper tubing properly in place (Fig. 15-13). For horizontal runs, use an electrician's snake instead of the rope. Tie it to the copper tubing and have another person pull from the other end as the pipe is fed into the hole on the upper floor.

STANDARD PIPE SIZES

In areas where no codes exist, the following figures will be of aid in providing adequate size piping for each section of the system:

Drainage Pipes

Sewer line	4 inches
House drain	3 to 4 inches
Soil stack	3 to 4 inches
Roof vent	3 inches or more
Branch Drains	1½ inches

Water Supply Pipes

Service entrance	1 inch
Branch to water heater	¾ to 1 inch
Hot and cold mains	¾ inch

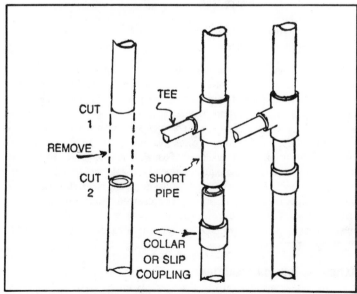

Fig. 15-11. Connecting a new drainpipe to a copper or plastic stack.

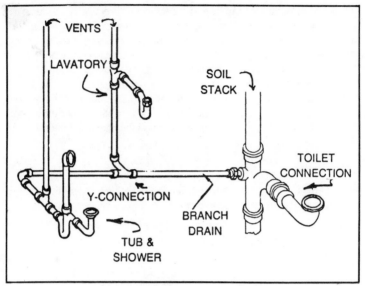

Fig. 15-12. Connection of a lavatory through a Y fitting.

Water Supply Pipes

Branch lines to dish-washers, laundries, tubs, showers, and sinks	½ inch
Branch lines to toilets and lavatories	⅜ inch

If it is necessary to join pipes of the same material but of differing size. A *reducer* is used. These are available in almost any form including reducing elbows, tees, and couplings. If the pipes to be joined are the same size but are made of different materials, it will be necessary to use an adapter. These also are available in common configurations such as adapter elbows, tees, and couplings. When joining copper DWV pipe to cast-iron pipe, transition fittings are used. To make the connection, the fitting is caulked into the cast-iron pipe at one end and soldered into the copper pipe on the other end. When connecting plastic pipe to an existing cast-iron line, the spigot end of the plastic pipe is inserted into the socket of the cast-iron pipe. The seal is made with a special compound designed for these purposes. This compound is easy to obtain and can be found at most plumbing supply stores.

FIXTURE DESIGN CONSIDERATIONS

The actual installation of new or replacement fixtures has been discussed earlier and will not be repeated here. Depending upon the situation, the amount of work required may be either quite simple or, on the other hand, quite extensive and time-consuming. Plan the additions in such a manner that if possible, the new fixtures can be connected to the existing lines in the most direct route possible. If a whole bathroom is being added, try to plan it so that its plumbing will be back-to-back with an existing fixture, such as a kitchen sink, to avoid running new distribution and drainage lines. Careful planning can save a good deal of both time and money.

When shopping for new fixtures and their appropriate trim, take the time to shop around. The higher the price tag, the better the quality. There is also a quite diversified selection as to the materials used in the manufacture of both fixtures and faucet assemblies. Generally, the more expensive faucet attachments may have a polished gold or chromium surface, or possibly a brushed satin finish. These are highly resistant to rust or any other type of corrosion and are also the least likely to break down. For those new additions where cost is a factor, try to stick with the middle-priced fixtures rather than the least expensive ones. These will perform satisfactorily, while the low-priced fixtures will soon show signs of de-

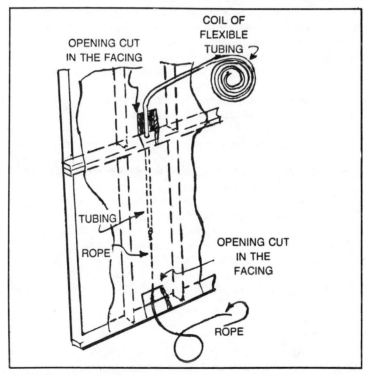

Fig. 15-13. To run copper tubing, simply tie a rope with a weight attached to one end onto one end of the tubing.

terioration and should be avoided if possible. When visiting a plumbing supply store, ask the dealer to show you the various qualities of faucets and to point out the differences between the cheaper and more luxurious items. The good quality items are usually made of heavy brass with a copper-nickel-chrome finish, while the cheaper brands are made of either zinc or aluminum die castings. The best proof of good quality is a well-known manufacturer's name stamp.

Shower nozzles are available in a number of design configurations. Some are equipped with such features as a volume spray control and a spray direction control, both of which are controlled by means of a ball joint inside the actual assembly. Nozzles can also be had with self-cleaning heads, which are quite useful in areas where the water supply is quite hard. Obviously, the cheaper shower nozzles will be more likely to break down or clog than their higher priced counterparts.

A number of safety devices are now available on shower heads which can prevent accidental burning. The first is a temperature control valve which maintains the shower water at a constant temperature. Almost everyone has had the experience of the water turning either hot or cold very suddenly while taking a shower. This is most often caused by some other member of the family using water at another fixture in the home, thus diverting some of the water from the shower. These valves can be had in two types, each of which performs the same basic function. One is a thermostat mixing type and the other is a pressure control type. The pressure control valve is a bit more popular and preferred due to the method in which it operates. A local plumbing dealer can provide information on both types and help you in selecting the type most suitable for your installation. The unit will usually come with complete instructions for ease of installation.

The other safety device available for showers is an automatic divertor control which automatically diverts the water supply back to the tub spigot after the shower is turned off. If the last person to use the shower forgets to pull down the shower lever that diverts the water to the tub, the next user will experience a sudden spurt of water from the shower nozzle when the faucets are opened. This is most unpleasant and aggravating and can be avoided completely by installing one of these automatic divertor controls. This type of device is not usually included with new fixtures, since most manufacturers provide a manual control with the assembly. However, these are available as individual units at most plumbing supply stores and are quite inexpensive and easy to install. Again, be sure to follow the instructions included with the unit.

SUMMARY

It is not really very difficult to improve a residential system that is either not performing adequately or is simply outdated in design and appearance. Yet it is not uncommon to find a great majority of people living with their system simply because they feel the job is either too costly or extensive a task. Both of these common reasons for not making the necessary modifications can be taken of quite easily.

Careful planning early on with regard to location will reduce both cost and time considerably. Examine the existing system, noting the location of both distribution and drainage lines, and try to determine the best possible place for the new additions in relation to convenient access points to these lines. If this is impossible or impractical and it is necessary to create either new stacks or branch lines, choose the route that will be as direct as possible and will require the least amount of construction. Any new additions will require some carpentry work and may necessitate the running of new lines. If all is kept to an absolute minimum, the work load will be considerably reduced along with overall costs.

Since some cutting into walls and/or floors will probably be required, this is a good time to consider making some changes in the appearance of the existing room. Floor tiles, coverings, fixtures, and their fittings come in a wide variety of colors and designs. By simply changing the materials or colors in a room, a whole new appearance can be achieved. These improvements don't have to be lavish or very expensive to bring about the desired result. Timing and planning done before beginning will provide for not only the necessary improvement, but may also result in a room or rooms with a much more modern appearance.

16 Preventive Maintenance

A home plumbing system which is installed properly and is periodically checked for malfunctions or inadequacies will provide a homeowner with many years of satisfactory service. Obviously, since most of this system will be located within walls and under floors, it may sometimes be difficult to perform checks on this portion of the installation. There are ways to maintain the system and use it properly over its entire lifetime to insure that no damage is done to the structure of the building inadvertently. Likewise, if all fixtures and mechanical parts of the system located outside the walls and floors are treated with respect rather than abused, they will last for many years to come. Proper treatment and preventive maintenance procedures discussed in this text will provide for a long life and a few repairs for a home plumbing system.

SEPTIC TANK AND DISPOSAL FIELD

Generally, a septic tank will need to be cleaned at regular intervals ranging anywhere from two to five years. Mineral sludge collects at the bottom of the tank during normal operation, while the other materials are discharged into the leaching field. The septic tank should be opened at least once a year in order to measure the dept of the top scum, the bottom sludge, and the liquid in between. When the combined depth of scum and sludge increases to half the total depth of the tank, this is an indication that the tank should be cleaned. This procedure is almost always done by professional septic tank cleaners with the necessary know-how to perform this operation properly. The best time to have this done is during summer or spring. If done in winter, the new bacteria will take some time to grow unless a small amount of old sludge and liquid is left behind for seeding purposes. Some people use special septic tank cleaning chemicals, but their usefulness is somewhat doubtful. These chemicals cannot really take the place of proper care and periodic cleaning.

If it becomes apparent that heavier demands are being placed upon the present system due to either an increase in family size or the addition of appliances, it is probably time to redesign the system and install a few more lines of trenches. If the original field and tank were planned taking into consideration future demands, this will not be necessary. As can be seen, this points out the need for careful planning before installing a septic system. If it is necessary to do some modifying of the present installation, be sure to consult a professional before beginning. He may be able to give some advice on the easiest and less costly approach to take, and he may even have some ideas on how to accommodate the increased demand using the present system as it is.

It is perfectly acceptable to plant grass over a leaching field and even over the septic tank and sewage lines leading to and from the field. Keep in mind that any bushes or trees with extensive root systems should not be planted too near the system, since these roots have a tendency to grow down through the joints and will cause blockages in the pipes. The field should be exposed to an adequate amount of sunlight, which aids in the absorption process. Any tall trees which shade the field should be removed, if this has not already been done at the time of the original installation. If a field is not functioning properly, this is probably one of the first things to check and also possibly the simplest to correct, since no digging up of lines and correcting the plumbing portion of the system will be necessary.

Some simple guidelines are listed here as practices which, if followed; will extend the lift of a septic system:

■ Do not overload the system by running several appliances at the same time.

■ Do not use strong bleaches and detergents in your washing machine. They slow down the bacteriological action in the septic tank.

■ Do not wash several loads of clothes all at once without giving the machines a break and the system a chance to absorb the water.

■ To prevent grease from entering the septic tank, install a grease trap on the kitchen line.

■ Do not discharge any strong chemicals, acids, paints, varnishes, etc., into the drainage system.

■ Because the capacity of the system is designed for the flow from the house, no rain water from the roof or yard should be connected to the field.

■ Do not discard such items as wrapping paper, paper towels, cooking oils and fats, coffee grounds, cigarette butts, and old rags into sinks or toilets.

If you find that the action of the system is a bit sluggish and no blockages can be found anywhere in the lines, here is a method which is extremely easy and produces satisfactory results. The most common cause for sluggish action that cannot be attributed to clogging is an insufficient amount of bacteria in the tank. The simplest solution is to flush some yeast down one of the toilets in the home at regular intervals, possibly once a week. This will serve to increase the bacteria population, since yeast is a form of bacteria. This is a very popular method which has been proven to work quite well and is also very inexpensive. The best type of yeast to use is brewer's yeast which can be purchased at any health food store.

If you find that an unpleasant odor is present at the site of the leaching field, this is a sign that the effluent is not being absorbed satisfactorily and can be attributed to many things. Check to see that the field is receiving an adequate amount of sunlight each day. Make the necessary corrections if this is the case. Some codes do not allow washing machines to be connected directly to the system, but that they must bypass the septic tank on their way to the leaching field. With either type of connection, this same problem may be encountered. If the washing of clothes places too great a demand on the absorbing capabilities of the field, it may be necessary to provide a separate drainage system completely independent of the septic system to take care of the water load from the washer. This can be provided either by a seepage bed or possibly a dry well.

WATER METERS AND SHUTOFF VALVES

Although not everyone has had the experience of a burst pipe or broken joint in their plumbing system, those who have can attest to the confusion trying to figure out where and how to shut down the system before any flooding damage occurs. This can be quite frustrating. Certainly anyone who has had this experience can look back and wish that he or she had known which valve controlled what at the time of the emergency. Even more frustrating can be knowing which valve must be shut off and finding that the device, through lack of a periodic inspection procedure, is not in working order and will not close properly. All valves should be checked regularly to insure that such a situation does not occur.

The information provided earlier should give you sufficient knowledge to go down in the basement and identify each valve in your system. With a marking pen and some tags, label each valve clearly as to which is connected to what line and so on. This will enable you and any other family member to shut off the right valve if the supply has to be turned off in a hurry. Everyone in the home should be shown where each valve is and how to operate the valves in the event of an emergency.

In cities and towns where the home is connected to a utility or privately owned water supply, the water meter will be the property of the company. If it becomes apparent by reading the dials on the meter that it is not operating properly, it is a simple matter to report this to the company. Regardless of who actually owns the meter, be sure you know where it is located and how to read it, as well as the location of its appropriate shutoff valves.

HOT WATER HEATERS

A water heater is really a safe device which, if properly maintained and equipped with some simple safety features, will last for many years. Up until the energy crunch of recent years, the average thermostat setting for a hot water heater was somewhere between 140 and 150°F. Also, most dishwashers required hot water at a temperature of about 150°F to clean the dishes satisfactorily. Today, the manufacturers of these units are producing energy-saving dishwashers which will perform at lower temperatures, thus providing their owners with a way to cut down on utility bills. Hence, the average temperature for hot water in the average home is decreasing somewhat. Not only will this bring energy costs down, but it will also increase the life of the hot water heater, since it doesn't have to work quite so hard to maintain the higher temperatures. The savings is really worth the loss of hotter water, although a setting of 120°F will produce sufficiently hot water for most uses.

The only real work that need be done to a water heater on a regular basis is to drain the water through the drain valve at the bottom of the unit to clean out any mineral deposits which may have accumulated. If this is not done, the unit will tend to perform less than adequately. All safety features should also be checked periodically to make sure they are in perfect working conditions. If they are worn or defective, their whole purpose in the installation is defeated.

PIPELINES AND FITTINGS

All types of piping and their appropriate fittings are meant to last for the entire life of the home. The materials are quite durable, as they should be, and are also corrosion-resistant. If respected and maintained properly by means of periodic inspections, they will endure normal use for years. The majority of the pipelines and fixtures will be concealed within the walls and the floors of the home. The major sites of exposure are in either the basement or attic. Since these lines are in the open, they have a greater chance of being damaged. If they are located in an area where it is expected that they may be in danger, either from furniture stored nearby or perhaps a carpentry workshop, the best solution is to "box" them in some appropriate manner. When providing this sort of protection, make sure sufficient clearance is given to the lines themselves and that any valves be easily accessible. If this is not possible or practical for any reason, the easiest alternative is to just be careful near the site of these lines and valves. If the family consists of children, they should be made aware of the locations of the pipe in both the attic and basement. Some types of pipe become quite warm and cause burning, so children should be warned in any case to stay away from any exposed plumbing. If it becomes necessary to move furniture, appliances, or any other heavy objects anywhere near the exposed lines, exercise caution in order to prevent damage.

FIXTURES AND FAUCET ASSEMBLIES

Sinks, showers, tubs, and toilets are constructed of heavy-duty materials designed to provide good resistance to staining and corrosion. They are not meant to be abused. If the guidelines given here are followed, they will keep their original shine and luster forever. Harsh and abrasive cleaners will reduce this shine if used regularly and are not recommended. If a cleanser is used, make sure it is a nonabrasive type.

Do not cut or chop vegetables and other foods on the surface of a kitchen drainboard. This will leave scratches which cannot be easily removed, if at all. Also, avoid scraping metallic objects such as spoons, forks, or utensils against these surfaces. Although some kitchen sinks are coated with materials which resist staining, continued contact with citrus fruit pulp and juice, vinegar, tomatoes, mayonnaise, yogurt, ketchup, cottage cheese, tea leaves, coffee grounds, and the like will tarnish the enameled surface. Remove such items promptly by throwing them in the garbage can or washing them down into the garbage disposer, if one is present in the system.

The kitchen sink drain, if not fitted with a disposer, is likely to get clogged if pieces of food, oils and greases, tea leaves, and coffee grounds are discarded into it. As dishes are being washed, it is almost impossible to prevent some of these materials from going down into the drain. These items may lodge on the interior walls of the drain and ultimately cause clogging or possibly a backup into the sink. If drain cleaners are used somewhat regularly, the drains will have a better chance of staying clear and will not require large-scale unclogging procedures. Care should be taken when using these cleaners. Two types of chemical cleaners are available: *acidic* and *alkaline*. Acids are more effective but should be used with extreme care. If acids come in contact with the skin, they may cause burning. The more reputable brands are recommended. Both acids and alkalis will dissolve grease, but only the acid type will prove successful if hair is causing the clog. Another thing to be careful of when using these cleaners is not to mix the acid and alkaline types together in the drain. This can cause a quite violent eruption. Unless it is possible to drain out one type completely, do not pour in the other kind.

A bathroom sink has a slight advantage over a kitchen sink in that it has a pop-up drain plug which prevents large objects from going down the drain. This should be lifted up occasionally and cleaned with a brush or cloth. Articles in the medicine cabinet should be arranged properly and carefully so that they are in no danger of falling down into the sink. If any type of corrosive substance does happen to spill on the surface of the lavatory, wipe it up immediately and wash the sink out with an appropriate cleaning solution. One simple maintenance procedure that should be performed periodically is to fill the lavatory to the brim after plugging the drain opening and then to suddenly release this water. If this is done on a regular basis, the flushing action will help to keep the drains clean and free of lodged materials.

The faucet assemblies on sinks, tubs, showers, and the handles on toilet tanks are normally chromium plated and will retain their shine for a long period of time. Wiping with a wet cloth or washing with soap and water is all that is required to maintain them in their original condition. Special liquid cleaners are available which work well on these types of fixtures. If the surface of a fitting appears to be damaged from misuse or accidental scraping, this may cause the entire surface to corrode, which will be apparent by the appearance of some green spots. To prevent further deterioration, remove the spots by rubbing them with very fine emery paper or a powder cleanser. Apply a thin layer of wax.

The best type of preventive maintenance for sink and faucet assemblies is a periodic check for any leaks either at the taps themselves or under the sink assembly. Leaks which go unnoticed or are ignored for any length of time may cause damage to the fixture and possibly even surrounding walls and/or floors. Although a minor drip does not seem to be any real cause for alarm, much water is wasted over a period of time.

Some types of leaks are not apparent simply by checking the faucets and looking around under the fixture for wetness. With regard to toilets, sometimes there is a continuous leakage from the toilet tank into the bowl which will go undetected. To check for this type of leak, get a dye from a hardware store and drop some of it in the tank so that the water takes on the color of the dye. Watch for a few minutes without flushing the toilet. If the colored water appears in the bowl, this is a sure sign that there is a leak somewhere. To repair this type of leak, refer to the information in Chapter 10.

If any outside faucets are present in the system, these should be drained and shut down at some point inside the house, probably in the basement, every fall before freezing becomes a danger. If this is not done, the pipes may burst. The faucet assembly will likewise be damaged. If there is a garden hose attached to this assembly, this should also be drained and stored away in a safe place. Hoses are made of some type of plastic or vinyl and are very prone to cracking if left outside, especially if there is any water left in them.

GUTTERS AND YARD MAINTENANCE

Almost every home is equipped with some sort of gutter system which serves the purpose of draining off excess water from the area close to the house to a point where it will not do any damage to the structure itself. The gutters are connected to vertical downspouts which send the water away from the foundation of the house. These downspouts should be inspected periodically, and in particular after a heavy downpour, to insure that they do not have any leaks and are properly conveying the roof water away from the house or into a dry well provided to receive this water. If they connect to ground pipes or to a dry well, open up the connection at ground surface and inspect the ground pipe. It is not unusual for this pipe to become either partially or completely clogged with leaves, dirt, paper, and other types of debris. To clean the pipe, simply direct a stream of water from a garden hose into the pipeline. Make sure the water has some force behind it to insure all debris are pushed out satisfactorily. If this is done regularly, there will be no danger of the water backing up into the ground and the substructure of the home, which will probably cause flooding in the basement. If the blockage is extreme, it may be necessary to open the line completely to clean it out. In any case, do a thorough job each time this procedure is done.

In areas where there are large trees extending higher than the point on the roof where the horizontal gutters are located, leaves may fall into the drainpipes and clog them, particularly in the fall. At the end of the season and before winter sets in, climb up on a ladder and inspect the whole horizontal gutter structure. Remove any leaves or other debris which may have collected; otherwise the gutters, will not perform the job for which they were intended. This task may be eliminated completely by installing special wire screens over the gutters. These screens are available at most hardware and plumbing stores and are manufactured solely for the purpose of protecting gutters from leaves and debris.

Most people do not realize that proper grading of the area around the house is a form of preventive maintenance, especially if a septic system is a part of the home's plumbing. If there are any depressions or low spots, particularly over the septic tank or the disposal field, these should be filled and the land graded so that all water flows away from the property along natural drainage lines. Water standing anywhere close to your house, particularly next to the foundation, may seep down to the basement, making it very damp and unpleasant smelling. The ground all around the outer walls should slope away so that no rain water is allowed to collect near the house itself. Another practice which is not recommended is the planting of flower beds in close proximity to the outer walls of the structure. The reason is that when these beds are watered, the excess water may seep down to the foundation and cause dampness in the basement.

PREVENTING FROST DAMAGE

Chapter 13 covered common methods of providing insulation for pipelines which may possibly be exposed to freezing weather. This applies to those lines which run outdoors to another building such as a garage. However, all water supply pipes located in the attic or in an unheated basement may need some sort of protection against frost. If this type of protection has already been provided, it should be checked periodically. Pipe insulation may have been damaged in some spots, or lines running close to an outside wall with no insulation in it may have been overlooked during the installation of this type of protection.

In situations where no insulation has been provided and weather forecasters are predicting a deep frost, there is a way to take care of the lines in a hurry. One way is to keep water running continuously in the uninsulated line, letting it go to waste. This arrangement, although expensive, will reduce the likelihood of a pipe freeze, which is more expensive. Another method is to place several electric bulbs along the exposed pipeline and turn them on. To avoid losing the heat, enclose both the bulbs and the line within a covering of aluminum foil, taking care to provide the necessary precautionary measures against electrical hazards.

PLUMBING SYSTEM SHUTDOWN PROCEDURES

A home or summer cottage may be left vacant for a period of time. If this period is expected to include the winter season and the heating system will not be running, it is important to drain the plumbing system entirely of its water to avoid permanent damage to any portion of the installation.

The first step is to shut off the cold water supply to the house by closing the main shutoff valve in the basement. Proceed with the following steps, which will drain off almost all of the water contained in the various lines and fixtures.

■ Starting at the top floor, open all the faucets and valves. After water stops running, take out the plug from the main shutoff valve and collect any remaining water in a bucket.

■ Flush the toilets to empty their tanks. Remove any remaining water in the tanks by means of a sponge tied to the end of a stick.

■ Drain out water from all the traps under tubs, lavatories, and sinks by opening their cleanout plugs. Those not provided with plugs should be dismantled and, after all the water is drained, reassembled.

■ Pour a liquid such as kerosene into the toilet bowls and in all other fixture traps, including the one in the basement floor drain. The seals will prevent sewer gases from entering the house. Add automotive antifreeze solution to the seals to prevent them from freezing.

■ Drain hot water radiators.

■ Drain the water from the boiler, the pneumatic tank, floor tanks, expansion tanks, and other containers.

■ Empty the hot water tank by opening the bottom drain valve and the cock in the pipe at the top which lets air in. Check to be sure that it is completely emptied, including all horizontal pipes. If necessary, pipes may be opened at joints to drain the water.

■ Washing machines, dishwashers, pumps, and any other equipment that may have some quantity of water inside should be completely drained.

■ To drain the insides of all the valves, open the small-plugs on them.

■ Shut off the curb stop valve and open the service pipe just close to the basement. Uncouple it so that it is open to the atmosphere.

■ Uncouple the water meter from the line, and tilt it to drain off any water.

■ Wrap the main shutoff valve and as much of the main supply pipe as possible with insulating tape.

This should cover everything involved in shutting down the system for any length of time. If you have your own individual well water, a few more steps will be included. First, let the water from the lines run back into the well by opening all the valves. The pump should be left dry, or water contained in it will freeze and cause damage. If the pump is a submerged type, don't do anything to it. If it is a reciprocating or rotary type, though, drain the water out of the casing. The water tank should also be drained completely.

When opening the building after if has been closed for some time, if all the outlets are closed and the water supply is turned back on, this will cause air to be trapped in the water lines and tanks. Also, when the faucets are opened, this air will rush out and water hammer will occur. To avoid this, first open all faucets and valves throughout the building and then open the main supply valve gradually. Air in the lines will be pushed out slowly through the faucets, which should be closed progressively upwards as the water rises in the pipes and flows out each faucet for a few minutes. Air should also be allowed to escape from the storage tanks and any other containers used to hold water. Air chambers for preventing water hammer should be inspected to make sure they are working properly. If air has been sucked out of them, open up the caps, let them fill with air, and then replace the caps. If the whole procedure has been done properly, no problems should be encountered either in shutting the plumbing down or in turning it back on.

SUMMARY

A regular periodical maintenance program is a must for every home, but many people neglect this simply procedure. The result is usually a good deal of frustration and money to make major repairs which probably would not have been necessary if the system had been inspected regularly. A minor defect of worn-out part, if undetected or ignored, will have the effect of a chain reaction, turning a minor problem into a major breakdown in a very short time. This is not always the case, but it can happen and the point is that it can be prevented through a regular maintenance program. If you were the one to do all the original work on the plumbing system, you probably know the amount of work and money that went into the original installation and appreciate the value of this type of preventive maintenance. If everything was installed and operational before you moved in, however, you are more likely to be inattentive to this type of detail. Whichever category you fall into, do not overlook this vital step in maintaining a plumbing system in its original condition. The procedures discussed in this text are very simply and not too time-consuming. Regardless of the amount of time involved, it is well worth the effort in the long run. Those who have ignored their plumbing system until it was too late are the ones who wished they had been conscientious before major repairs were necessary which could have been avoided. Take the time to inspect your plumbing system regularly and thoroughly. You will never be sorry you did.

Municipal Water and Sewage Treatment Plants

A proposed residence which is near enough to a water and sewer main provided by a municipality will probably be required to use these facilities in lieu of the installation of a new septic system or drilling of a new well. In areas where water and sewer systems are not provided, such as in rural areas where it is not economically feasible to extend lines, homeowners must install individual sewage disposal systems and provide for their own water supply. If sewer and water lines are extended into these areas at a later date, these homeowners will usually have the option of tapping in to the lines if desired or remaining with their private water and sewer systems as long as they continue to meet health department requirements. Should either the well or septic system become defective, these same homeowners would probably be required to tap into the municipal system rather than to rebuild the private facilities.

Both water and sewage treatment plants perform a vital service for those who tap into them. Millions of dollars will be spent by the municipality in order to effect the study, development, and construction necessary to complete a typical facility. The main objective of these expenditures is to provide a supply of fresh water which is purified and a safe and effective method of disposing of solid and liquid wastes. The homeowner pays for these services, both on a monthly basis and through real estate and personal property taxes. Generally, these services are not tremendously expensive and are quite cost effective when compared to the installation of private systems. To offset the multi-million dollar cost of even small facilities, many municipalities receive direct partial financing from the federal government. This helps to lower the portion which must be assumed by homeowners.

WATER TREATMENT SYSTEM

For the purpose of this discussion, the water and sewer facilities in Front Royal, Virginia, were used as examples. Front Royal is a growing

365

community with a population of approximately 12,000. Its water treatment plant was completed a little over 20 years ago and has a storage capacity of approximately 1.2 million gallons. While this plant was over-designed at the time it was built, it is being sorely taxed due to the recent growth in this area approximately 75 miles from Washington, DC. The average daily needs of the town in summer months are about 1.2 million gallons, which pretty equally matches the storage capacity. Fortunately, facilities such as these can be added to, and the municipality is in the process of installing an additional 3 million-gallon storage tank.

Figure 17-1 shows a block diagram of the water treatment system. Water is taken from the Shenandoah River a little over a mile away and from a mountain stream. Two large 250-horsepower pumps are used at the river site to draw the water to a large holding pond located in close proximity to the treatment plant. This pond is shown in Fig. 17-2. The water from the mountain stream is gravity fed. Through a continual process, the water is delivered from the holding reservoir to a lower level in the treatment facility, where chemicals are injected. This process is done automatically by individual machines for each different chemical. See Figs. 17-3 and 17-4. The water is tested to determine what quantities of chemicals are needed and when to add them.

INJECTION PROCESS

Each chemical serves to produce a purifying effect on the water which will ultimately be delivered for human consumption. *Copper sulfate* is added at the storage reservoir and serves to minimize the growth of algae. During the injection process, alum, lime, carbon, and chlorine are added along with fluoride to complete the purification process. The alum is used to provide a settling action of the mud and dirt present in the water. Lime controls the pH, or acidity, of the water. Taste and odor control is accomplished by the injection of carbon. Chlorine is used to purify the

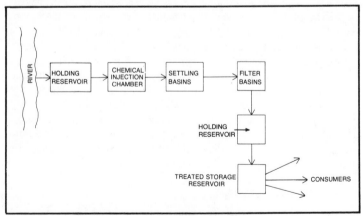

Fig. 17-1. Block diagram of water treatment facility.

water. Most localities also inject fluoride into the water for dental hygiene purposes. This is accomplished in the same manner as with the other chemicals.

The injection process is accomplished by electromechanical hoppers, each of which is loaded with a different chemical. These hoppers hold the chemicals until the machinery on the lower levels automatically trigger an injection sequence. The triggering machinery contains gauges which are set by a lab technician based upon the condition of the untreated water arriving at the plant. Obviously, a large quantity of chemicals are needed for this treatment process, which goes on 24 hours a day. Figure 17-5 shows the upper level of the treatment building where chemicals are both stored and loaded into the hoppers which are mounted nearby. A loading ramp allows for easy access to the chemical storage area, which is kept separate from the main treatment facility to avoid accidental chemical contamination. Figure 17-6 shows the opening in the hopper into which the chemicals are loaded by hand. The hoppers are monitored on a regular basis to make certain that they are in perfect working order. This assures that the correct amount of chemicals are added during each triggering sequence.

SETTLING BASINS

Once the injection process has been completed, the water is then pumped to an area outside the main treatment building, which contains a series of settling basins. These basins are series-connected with the first feeding into the second, the second into the third, etc (Fig. 17-7). In this particular facility, matched pairs of basins are used to handle the quantity of water which is processed daily. A main control valve inside the treat-

Fig. 17-2. Holding pond receives water pumped from river and streams.

367

Fig. 17-3. Chemical injection system.

ment facility controls the water level in the basins, making certain that they are always full.

The first pair of settling basins which, like the others, are approximately 12 feet deep, contain motor-driven agitators. The slow turning motion serves to provide a means of evenly mixing the injected chemicals into the water. A low rpm motor is coupled to the agitators by a chain drive slowly turning the blades around the central shaft (Fig. 17-8). The wooden agitator arms run almost the entire length of the settling basin and extend to a depth of approximately 10 feet. Figure 17-9 shows one of the wooden agitator arms as it turns just beneath the surface of the first basin.

From this point, water and mixed chemicals are fed to the second series of basins where sediment begins to fall from the water and onto the bottom of the cement enclosure. The alum crystals cause the earth and sediment to form globules which are heavier than the water, thus causing them to sink to the bottom. A close inspection of the water in the second and third pairs of basins will easily reveal these globules. As the water continues to travel through the series, more and more of the sediment is removed. Obviously, there is a sediment buildup on the bottom of the basins. After a period of time, the depth can easily approach 5 feet. For the

Fig. 17-4. Injection process.

Front Royal plant, this depth means that it is time for a cleaning process which involves the draining of the basins and the removal of the sediment by hand. The second and third pairs of basins tend to have faster buildups than do the fourth and fifth pairs because of the higher levels of sediment in the early stages of the settling process.

Fig. 17-5. Stocked chemical storage room.

Fig. 17-6. Chemicals are loaded into the hopper.

FILTERING OPERATION

When the water has traveled through the entire series of settling basins, it finally arrives to a point inside the main treatment facility. Here it is contained in four large basins where it prepares to undergo the final purification process. It is pumped to storage reservoirs and in turn to the

Fig. 17-7. Series-connected settling basins.

Fig. 17-8. An electric motor is connected to large agitators.

town. Final purification involves passing the water through a series of large filters. This removes any microscopic contaminants which were not eliminated during the settling process. The filtering is controlled by means of central panels located at each of the four filtering basins. This panel contains several valves and two chart recorders to allow the treat-

Fig. 17-9. Agitator can be seen just beneath the water's surface.

Fig. 17-10. Central control panel contains several valves and graph monitors.

ment plant operator to constantly monitor and control this portion of the purifying operation (Fig. 17-10). Each control provides a different function. The *influent valve* is opened to receive water from the settling basins, although in normal operation the flow is usually continuous. Any remaining wastes which may still be contained within the water are drained off by means of the *waste control valve*. This water is disposed of by pumping it to two large lagoons or settling ponds which are located about ½ mile from the treatment facility proper. The water is allowed to stand for a period of time and is then allowed to drain into a natural creek. The water which is channeled into this natural creek is of better quality than the water already there; thus, no contamination of a natural stream takes place. The actual passing of the water from the large basins within the plant through the purifying filters is controlled by the *wash* and *rewash levers* on the central panel. During the wash process, the water is backwashed by means of a large pump. The backwashing process involves the passing of the water through layers of sand and gravel. The rewash process provides a second washing without the use of a pump.

The purified water is now fit for human consumption, having gone through the entire filtering process. It is pumped into a large holding reservoir beneath the main building. The next step in the chain is to pump from this location to the treated storage reservoir (Fig. 17-11). The water is now ready to be pumped to the town residences where it is used in conventional ways.

Supplying water to even a small area involves a very complex process. There can be little room for error in supplying this type of service because of the importance of insuring the safety of those using the facility. The *State Water Control Board* regularly inspects and monitors these types of operation, and very rigid standards must be maintained.

MONITORING EQUIPMENT

The water treatment plant operators are staffed around the clock to take care of any contingencies or emergency situations that may develop. Since the Front Royal plant has a minimal holding capacity when compared with average daily usage, water main breaks could quickly deplete any emergency reserves. Monitoring equipment is contained at the treatment facility, which will indicate any unusual demands on the plant. A portion of this equipment is shown in Fig. 17-12 and is tied into an alarm system. Should a main break, the alarm will be sounded. The operator on duty can immediately notify a work crew to locate the problem area and make the necessary repairs. In some instances, the treated water may be temporarily shut off to maintain holding capacity until the repairs are effected.

Tests are constantly run on water quality in the small laboratory, which is housed at the Front Royal treatment plant. This facility contains all of the equipment necessary to perform the quality measurements required by the State Water Control Board (Fig. 17-13).

PROBLEMS AFFECTING DRINKING WATER

Certain environmental occurrences can have direct effects on the taste and quality of the drinking water supplied. Obviously, during dry periods when water tables are low, the plant may not be able to pump as

Fig. 17-11. Enclosed treated water storage reservoir.

Fig. 17-12. Master monitoring panel reads overall demand of treated water.

much water from the river and streams as at other times. It is sometimes necessary to curtail water usage by prohibiting the washing of automobiles and watering of lawns until the supply becomes nominal again. Since the water is pumped from natural rivers and streams, the treatment operators must constantly test the raw water, looking for foreign contaminants. An early snow fall one October caused a weight buildup on the branches of many trees which had not shed their leaves. This caused an unusual amount of broken limbs to be desposited on the ground. As these began to decay, a high amount of *tannic acid* was deposited in the natural streams. This was immediately discovered at the treatment facility, and steps were taken to modify the treatment process. Town residents were advised that their water might have a slightly unusual taste, but that it had been tested and was safe for human consumption.

One problem of concern to all municipalities which run water treatment plants is that of contamination through backflow into the water lines. When a major water main breaks, it is possible in some instances for a suction to occur within the water system and for contaminants to be drawn through residential spigots into the water lines. For example, a garden hose left in a bucket of water might serve as a channel for this water to be drawn back into the water lines should a main break and a backflow be created. If there were chemical agents such as insecticides in the bucket, they could be introduced into the main system. Many municipalities are now requiring the installation of backflow valves on residential hookups to prevent this condition from occuring.

Municipal water systems were used by almost every business and residence within the areas they serve. When many persons are dependent upon a single source of water, it is essential that the facility be maintained in strict accordance with the regulations governing their operation and use. The main objective of any such facility is to provide the adequate amount of pure water needed by the community in as efficient a manner as possible. Since most communities continue to grow, water treatment plants must be designed to handle two, three or even four times the current demand. The planning stages alone for a new facility may cost hundreds of thousands of dollars before the first construction step ever takes place.

WASTEWATER TREATMENT FACILITIES

For most municipalities, wastewater treatment facilities are the most expensive service rendered to the community as a whole. Even small communities are looking at million dollar-plus expenditures when considering the installation of sewer lines. Treatment facilities, depending upon the area's size, can cost upwards of $10 million or even $100 million or more for the larger municipalities.

While writing this book. I was fortunate to have access to the secondary sewage treatment facility which was still under construction within Front Royal, Virginia. At the time of this writing, the plant is about 90

Fig. 17-13. Laboratory test instruments are used to indicate water quality.

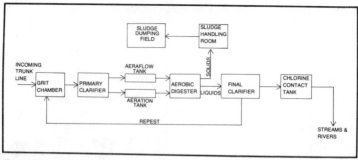

Fig. 17-14. Block diagram of secondary sewage treatment facility.

percent completed and will be ready for activation in a few months. It is easier to see the many processes which take place shortly before the facility is to be used than when it is actually in operation.

Figure 17-14 shows a block diagram of the Front Royal plant. Raw sewage from the various lines enters the plant by gravity feed where the processing takes place. Some outlying areas may require sewage pumping stations to deliver the wastes to a point which will allow for this gravity feed. After all of the processing has been completed, the conditioned liquids are allowed to flow back into area streams and rivers. This output does not present a contamination hazard, as all foreign elements are removed during the processing operation.

Tracing the flow of sewage from the lines and through the plant begins at the *grit chamber* (Fig. 17-15). Sewage from the lines is fed to the input of

Fig. 17-15. Grit chambers remove sand and gravel.

Fig. 17-16. Primary clarifier allows for larger contaminants to settle from sewage.

this facility, which removes sand and gravel from the wastes. Screens are located within the grit chamber that filter out this larger matter and then deposit it on a conveyor for loading into trucks for disposal.

Fig. 17-17. Aeration tanks oxygenate water, making it a suitable environment for bacteria.

Fig. 17-18. Aerobic digester allows bacteria to grow, further removing solid contaminants.

Fig. 17-19. Overhead view of aerobic digester showing control lip which allows for liquid to flow through the trough on left for further treatment. Solids remain in the basin.

Fig. 17-20. Final clarifier.

After passing through the grit chamber, the sewage enters the *primary clarifier* shown in Fig. 17-16. This is an arrangement of settling basins which allows for the initial settling action to take place. The heavier solids in the sewage sink to the bottom. This includes the matter which was too fine to be screened in the grit chamber.

After the initial settling process has been completed, the remaining sewage is transferred to *aeration tanks* which are shown in Fig. 17-17.

Fig. 17-21. Slowly revolving agitator is located in the bottom of the basin.

Fig. 17-22. Pumps channel treated water back to area streams.

Agitators within this unit mix oxygen with the effluent, making it habitable by bacteria which will further break down the microscopic waste materials within the water.

The next step in the treatment process channels the effluent into the aerobic digester. Bacteria are allowed to grow here. The digester, shown in Fig. 17-18, closely resembles some of the earlier settling basins. The liquid effluent is drained from the top of the digester by means of a lipped trough (Fig. 17-19). As liquid flows over the top, it is channeled off to another area of the plant, leaving the solid wastes in the basin. These solids or sludge are transferred to the sludge handling room, where it is

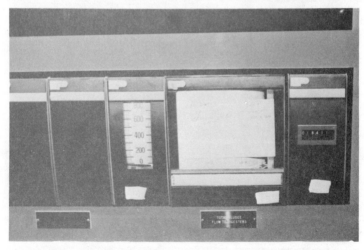

Fig. 17-23. Graph recorders constantly monitor treatment operations.

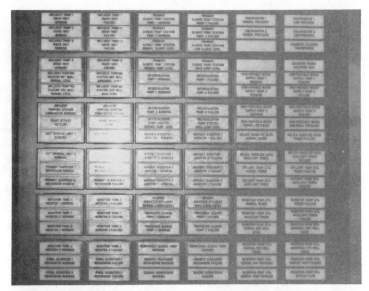

Fig. 17-24. An alarm panel is activated should a breakdown occur at any point in the treatment process.

further processed and will eventually be trucked to a sludge dumping field where it will be allowed to mix with the soil.

Meanwhile, the liquid effluent is pumped to the final clarifier which contains a slow-moving agitator to gradually settle out any remaining

Fig. 17-25. Standby electrical generator is automatically activated in the event of a power failure.

Fig. 17-26. Overall view of a portion of the secondary sewage treatment facility located in Front Royal, Virginia.

impurities (Figs. 17-20 and 17-21). Following this process, the liquid effluent is channeled through all of the former stages again where particles are continually removed. The final stage in the processing is the chlorine contact tank which chlorinates the water. The chlorine feeders inject the chemical into the water, purifying it to a point where it is safe to be deposited into area streams and rivers where it is then pumped. These pumps are shown in Fig. 17-22.

The operation of a secondary sewage treatment facility is vital to the health and safety of all users within the municipality. Stringent monitoring requirements are used throughout each step of the operation. Figure 17-23 shows one of many graph recorders which are part of this monitoring process. This is a complex operation with every stage dependent upon all the others. If one link in this chain breaks down, the process must be halted until repairs can be made. A central panel at the main control point is shown in Fig. 17-24. Should a problem develop in any of the many stages, an indicator light is activated and an alarm sounded.

With all of the many large pumps and motors used in this secondary treatment facility, the plant is obviously highly dependent upon electrical power and must be prepared to operate when this power is temporarily interrupted. Figure 17-25 shows auxiliary generators which can drive the plant during such an outage. The motors run on diesel fuel and offer a backup system which can run the plant for an indefinite period of time. Two 12-cylinder diesel engines provide adequate redundancy.

The plant is rather large, covering several acres (Fig. 17-26). Like the water treatment facility discussed earlier in this chapter, this secondary sewage treatment plant is designed for many years to come. The completed facility will cost in the neighborhood of $5 million, but the price of installing new sewer trunk lines to various parts of the town and country increase the overall price to in excess of $16 million.

SUMMARY

Many home plumbers become discouraged when faced with many of the rules and regulations a local municipality may place upon them when they are planning a modification to an old plumbing system or the building of a new one. By understanding the complicated procedures and the tremendous costs involved in building, operating, and maintaining municipal water and sewage treatment facilities, the reasons behind many of these regulations become apparent. You should always be in the strict accordance with state and local regulations during these installations. What you may feel is perfectly satisfactory for your home plumbing needs could mean disaster for the entire system if every other plumbing system was designed in the same manner.

A Solar-Derived Hot Water System

18

The plumbing field itself is a highly regulated profession for many reasons, the major one being the safety of those using the system. Before any changes to rules and regulations are made much time goes into testing the proposed modifications and making sure they will perform as adequately and successfully as the old tried and true methods.

The last real addition to the plumbing field was the introduction of plastic piping some years ago. Although it does meet with approval on a national scale, some localities still do not permit its usage in home plumbing applications. There are probably some people who may find this frustrating. They may be justified in this feeling, but keep in mind that there will always be those who feel just as justified in sticking with what has proven successful and safe in the past.

The introduction of solar methods of heating hot water has also met with some resistance on a national scale, but as the energy crunch strikes home, those governing the plumbing industry have begun to give their acceptance to the tapping of the sun. This is not a new method at all; people have been using the heat of the sun for a multitude of purposes for many years. It is only recently that this practice has gained widespread approval and popularity due to the need to find alternative sources of energy. Solar energy utilization is very cost effective once installed, although the original installation expense may still not be within reach of some people. As more and more people opt for solar power and the demand increases, costs have dropped considerably and should continue to do so.

As with any new products to hit the market, you should proceed slowly and carefully. It is not within the scope of this book to go into much detail with regard to the principles involved in tapping the sun's energy. However, there is a wealth of information available on this subject, and it is wise to obtain an understanding of these principles in order to be able to make a determination as to whether they can be applied to your home in particular. It is also wise to check the reputation of the company being considered as a dealer and have some knowledge of the working mechanisms of the products being purchased. This is not to say that all people involved in the solar field are not to be trusted; it is merely a warning to know what you are buying and whether the product will perform adequately for your intended purposes.

The economic attractiveness and feasibility of each individual solar application will vary widely, depending upon local setting, geographical location, climate conditions, and the price of competing energy sources. People living in areas with greater temperature extremes who use large amounts of energy will most often realize more immediate savings than those living in more temperate climates. The government is now allowing tax credits to those who spend part of their income on converting portions of their home to allow for the use of some forms of alternative energy. Likewise, some types of improvements or investments in a new home which will utilize solar energy can be amortized as part of an increase in the purchase price. Since solar installations are part of the real value of the property, they appreciate at the same rate as the rest of the property.

THE HISTORY OF SOLAR ENERGY

Three basic methods of collecting solar energy directly have been developed by man over thousands of years. These are:

■ *Flat collector method*—uses unfocused or diluted solar radiation.

■ *Concentrating collector*—uses concentrated or focused solar radiation.

■ *Photovoltaic or photogalvanic cells*—use diluted solar radiation.

The first two methods convert sunlight directly to heat; the third converts it directly to electricity. Concentrators, convex lenses, or concave mirrors (*parabolic*) are the oldest known types of solar collectors. Evidence of both convex and concave mirrors has been found in ancient civilizations. It is assumed that they were used to start fires.

In 212 B.C. *Archimedes* used a large battery of mirrors to set the sails of invading ships on fire. Serious studies of the sun and its energy potential began in the 17th century. By that time, diamonds had been cut and melted using the sun's heat. *Galileo* and *Lavoisier* utilized the sun in their studies. In 1774 Lavoisier used in his scientific studies a 52-inch lens to heat materials. The heat produced was theoretically purer than other forms because it contained no contaminants such as ashes or soot. By the early 1800s, heat engines were operating with energy supplied by the sun. Small solar furnaces have been used to power steam generators since 1870. By 1878, ice was produced using solar energy.

In the early 20th century, solar energy was used to power water distillation plants in Chile and irrigation pumps in Egypt. In 1929 Robert Goddard applies for patents on five different solar devices to be used to send a rocket to the moon. Most of these projects were considered to be ridiculous because they were so far ahead of their time.

By the end of 1930, many Californians were making practical use of solar energy. "California hot water heaters," as they were known, produced hot water in many homes. Other devices similar to present day collectors were used to heat homes. Many passive systems were located throughout California as many architects began to design buildings that incorporated solar design to some extent.

Frank Lloyd Wright was one of the first architects to recommend that solar energy design be employed in the construction of new buildings. Wright lived long enough to see some of his ideas and recommendations become reality. The first building to be heated by solar energy-generated hot water (through a circulating process) was built at the Massachusetts Institute of Technology in 1938. Some 18 other experimental buildings employing solar design were built between 1938 and 1960. These experimental projects employed solar energy as their total heating energy source. Performance data recorded on those projects is used as engineering design data today.

Twentieth century space age technology gave solar energy its mightiest boost National Aeronautics and Space Administration (NASA) engineers took a small amount of knowledge and developed enough energy to enable spacecraft to be self-sufficient once they reached outer space. Much of this achievement was directly related to solar energy utilization.

ECONOMIC CONSIDERATIONS

The solar payback period will depend upon each individual application. If solar can be used year round, such as for domestic hot water service, then the payback time is much shorter than if it were used only during the winter months, such as for space heating. In fact, domestic hot water heating is probably the most attractive solar application because of the short payback period, and also because it requires such a small amount of collector area. In most areas of the United States, only 40 to 60 square feet of collector is required to meet 70 percent of the annual hot water demand in the average home. As fuel costs continue to rise, the payback period will become even shorter.

There are several reasons why solar equipment is more expensive than that of conventional systems:

■ The cost of the collectors, the heat storage system, and the piping and controls that must be added. Add these items to the backup system and the labor factor, and it is easy to see why the cost is high.

■ The solar energy industry is really in its infancy; production volumes are still low. Therefore, the cost of the solar equipment is higher now than it will be as production increases. However, the cost is still a relatively fast payback. In considering retrofit, the backup system cost can virtually be eliminated. Also, if the homeowner is handy with tools, most of the labor costs can be eliminated or reduced.

THE SOLAR COLLECTOR

A solar heat collector is a device which absorbs the sun's energy and converts it to heat. Flat plate collectors are the most practical and least expensive for residential use. Obviously, this collector is the chief component of the system. Once the sun's energy has been converted into heat energy, it is transported into the home by a working fluid. The flat collector may be mounted in the roof, on the roof, directly to the roof, on or in the walls of the home, or mounted on its own support structure at the proper tilt angle in the backyard.

The basic principle of a collector is not all that complicated. The sun's energy is absorbed on a flat plate surface. The plate is usually constructed of copper, aluminum, or steel. There are two categories of flat plate collectors: air and liquid types. They are utilized according to the type of working fluid that is circulated through them to affect and transport the captured solar heat.

The basic components of a flat plate collector are shown in Fig. 18-1. The absorber stops and collects the sunlight, converts it to heat, and transfers the heat to the circulating fluid, designated the working fluid. The absorber surface is usually painted black or dark green to improve its collection efficiency. In order to minimize heat loss out the front of the collector, transparent cover plates are placed above the collecting surface of the absorber. Heat loss out the sides and back are reduced by insulation. The components are enclosed and sealed for protection against moisture.

Although it is altogether possible to construct a collector yourself, manufactured collectors are superior for several reasons. In examining the primary component of the collector, the absorber, it is important to know that this component is the one that gets the hottest. It is darkened for maximum absorption of the sun's energy. Three types of color are used: one is a flat black; another is a flat dark green; and the other is called a selective black. Flat black and flat dark green paints come in spray cans and

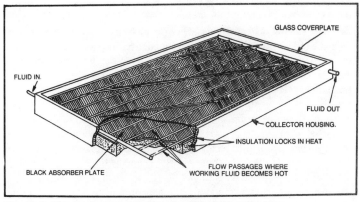

Fig. 18-1. The basic components of a flat plate collector.

have long been used to darken collector absorbers. These colors have an absorptivity in the solar radiation region of the spectrum of approximately .95 and an emissivity in the thermal radiation region of the spectrum of about .95; that is, heat is reradiated at a significant rate. A selectivity darkened absorber has an absorptivity in the solar region of the spectrum of approximately .90 and an emissivity in the thermal region of the spectrum of approximately .10 or less. Thus, a selective surface improves the efficiency of the collector significantly because it greatly reduces the reradiation of the heat from the absorber. For example, when operating at a 100°F temperature difference between collector absorber and outside air, a collector using a selective surface and single glass cover delivers 25 to 50 percent more heat daily than a collector using a flat black or flat dark green absorber surface and a double glass cover. Therefore, if efficiency is of prime consideration, which it should be, a manufactured collector should be the choice.

SHOPPING FOR SOLAR EQUIPMENT

Many different designs are being sold today commercially with considerable variation in appearance as well as performance. Be very selective in shopping for this equipment. A buyer of solar equipment should be looking for a product that delivers the maximum amount of heat per unit area and has the maximum durability. Both objectives should be sought at the lowest possible cost. Traditionally, collectors have been compared simply on a cost per square foot basis. When there was little difference in efficiency among available collectors, this was an acceptable basis for comparison. However, with the availability of high performance collectors, a different basis for cost comparison should be used. It make little sense to compare two collectors on a cost per square foot basis when collector A may deliver 25 to 50 percent more heat per month than collector B. Neither does it make any sense to purchase a collector that has a short life expectancy, as the collector may fail before it can pay for itself.

The amount of heat delivered per month by a given collector depends upon the temperature level of the heat needed for its particular application, the efficiency of the collector at the required temperature level, and local climatic conditions. Different applications require different temperature levels of heat. For example, pool heating requires a fairly low temperature level of heat. Domestic water heating will also require a moderate temperature level, on the order of 60°F to 120°F above ambient.

The performance of a given manufacturer's collector should be determined by performance tests made by an independent testing laboratory. Test procedures should be those as established by the National Bureau of Standards. If a manufacturer cannot or will not furnish you with a copy of the test results relative to the equipment in question, and a warranty clearly establishing a guarantee, you should take you business elsewhere.

Several additional cautions should be kept in mind when shopping for solar components. Manufacturer's price sheets vary widely in costs and in what you receive for your money, aside from differences in efficiency. Some manufacturers base their price on cost per square foot of gross collector area, while others base their price on cost per square foot of net absorber area. The gross collector area can be as little as 10 and as much as 50 percent more than the net collector area. Also, some manufacturers include no provisions for mounting the collectors; thus, such provisions must be provided by the installer. Often it is desirable to hide the pipes or ducts at the top and bottom of the collector. These are fairly insignificant items, but some manufacturers sell a kit that can make the installation easier. This is very handy for the do-it-yourselfer, and it can also save some money. Where a limited area is available for mounting a collector, the criteria for choosing a collector should be the one that produces the greatest amount of heat from the total area available for mounting.

OTHER COMPONENTS

Since the sun does not shine all the time, it will be necessary to provide for some sort of storage area for the water once it is heated. You need a storage tank along with the necessary piping to connect it to the collector. The solar system will provide somewhere around 70 percent of the home's need, so a backup water heater will probably be a part of the system, too. The costs will be significantly reduced since this backup heater will not be in use as much of the time.

In Japan there are thousands of solar water heaters in use, many of which are no more than a black box on a roof filled with about 30 gallons of water. In the morning the box is filled with tap water, and by late afternoon the water is heated to over 100°F. This will obviously only be useful in direct relation to the amount of sunlight available.

A simple solar water heater can be built from an old water tank. To make it into a solar water heater, paint it black, connect it to the hot water system, fill it with water, and expose it to the sun. To step up the efficiency, put the black tank in a box with glazing (for heat retention), add reflectors around the tank, and enclose it with insulation during the night or on cloudy days. See Fig. 18-2.

THERMOSIPHONS

Another solar water heating system is called a *thermosiphon*. It operates on the principle that cold water will sink in a water tank and thereby cause hot water to rise. The water heated in the solar collector will flow up into the storage tank because warm water rises, just as warm air rises. The bottom of the storage tank must be at least 2 feet above the top of the collector. This will prevent circulation of water in the wrong direction. The thermosiphon system eliminates the need for pumps and controls. If freezing is a factor, a closed system with a heat exchanger may be needed. See Fig. 18-3 and 18-4.

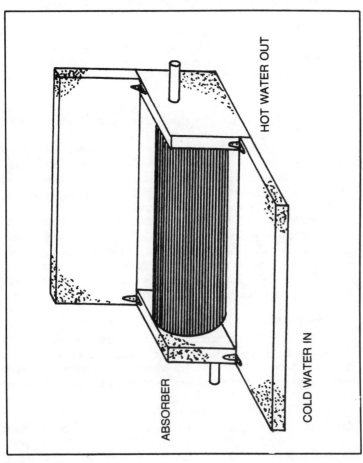

Fig. 18-2. A bread box solar water heater.

HOT WATER OUT

COLD WATER IN

ABSORBER

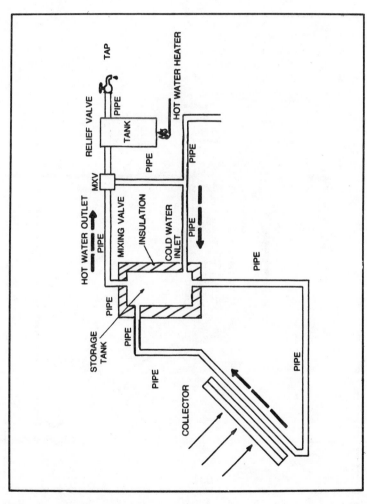

Fig. 18-3. In a thermosiphon, hot water flows up to a storage tank.

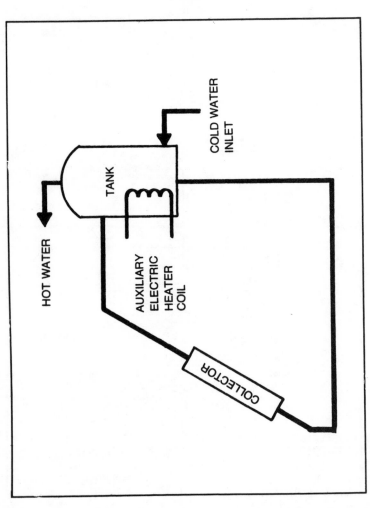

Fig. 18-4. A variation of a thermosiphon in which the tank has an auxiliary electric heater coil.

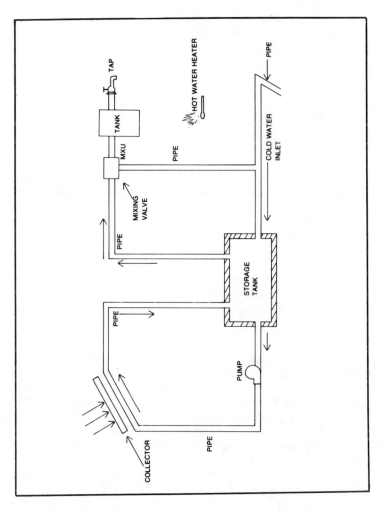

Fig. 18-5. If the storage tank cannot be placed above the collector, a small pump may be used to circulate the water.

PUMPED SOLAR WATER HEATING

Where it is not feasible to place a storage tank above the collector, a small pump may be used to circulate the water. Combined with a heat control system, the pump will move water through the collector only when the water in the collector is warmer than the water in the tank (Fig. 18-5).

In locations where freezing is common, a dual system with a heat exchanger is recommended. The heat exchanger system requires only a single tank with an auxiliary heater. The collector system is filled with antifreeze, and the heat exchanger is isolated from the hot water. The heat is transferred to the water through an immersed heating coil (Fig. 18-6).

ADDITIONAL CONSIDERATIONS

When considering the installation of a solar water heating system, it will be necessary to determine a number of things such as cost, estimated usage, and convenience of location of the components which will be involved. The first thing to do is determine your current daily or monthly hot water usage and how much this usage is costing using conventional sources of energy. The higher the approximate usage, the higher the potential for realizing a significant savings. The average cost of a residential solar water heating system will be between $1,200 and $2,000. This cost can be cut by providing much of the labor yourself. On the average, a family of four will need a storage tank of about 80 gallons and from 40 to 60 square feet of solar collector to supply their domestic hot water needs.

The roof is usually the most practical location for collectors, although they can be mounted almost anyplace: on the ground, on the garage, on a storage shed, or on exterior walls. Keep in mind that the more structure required to support the collector, the more the total system will cost. The degree of tilt for the collector is approximately equal to the latitude where the building is located. Under certain conditions, it may be practical to vary the pitch of the collector by a few degrees. In order to optimize for winter, the collector can be mounted at a steeper angle. If the pitch of an existing roof is only a few degrees more or less than the optimum, mount the collectors on the roof.

SUMMARY

This chapter has touched lightly on solar-derived residential hot water systems which are becoming more and more popular today. You will find that there are numerous commercial and homebuilt designs on today's market with each offering specific advantages and disadvantages. Many people find that a solar hot water system makes an excellent adjunct to the more standardized systems which have been in use for many years. Specific geographical location will be a major determining factor in deciding to go with a solar design.

Solar water heating systems are much more practical for most individuals than are complete solar heating facilities. as the former are not

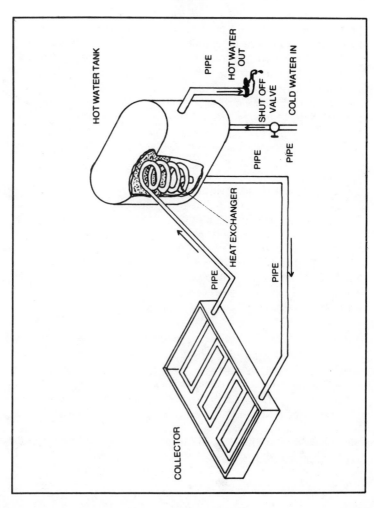

Fig. 18-6. Typical hot water heater with heat exchanger inside the tank.

nearly so complex and can be built using home construction methods. Many are quite cost effective and pay for themselves in a few years. Small commercial systems are also available and may be chosen to suit individual needs and to fit the average climate conditions of a certain area. Rarely is a solar hot water system used exclusively to supply the needs of a residence. The owner must anticipate the need of an electric or gas hot water heater as a main system during certain periods with the solar unit serving as a backup unit. During certain times of the year, however, a solar water heating system may be able to supply 100 percent of the user's needs.

Index

Edited by Robert Ostrander